GRAPHICAL TECHNIQUES

for

ENGINEERING COMPUTATIONS

SCALES—TOOLS OF SCIENCE AND INDUSTRY

GRAPHICAL TECHNIQUES FOR ENGINEERING COMPUTATIONS

by

WALTER HERBERT BURROWS, A.B., M.S.

Principal Research Chemist, and
Head, Industrial Products Branch,
Engineering Experiment Station,
Georgia Institute of Technology

CHEMICAL PUBLISHING COMPANY, INC.
New York
1965

CHEMICAL PUBLISHING CO., INC.

New York

Printed in the United States of America

To My Wife

CATHERINE

PREFACE

There is a steadily growing interest in the capabilities of graphical methods in the field of computation and an increasing demand for applications of these methods to a broad spectrum of scientific and engineering formulas—scientific principles or laws expressed in mathematical symbols.

Over the years, man's scientific endeavors have resulted in the accumulation of ponderous volumes of these formulas involving computation for their application to engineering problems. At the same time, man has developed a number of devices for reducing the labor of these computations—numerical devices such as the abacus and tables of logarithms, mechanical devices such as adding machines and desk calculators, electronic devices such as the modern computers, and graphical devices such as the slide rule and the nomograph. It is this last class of devices with which we are now concerned.

In a day that is witnessing the lightning invasion of high-speed desk calculators and electronic computers into the research laboratories, engineering departments, and business offices of industry, university, and government, one might well expect the comparatively simple graphical device—engineering workhorse of a decade or so ago—to dissolve gracefully into oblivion before the onslaught. Instead, the interest continues to grow.

The reason is quite simple. The calculation of a series of values required for the solution of an engineering problem (e.g., the design of a column for the fractionation of a hydrocarbon mixture) can be quickly performed with all required accuracy by the use of charts contained between the covers of a handbook at the engineer's fingertips. To use the electronic computer with equal ease and speed, it would be necessary that the engineer have at his fingertips both the computer and a program on punch cards or tape for the specific problem, and that he be able to enter his data without delay and receive his results immediately. Such, of course, is not the case. The computer is not at his fingertips, but in another room or possibly another building; the program for his problem, if it exists, is in a card file or tape repository, and must be fed separately to the machine; his data must be prepared in the proper form for entry; and his problem must await its turn in the schedule of a busy computer. Computer time and punch card programs are expensive in comparison with nomographs, many of which are available at little or no charge. For the computation of an extensive table of values based upon a single formula, the computer is economical, efficient, and remarkably fast, but in convenience and ease of operation it does not compete with the graphical device for many of the day-to-day computations of engineering. Indeed, by providing a high-speed method for processing and analyzing the mountains of data generated in today's unprecedented program of research, the computer is constantly spawning new formulas for which graphical methods offer the most satisfactory means of every-day application.

Many scientists and engineers that use graphical devices have little idea of the relative merits and applicability of the various types of devices, and virtually no knowledge of the underlying theory of their construction. Yet, the mathematics of this theory is so simple that mathematics advisers on projects for high school science fairs would do well to consider some of the methods described herein, such as construction of special slide rules, nomographs for formulas of current interest, and three-dimensional nomographs. There is no lack of technical literature in this field, but what is lacking is a systematic approach to the subject as a whole, from the standpoints of both organization and theory.

This book is an outgrowth of very earnest efforts towards unifying my own knowledge in this field. Having made a thorough study of nomographic methods and theory, I nevertheless found myself in poor shape to produce a series of nomographs based on certain polynomials describing the characteristics of flight of helicopters. A modification of existing theory greatly simplified the procedure for representing polynomials; however, it also pointed the way towards the development of a new theory of nomographic representation, the hyperbolic coordinate method from which have grown generalizations and extensions covering the entire field of nomography. At the same time, extension of the idea of the scale equation into the areas of graph papers, graphs, and slide rules has simplified the application of these devices in the field of computation.

Although some topics (e.g., the hyperbolic coordinate method of nomography) are treated in much detail because of the lack of thorough treatment elsewhere, other topics (e.g., graphical integration and differentiation) are recognized as being adequately covered in other sources and are here given only introductory discussions by way of recognizing their family relationships. Much of the material has been expanded from my original publications in this field; however, it will be obvious to the reader that I have also leaned heavily upon the works of such outstanding authors as d'Ocagne, Allcock and Jones, Lipka, and Mavis.

The illustrative examples and the problems that close each chapter represent many diverse disciplines, and they are derived largely from private correspondence with many that have discovered the value of graphical methods in their own fields; to these contributors I am pleased to express my appreciation. I am also deeply grateful to the many that contributed kindly advice and assistance as this work progressed, especially Mrs. Margaret Looper Weill and Mrs. Caren Burrows Yarbrough, who did some of the freehand sketches; Mr. Howard G. Dunlap and Mr. Edward Foster, who read portions of the manuscript; and the Engineering Experiment Station, Georgia Institute of Technology, which contributed generously to the support of the original research as well as to the preparation of the final manuscript.

November 1964 W. HERBERT BURROWS

CONTENTS

chapter 1

SCALES

All graphical methods of computation, indirectly or directly, are based upon scales. This fact makes it imperative that a clear and concise conception of the nature and properties of scales be acquired from the outset. To be sure, readers of this book have been acquainted with scales of one sort or another since infancy, and have used them to measure the dimensions of a room, to read the time of day, and to ascertain the temperature or the velocity of an automobile. But the background of experience thus acquired is of greater value in the application of graphical devices to the solution of problems than in the design of the instruments of computation themselves.

NATURE AND PROPERTIES OF SCALES

1.1 Some Familiar Scales

The familiar scales in the frontispiece display several features characteristic of all scales. Each is a line segment, whether straight or curved, with subdivisions marked by points or dashes along its length. The scale thus marked is said to be graduated; the points or dashes are called graduations. Each scale has a point (which may or may not appear on the segment shown) which is denoted as the origin of the scale. The positions of remaining points on the scale are determined by a functional relationship of the type,

$$s = mf(x) + l_0$$

in which s is the length of the line from the origin to the

said point, m and l_o are constant, and f(x) is some function[1] of the variable x represented by the scale. This relationship is commonly known as the defining equation of the scale.

One scale shown is a foot-rule; divided into inches and sixteenths of inches, but showing the useful half-inch, quarter-inch, and eighth-inch subdivisions by dashes of different lengths. The defining equation of this scale is s = x, which is to say, the length of the segment of the scale is equal to the distance it measures. Here f(x) is x, itself. The value of m is unity, and that of l_o, zero. A later discussion of the meanings of m and l_o should make the reasons for these values clear.

Another scale shown is a thermometer scale, for which the defining equation is

$$s = mt + l_o$$

in which t is the temperature in degrees Fahrenheit, m is a proportionality constant, namely, the increase in length of the mercury thread for each degree rise in the temperature, and l_o is the length of the mercury thread when the temperature is 0°F.

Still another scale is a radio dial, graduated in kilocycles per second, a unit of frequency. Rotation of the dial in a clockwise direction decreases the capacitance of a condenser in the radio circuit, which, in turn, increases the frequency of the tuning circuit. Neither of these relationships is linear (i.e., in direct proportion), for the condenser is shaped in a manner such that equal angles of rotation do not produce equal changes of capacitance; furthermore, frequency is inversely, not directly, proportional to capacitance of the condenser. The nature of f( ) is determined by this inverse relationship and by the shape of the condenser blades, which differs in various makes. Actually, the positions of the dial corresponding to the frequencies shown were probably ob-

1 By "function of x" is meant a quantity whose value is fixed for each fixed value of x. In a pure sense, f(x) may be a constant; i.e., it may have the same value for all values of x. In all of the present cases, f(x) is a variable, taking on different values for the different values of x.

tained by calibrating an identical condenser against standard frequencies. The exact nature of $f(\nu)$ is of no importance, and the scale may be quite satisfactorily defined by the relationship

$$s = mf(\nu) + l_0$$

The frequency is represented by ν and m is a proportionality constant, the increase in length of scale for each increase of unity in the value of $f(\nu)$. Because of the inverse relationship of $f(x)$ to x, l_0 is an infinite length, the length of the scale when $f(x)$ is zero.

1.2 Modulus

In the defining equation of the temperature scale, the term m is identified as the increase in length of the mercury thread corresponding to an increase of 1°F in the temperature, for in this case the "function of the temperature" is the temperature itself. In the defining equation of the scale of a radio dial, the term m is not a length of scale corresponding to an increase of unity in the frequency, for this length varies over the entire length of the scale. It is, instead, the length of scale corresponding to an increase of unity in the value of the "function of the frequency." This term, which is the length of scale corresponding to an increase of unity in the value of the function represented, is known as the modulus of the scale.

The scale is constructed by plotting the positions of its graduations on a reference scale, or reference frame. The reference frame may be an axis of the cartesian coordinate system—another scale, or simply a straight line having on it two points with assigned values. The point on the reference frame having the assigned value of zero is known as the origin. The origin of the reference frame does not necessarily correspond to the zero point of the scale, which is the position of the point $f(x) = 0$ on the reference frame. Figure 1:1 shows three logarithmic scales all having the same modulus (2.5 inches) but differing in the position of its zero point on the reference frame. The zero point of scale A coincides with the origin of the reference frame. The zero point of scale B lies 1.5 inches to the left of the origin; that of scale C lies four inches to the right of the origin of the reference frame. These positions are defined by the values of l_o in the defining equations, which are, respectively, $s = 2.5 \log x$, $s = 2.5 \log x - 2.5$, and $s = 2.5 \log x + 4$.

1.3 Some Useful Types of Scales

Some of the more useful types of scales are shown in Figure 1:2. These scales are extensively used in the con-

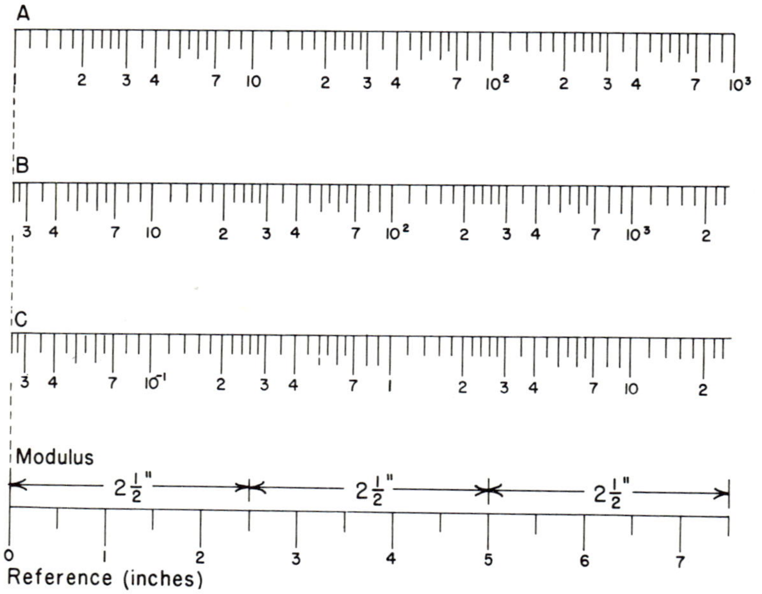

FIG. 1:1 LOGARITHMIC SCALES, MODULUS 2.5 INCHES, WITH DIFFERENT ZERO POINTS
A) $s = 2.5 \log x$; B) $s = 2.5 \log x - 1.5$; C) $s = 2.5 \log x + 4.0$

struction of graph papers, nomographs, and slide rules. The characteristic defining equations of these scales are

Linear: $x = mx + l_o$

Logarithmic: $s = m \log x + l_o$

$$\text{Reciprocal:} \qquad s = m/x + l_o$$

$$\text{Hyperbolic:} \qquad s = \frac{mx}{x + r} + l_o$$

The last of these is shown for the case r = 1.

1.4 Errors in Absolute and Relative Scales

It must be recognized from the beginning that accuracy in the use of a scale depends upon one's locating the exact point

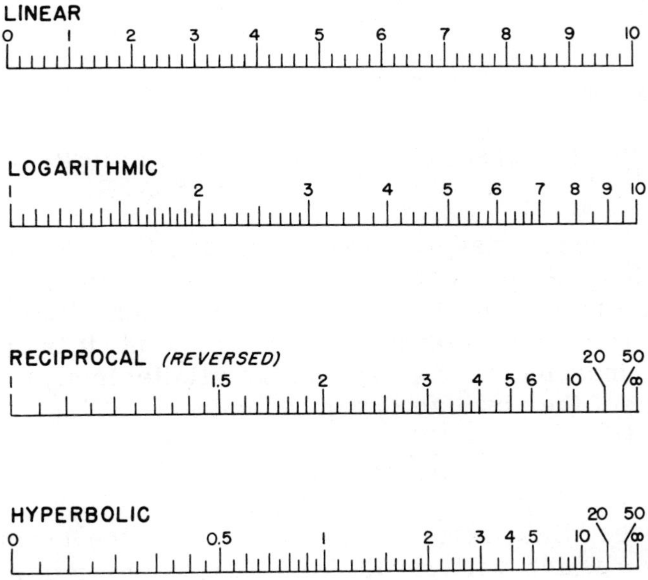

FIG. 1:2 TYPICAL LINEAR, LOGARITHMIC, RECIPROCAL, HYPERBOLIC SCALES

representing the desired value of the variable that the scale represents. As this is virtually impossible, there is always an error, however slight, represented by the difference in the position of the point actually located and the position of the desired point. If the value of x at the desired point is x_1, and that at the point actually located is x_2, then the difference $x_2 - x_1$ is the error in reading the value of x at that point.

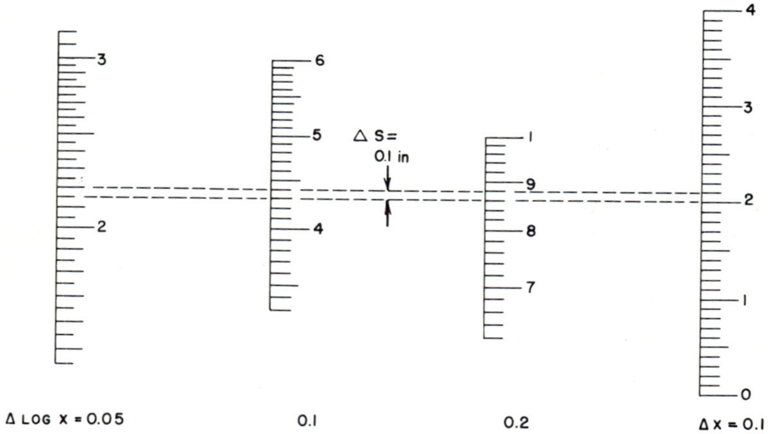

FIG. 1:3 ABSOLUTE AND RELATIVE ERRORS OF LINEAR AND LOGARITHMIC SCALES

This difference may be designated Δx; then $\Delta x = x_2 - x_1$. The absolute error of the scale is the ratio of the error in the value of the reading to the error in position of the point, or $\Delta x/\Delta s$. To confine attention to a single point, it is necessary that the increments Δx and Δs be infinitesimal, whereupon the value of the absolute error becomes

$$\text{Absolute error} = \frac{dx}{ds}$$

In essence, this factor dx/ds determines the magnitude of the error which will result from a given error in locating a scale point.

In financial matters, virtually all computation is performed with the same absolute error; regardless of the size of the budget, it must be accounted to the penny. This is not the case in numerous other fields, wherein errors of measurement are related to the size of the measurement. In such cases, the relative error has greater significance. The relative error is defined as the ratio of the value of the absolute error to the value of the variable being measured, or

$$\text{Relative Error} = \frac{dx/ds}{x}$$

Absolute and relative errors of the linear and logarithmic scales at various positions are shown in Figure 1:3. Δs is taken as 1/10 inch, which is large compared with the infinitesimal, ds; consequently, the estimates given for the values of the errors on the logarithmic scale cannot be exactly equal to the true values obtained from the formulas. They are close enough, however, to show the significance of these errors. It may be noted that the linear scale has the same absolute error throughout its length. That is, the absolute error of the linear scale is independent of the variable represented. This

TABLE 1:1

ERRORS IN SCALES

Type	Defining Equation	Absolute Error	Relative Error
Linear	$s = mx + l_o$	$\frac{1}{m}$	$\frac{1}{mx}$
Logarithmic	$s = m \log x + l_o$	$\frac{x}{m}$	$\frac{1}{m}$
Reciprocal	$s = \frac{m}{x} + l_o$	$\frac{-x^2}{m}$	$\frac{-x}{m}$
Hyperbolic	$s = \frac{mx}{x + r} + l_o$	$\frac{(x + r)^2}{rm}$	$\frac{(x + r)^2}{rmx}$

is not true on the logarithmic scale, where the absolute error increases with the value of the variable; the logarithmic scale has the same relative error throughout its length.

The absolute and relative errors of the scales shown in Figure 1:2 are given in Table 1:1.

1.5 Non-linear Reference Frames

The usual reference frame for the construction of scales for graphical computations is an axis of the cartesian coordinate system; these coordinates are linear. It is occasionally necessary, however, to construct a scale upon some other base, as, for example, an hyperbolic coordinate axis. In these cases, the scales are designated as such by name, using the name of the function followed by the name of the reference

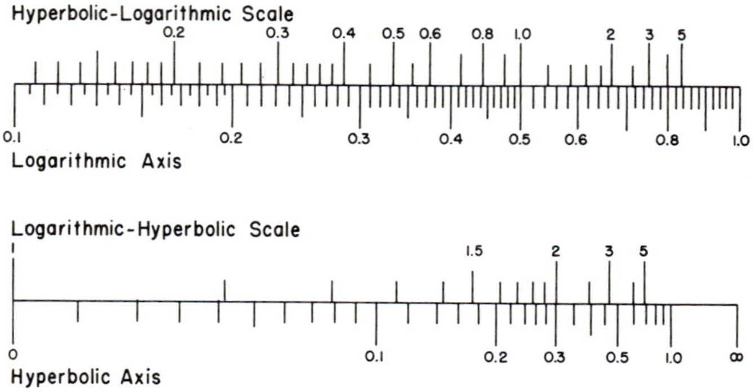

FIG. 1:4 LOGARITHMIC-HYPERBOLIC AND HYPERBOLIC-LOGARITHMIC SCALES

frame. For instance, a logarithmic-hyperbolic scale is a scale whose graduations are computed from a logarithmic function and plotted on an hyperbolic reference frame. Figure 1:4 shows a logarithmic-hyperbolic scale and an hyperbolic-logarithmic scale. Obviously, the two are not identical.

METHODS FOR CONSTRUCTING SCALES

1.6 Construction of a Scale from Data

If the value of s for each value of x that is to be plotted has been computed, or exists in the form of a data table, the scale may be constructed by locating each scale point on the reference frame. Table 1:2 shows the vapor pressures of water over a range of 0° to 100°C, which may be used as a basis for the construction of a scale of vapor pressures as functions of temperature.

As v.p. = f(t), the defining equation of the scale is

$$s_t = m(\text{v.p.})_t - l_o$$

l_o may be taken as zero; the modulus m is selected to accommodate the range desired within the length of the scale:

$$\text{modulus} = \frac{\text{length}}{\text{range of function}}$$

TABLE 1:2

VAPOR PRESSURES OF WATER
(in mm of Mercury)

t, °C	V.P., mm	t, °C	V.P., mm	t, °C	V.P., mm	t, °C	V.P., mm
0	4.58	58	136.08	73	265.7	87	468.7
10	9.21	60	149.38	74	277.2	88	487.1
20	17.54	61	156.43	75	289.1	89	506.1
25	23.76	62	163.77	76	301.4	90	525.8
30	31.82	63	171.38	77	314.1	91	546.1
35	42.18	64	179.31	78	327.3	92	567.0
40	55.32	65	187.54	79	341.0	93	588.6
42	61.50	66	196.09	80	355.1	94	610.9
44	68.26	67	204.96	81	369.7	95	633.9
46	75.65	68	214.17	82	384.9	96	657.6
48	83.71	69	223.73	83	400.6	97	682.1
50	92.51	70	233.7	84	416.8	98	707.3
52	102.09	71	243.9	85	433.6	99	733.2
54	112.51	72	254.6	86	450.9	100	760.0
56	123.80						

In this case the range of v.p. is approximately 800 mm. and the length of scale desired is eight inches. The modulus is then 8 inches/800 mm. = 0.01 inch/mm. It follows that 1 inch = 100 mm. (Fig. 1:5).

The reference frame is a scale divided into inches and tenths of inches, marked with the values of vapor pressure. Points corresponding to the values of vapor pressure for each value of temperature are marked off on this reference frame, and the temperatures to which they correspond are printed on the scale.

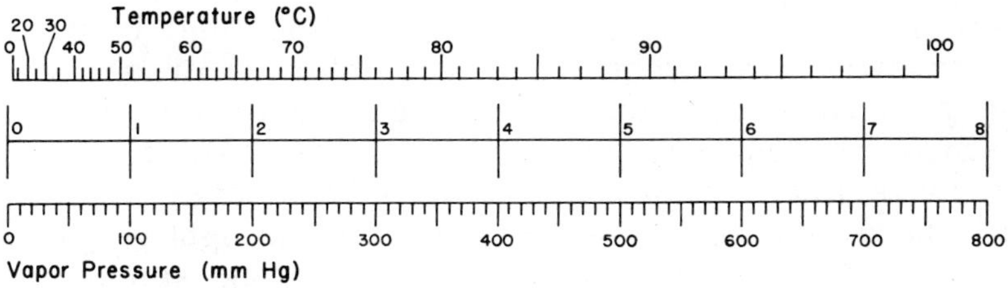

FIG. 1:5 SCALE SHOWING VAPOR PRESSURES OF WATER AS A FUNCTION OF TEMPERATURE, °C

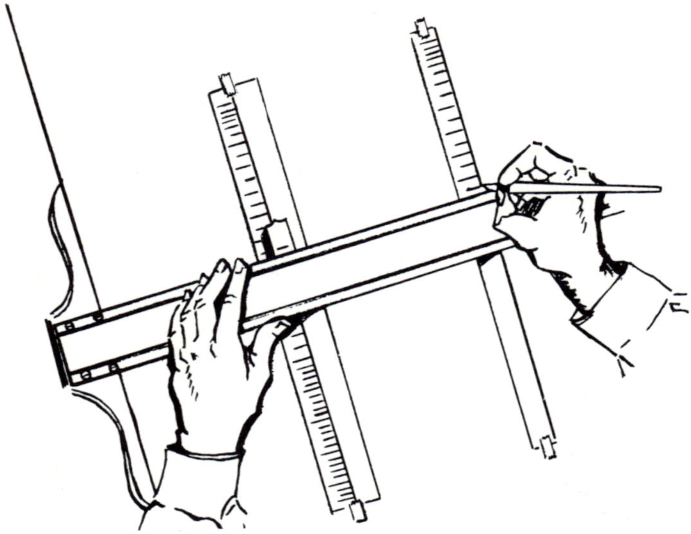

FIG. 1:6 USE OF T-SQUARE IN GRADUATING SCALES

It is not necessary to construct the scale directly on the reference frame, and in many instances it is considerably easier to achieve the desired degree of accuracy by laying off the scale on a line parallel to the reference frame. This is particularly true in the use of a T-square as a guide for the graduating pen. In this case, an index may be attached to the T-square so that it moves along the reference frame as the T-square is moved up and down on the drawing board (Fig. 1:6). A group of parallel lines is lightly ruled with pencil to serve as a guide for the different lengths of graduations to be used. If tracing paper is used, these lines may be ruled on the reverse side, where they may later be erased without damage to the graduations. Each graduation is then made with the index in position against the reference scale. The solid line (base line) running the length of the scale is not added until the graduations are complete and the light pencil lines are erased. Then the numbers designating values of the variable represented are inserted to complete the scale.

1.7 Construction of a Scale from a Graph of the Function

In many instances it is desirable to prepare a scale representing a function available only in the form of a graph, or in an incomplete set of data from which a graph may be prepared. The graph is made on suitable graph paper, having an ordinate whose modulus is equal to that desired for the

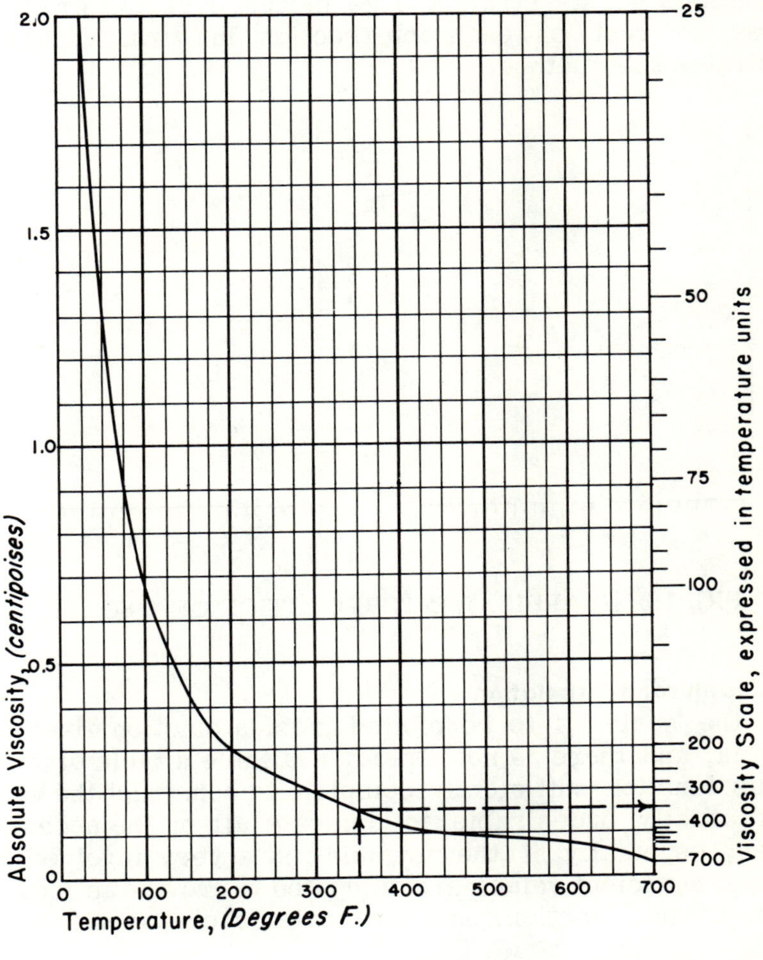

FIG. 1:7 METHOD OF CONSTRUCTING A SCALE FROM A GRAPH

reference frame. The abscissa of the coordinate paper may be linear, or logarithmic, or of any form suitable to the data being plotted. The curve is plotted and drawn in as smoothly and accurately as possible. The points of intersection of this curve with abscissa lines are then projected onto a line parallel to these abscissa lines, and the graduations are made in the same manner as described in the preceding section. The T-square is a useful instrument for making the projections as its straight edge moves over the graph and comes to rest on each intersection in turn. Figure 1:7 illustrates this method.

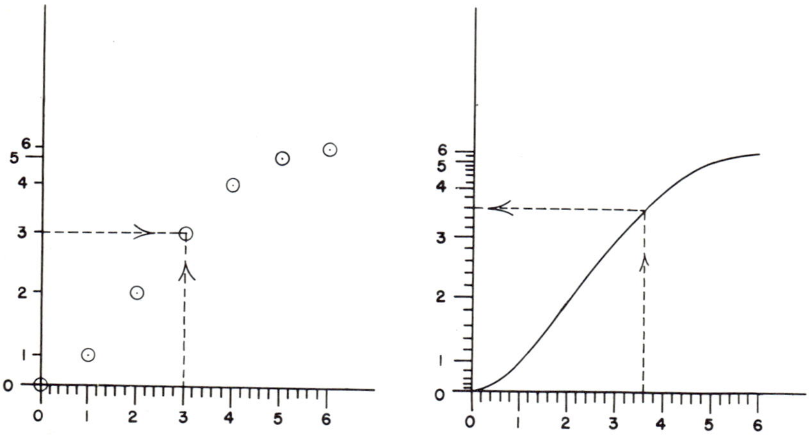

FIG. 1:8 GRAPHICAL INTERPOLATION OF A SCALE

1.8 Graphical Interpolation

If the scale is to be plotted from a function given by a formula, and there is not already available a table of values of this function sufficiently complete to construct the scale, it is usually undesirable to compute all of the necessary values, especially if the formula is a very involved one. Instead, sufficient values are computed to provide an accurate graph of the function, and the scale is prepared from the graph.

In the same manner, it is possible to complete a scale that originally contains only a few of the desired points. The incomplete scale is made the Y axis of a coordinate system, and a suitable X axis is chosen. Points are located horizontally from the graduations on the incomplete scale, and vertically from the equal values of x on the X-axis. A smooth curve is drawn through these points, and the scale is then completed in the same manner as that described for the construction of a scale from a graph. This method, often known as graphical interpolation, is illustrated in Figure 1:8.

1.9 Geometrical Construction of a Scale

In some cases a function can be plotted by means of a geometrical construction. Figure 1:9 shows the construction of scales representing the functions $x = p/(p + 1)$, $x = 1/(p + 1)$ and $x = 1/q$. No general rules can be given for geometrical methods of constructing scales, but a familiarity with plane geometry is an aid in devising methods.

METHODS OF ALTERING MODULI OF SCALES

In calculating the values of scale points from the defining equation, the modulus of the scale desired for a particular application may be included in the computation. If a scale based upon the same function is already available, however, it is unnecessary to make the computations again just to effect the change in modulus. Several methods of altering the modulus of a scale are described in the following sections; the one chosen in a particular case will depend very largely upon the materials available, the urgency of the situation, and the degree of accuracy demanded.

1.10 Projection through a Point

Figure 1:10 shows a method of producing a scale of desired modulus from a similar scale of greater or lesser modulus. A second line is laid off parallel to the original scale. The range of the new scale is marked off on the second line. Straight lines are drawn through the extremes of both, intersecting at a point P. Through this point additional lines are drawn to the points on the original scale. The points of intersection of these lines with the second parallel line determine the positions of the graduations on the new scale. For either

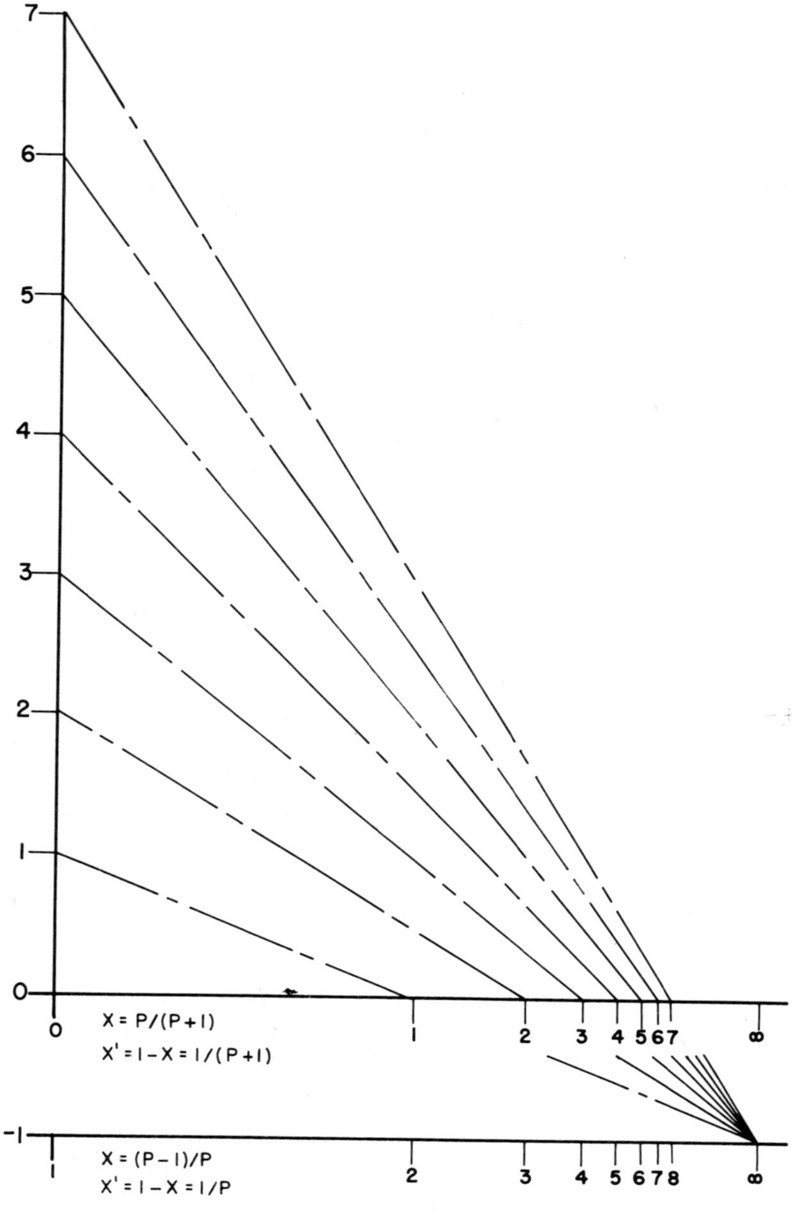

FIG. 1:9 GEOMETRICAL CONSTRUCTION OF SCALES

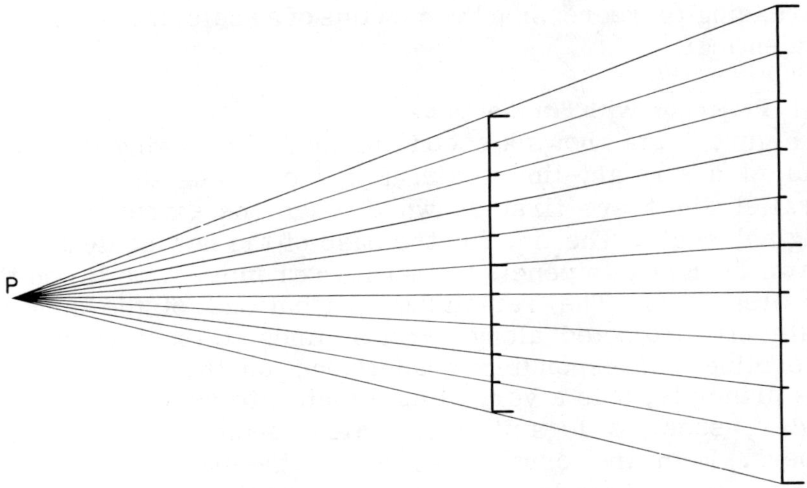

FIG. 1:10 CHANGE OF MODULUS BY PROJECTION THROUGH A POINT

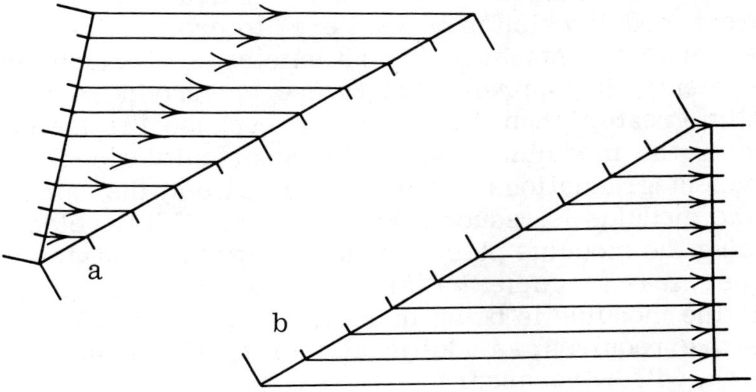

FIG. 1:11 CHANGE OF MODULUS BY PROJECTION WITH PARALLEL LINES

increasing or decreasing the modulus of a scale, the procedure is identical.

1.11 Projection with Parallel Lines

Figure 1:11a shows a second method of changing the modulus of a straight-line scale. For increasing the modulus, parallel lines are first drawn through the extremes of the original scale. The line for the base of the new scale is then drawn in lightly in pencil so that its extreme points lie on the parallel lines. The remaining graduations of the original scale are projected along parallel lines to the new scale to determine corresponding graduations on that scale. Using this principle, it is a very simple matter to construct certain scales, such as logarithmic scales, from common graph papers. Both the original scale and the parallel lines are printed on the graph paper, and it is necessary only to construct a diagonal across these lines in such position as to give the required modulus for the new scale.

A similar method is used for decreasing the modulus, as shown in Figure 1:11b. Here, one has the added advantage that the graduations may be ruled in during the process of subdividing the scale, as it is possible to set the base line of the scale at right angles to the T-square with which the parallel projections are made.

In the event that the original scale is available in several different moduli which may be altered to produce the desired scale (as in the standard logarithmic scales included in the Supplement), it is preferable to use one having a modulus slightly greater than that desired, because the method of reducing the modulus provides the opportunity of making the permanent graduations of the revised scale at the same time that the modulus is reduced. Furthermore, if it develops that reducing the modulus places some of the graduations too close together, it is a simple matter to omit alternate graduations; but if the modulus is being increased and additional subdivisions are required, an additional operation, graphical interpolation, will be necessary.

In any event, it is undesirable to select for the original a scale whose modulus differs greatly from that desired, as the number of changes required in such a case makes the process as laborious as an original construction. Nor should any scale modulus be modified with lines of projection that meet the revised scale at an angle less than approximately 60°, because little accuracy can be attained in the location of the exact intersection of such lines with the base line of the scale.

1.12 Optical Projection

Figure 1:12 shows diagrams of two optical systems for altering scale moduli. The nature of the original scale and of

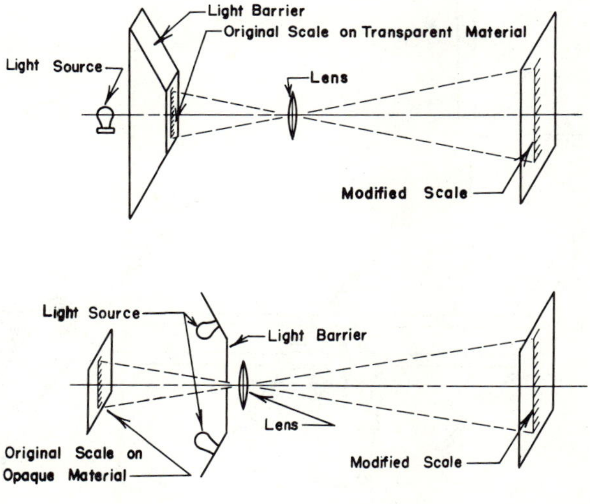

FIG. 1:12 OPTICAL SYSTEMS FOR CHANGING MODULUS

the material on which the revised scale is to be constructed determines to a large measure which of these systems should be employed in a given situation. The actual apparatus may be put together with simple materials from an elementary physics laboratory, or it may be purchased in the form of much more expensive, but more convenient units.

The original scale may be projected onto the drawing board in such size as to give the desired modulus. An index attached to the T-square in such a way as to move along this projected scale may then be employed as a guide in making the graduations on the finished scale, as shown in Figure 1:13.

1.13 Photographic Methods

An alternate procedure employing optical systems is the photographic process. There are several types of materials and methods that may be used. Of these, the most accurate

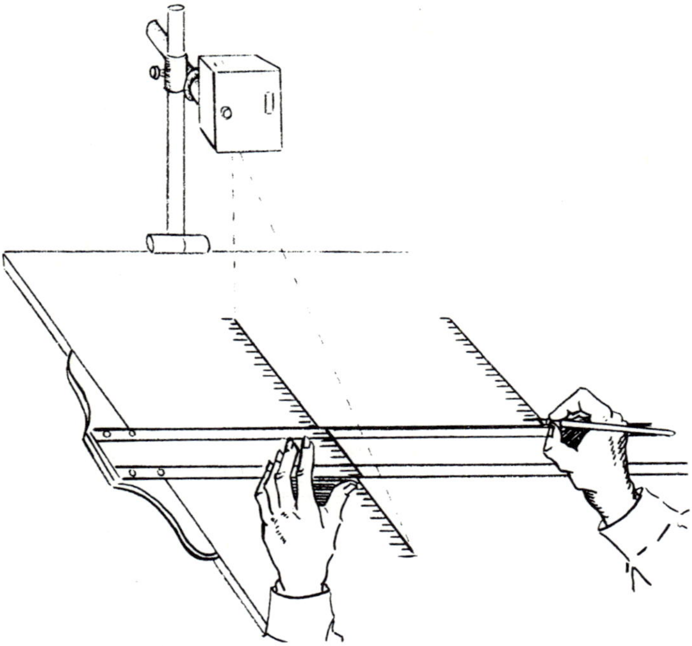

FIG. 1:13 RULING A SCALE FROM ITS OPTICAL PROJECTION

for producing black line prints on white paper is the photo-offset method. The original scale with the desired change in modulus is photographed onto film, then contact printed onto a photo-offset master plate, from which it is printed by lithography. If a rotary offset press is used, with the scales placed lengthwise on the master, there is a slight correction factor due to the "creep" of the impression paper as it passes through the rollers. This correction factor must be determined by experiment, but once it is established for a given machine, it may be applied to all subsequent cases. This process involves materials that are subject to only negligible dimensional changes during developing and fixing operations; hence, it may be used with very accurate results. It is, nevertheless, an expensive process for the reproduction of a few copies of the revised scale. This disadvantage loses its significance on large jobs, however, because a number of changes in scale may be photographed on the same or separate film, and then reproduced on the same master.

The least expensive method is the Photostat process. The size of an original scale may be altered by as much as fifty per cent in each step, and as many steps as required may be used. These steps provide alternate negatives and positives of the original scale, as in ordinary photography, but negatives and positives are all printed on the same material, opaque paper. There is some shrinkage of the paper during developing and fixing of the prints, a shrinkage that is not uniform for the same paper in successive prints or in different directions on the same print. One remedy for these unpredictable dimensional changes lies in multiple prints, enough prints being made at various angles to increase greatly the chances of obtaining one print with the desired modulus. Another remedy is the steam bath; immersing the paper in steam for a few moments relaxes the fibers allowing the print to be stretched enough to correct a scale that turns out to be slightly too short.

A third method is photography. In the first stage, the production of the negative, no appreciable dimensional changes occur; hence, the only drawback is the cost of film of the size usually required (scales sixteen to twenty inches long are frequently used in nomograph masters). This cost is still considerably less than that of the photo-offset method, but the scales appear as clear lines on black film, rather than black lines on white paper. The latter form is more desirable in most applications.

The photographic process may be carried one step further, the production of a contact positive print on photographic printing paper. The result is a scale having black lines on a white background, but the printing paper suffers unpredictable dimensional changes in the baths and driers. The same remedies used for the photostat process may be applied here, or a new plastic printing material which is not subject to dimensional changes, may be used to prepare photopositive prints.

1.14 Mechanical Methods

A type of dividing machine is designed to effect large or small changes in the moduli of scales; Figure 1:14. The original scale is passed under an index by the rotation of a roller, while the paper on which the revised scale is to be ruled passes under a ruling pen on another roller. The two rollers

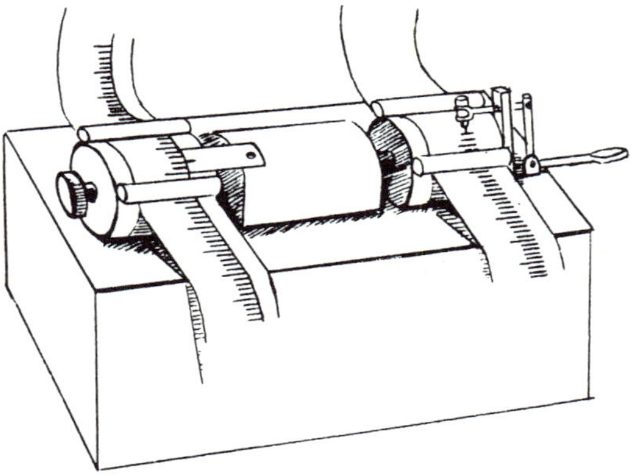

FIG. 1:14 DIVIDING MACHINE FOR ALTERING MODULUS

are connected through a system of gears of which the ratio of their rotations may be varied with great precision over a wide range.

A modification of this machine is employed for ruling scales from data tables. The first roller is connected to a counting device which records numerical equivalents of its rotations. It is rotated in turn to each computed value of the scale points to be plotted, the gear box establishes the modulus of the desired scale, and the scale is ruled as it passes over the second roller under the ruling pen.

STANDARD SCALES

In the construction of nomographs, special slide rules and graph papers, and other graphical devices, it is necessary to construct scales characteristic of functions appearing in individual formulas. Such scales can be used only for the particular device for which they are constructed. In addition to such scales, however, there are certain scales that appear repeatedly in graphical devices, and which should, therefore, be readily available in a wide variety of moduli and characteristic forms. Principal among these are the linear, logarithmic, and hyperbolic scales. An assortment of such scales in suitable use for application to graphical methods constitutes the laboratory's file of standard scales. An adequate file is a tremendous time saver; the method of accumulating and storing it should be devised with careful attention to the fields of application in which it is to be employed.

1.15 Logarithmic Scales

Logarithmic scales should be available in an assortment of moduli capable of accommodating the ratios of moduli frequently encountered in the construction of nomographs and slide rules, and of length characteristic of the size master generally produced in the laboratory to which the file is attached. The author's laboratory employs logarithmic scales, printed by photo-offset method, in sheets like the one in the Supplement of this book. The moduli of these scales are l/n, where l is the length of the scale and n has the values, 1, 3/2, 2, 3, 4, 5, 6, 7, 8, 9, and 10. An auxiliary file is a photographic plate made from this sheet. When additional moduli are required, this plate is projected by optical means to produce the desired modulus from one of these scales. Occasionally, these scales are modified by one of the graphical methods of projection; sometimes the scales of logarithmic graph papers are used in similar fashion.

1.16 Linear Scales

A similar assortment of linear scales is unnecessary; a few standard graph papers with divisions at 1/2, 1/4, 1/8, 1/10, 1/16, and 1/20 inch are sufficient, together with the methods described for altering moduli. Where graphical methods of projection are to be used, standard graph papers are preferable to standard linear scales, because they are more readily adapted to both increasing and diminishing moduli than are single scales.

1.17 Hyperbolic Scales

Hyperbolic scales may all be constructed with a single modulus, the modulus of the hyperbolic scale being identical with its length between the points $p = 0$ and $p = \infty$. However, the function represented is $p/(p + r)$, where r is a constant, and it is necessary to have at hand scales characteristic of the usual values of r. Fortunately, the scale for $r = n$ is identical, as far as graduations are concerned, with the scale for $r = 10^m n$, where m is any integer, positive or negative (see Chapter 10).

The only feature of an hyperbolic scale that is changed by multiplying its r factor by 10^m is the numerical values of its scale points; these must be multiplied by the same factor. In the principal application of hyperbolic scales—the construction of nomographs—the selection of the value of r is subject to several limitations which generally lead to a compromise value. Enough freedom of choice exists to permit the selection of a one-digit number (such as 80, 8, 0.08, etc.) for the value of r. This combination of circumstances makes it unnecessary to keep an extensive file of hyperbolic scales; scales of a single length with r-values of 1, 2, 3, . . . 9 (see Supplement) are quite sufficient. If other lengths are occasionally required, standard methods of projection may be employed.

1.18 Standard File

Scales for other common functions, such as squares, cubes, reciprocals, and the trigonometric functions are not encountered frequently enough in the author's laboratory to warrant inclusion in the file of standard scales. With extensive tables of these functions available, it is satisfactory to construct each from the computed values when the occasion demands it.

Whatever the scales considered standard by a given laboratory, in addition to other forms of storage used, it is strongly recommended that every scale be stored also on photographic plate, so that it may be reproduced at any time and with any modulus by optical projection methods.

SOME MECHANICAL ASPECTS OF SCALE CONSTRUCTION

1.19 Methods of Reproduction

The method of reproduction to be employed determines largely the method of ruling a graphical device. The processes most commonly used are (a) contact, (b) photographic, (c) photo-offset, and (d) offset.

The contact processes most commonly employed are Ozalid and blueprint. Both require an original ruled in black ink on a transparent medium such as tracing paper, tracing cloth, or acetate. Both methods employ a sensitized paper which is exposed to ultraviolet light through the original with which it is in direct contact; hence, a print is produced in the same size as the original. The Ozalid method produces blue or black lines on a white background, whereas the blueprint method produces white lines on a blue background. The Ozalid method is generally preferred today because of the closer resemblance of the print to the original drawing.

Blueprints are developed by immersion in water, after which they are dried on a hot drum. Ozalid prints are developed in moist ammonia fumes. As paper is very sensitive to moisture in any form, there is loss of dimensional stability encountered in both methods.

The photographic processes commonly employed are Photostat and photography. The original may be made on either opaque or transparent material; if on transparent material, however, it should be placed over a white background for reproduction. Both methods produce alternate positive and negative prints of the original. In the Photostat process these prints are all made in the same manner; the original is photographed with a special camera onto a piece of sensitized paper, producing, after developing and fixing, a negative. This negative is, in turn, photographed in the same way to produce a positive print. The dimensions of the original may be reduced or increased at any camera stage of the process. Chapter 7, on NOMOGRAPHY, describes the use of Photostat negative prints, as well as positive prints.

In the ordinary photographic process there is only one camera stage, that in which the original is photographed to produce a negative film or plate. The print may be made by contact exposure of sensitized paper, or by exposure using an enlarging device. The latter is the more economical if large prints are required. By either method, a positive print is obtained from a photographic negative.

The photo-offset method, of which Multilith is one example, involves three stages. The first is a camera stage, in which the original is photographed onto film to produce a negative with any desired changes in dimensions. This negative is then printed by contact exposure onto a photo-offset master plate. Finally, the master plate is mounted in an offset printing press, the run producing copies of the original with black (or any other color, depending upon the ink used) lines on a background the color of the paper employed. This method is popular for publications, for industrial nomographs in which a large number of copies are required, and as a means of reproducing special slide rules. A new photo-offset process by-passes the negative, and produces the printing plate directly from the original.

The offset method involves methods of preparing the original directly onto the offset master plate, using special carbon paper to make the impression. The carbon impression conditions the plate to accept ink for the printing process, consequently an ink impression is obtained only where the rulings are made.

1.20 Ruling and Lettering Pens

The preparation of originals by pen-and-ink methods requires considerable skill in drafting, because the graduations and base lines of the scales should be made with the finest lines possible. The area covered by intersection of two coarse lines can represent an error larger than what should be permissible in the reading of a nomograph or other graphical device. If a ruling pen is used to produce very fine lines, however, it is necessary to work rapidly to keep the flow of ink through the point of the pen at a level that will prevent or minimize clogging. Even so, it is difficult to obtain uniform fine lines, as the ink occasionally forms a crust at the point of the pen which increases the width of the line slightly.

Lettering pens are much less subject to clogging or encrustation, because the ink is exposed to the air only at its top surface. It is possible to use a fine lettering pen for a much longer period without cleaning or renewing it. Even the finest lettering pen, however, does not produce a line so fine as one that can be obtained with a good ruling pen, nor does it deposit so thick a layer of ink.

In view of these limitations, it is generally unsatisfactory to use pen-and-ink originals unless a photographic method of reproduction is planned; this virtually eliminates the Ozalid and blueprint methods of reproduction for work where a high degree of accuracy is desired. If a photographic method of reproduction is to be employed, the original may be ruled at one and one-half or two times the size of the print desired. During the photographic reduction, the lines are reduced to the desired thickness, giving the final print the appearance and degree of accuracy desired.

1.21 Styli

The offset method is one means of producing fine lines on an original without photographic reduction. A fine stylus is used to trace out the lines on the reverse side of a carbon paper which faces onto the offset master plate. The resulting carbon impression is almost as fine as an engraving. If the plate is given a very light inking during the printing process, excellent results are obtained.

The mark of the stylus on the reverse side of the carbon is, however, almost invisible; hence, the work becomes exceedingly tedious unless a dividing machine is used to hold the stylus in proper position and set the length of its stroke. The combination of dividing machine and offset method lends itself particularly well to the construction of a set of standard scales.

A stylus used directly on a piece of exposed film of the type used in photo-offset work removes a thin strip of the black gelatin emulsion. If the work is lighted from behind, all marks made by the stylus are clearly visible as the work progresses. The resulting master consists of an opaque sheet with transparent lines. This master may then be used as the negative for the production of a photo-offset plate. Scale values, legends, etc., are best added directly to the plate with pen and special ink, or with hand-stamping apparatus. Rulings on the film should be made in such a manner that the film can be placed with its emulsion side against the photo-offset master plate during the exposure of the master plate, otherwise a slight blurring of the lines may result from scattering of the light as it passes through the film.

The film may be printed on sensitized paper; if so, it is desirable to engrave scale values and legends in the emulsion with a sharp stylus and a reverse lettering template, so that the print may be made with the emulsion in contact with the sensitized paper.

One may obtain from the dealer in art supplies a type of scratch-board consisting of a black surface on white board. The stylus penetrates the black surface, removing a thin strip, and exposing the white surface underneath. Scale values and legends are engraved in the same manner, or pasted on in the form of cutouts from a previously prepared Photostat negative. The resulting master is then reproduced by the photostat method, giving a positive print, because the master is similar to a negative.

1.22 Distribution of Graduations

The length of stroke used in graduating a scale of a nomograph serves as a means of emphasis. The longer strokes, by emphasizing principal subdivisions of the scale, serve to facilitate the location of desired points. Emphasis, however, is largely a matter of personal taste in nomography as it is in rhetoric, and it may not be precisely dictated by rule. One may, however, quote accepted usage, and the following seems the order of the day:

1. Use as few different lengths as possible; usually, not more than three. If the principal subdivisions have numbers attached, that is emphasis enough, and a

single length may be used for both principal and secondary subdivisions.

2. Do not make even the shortest graduations too short. In a scale that is to be reduced to one-half size, the three common lengths of graduations are 0.2, 0.3 and 0.4 inch. If a fourth length is used, it is 0.1 inch. Upon reduction, this stroke becomes 1/20 inch, which is quite short enough; the longest becomes 1/5 inch. which is not too long for a principal or secondary subdivision. If many parallel scales must be placed close together, these lengths may be reduced proportionately on each scale, as is done on some slide rules.

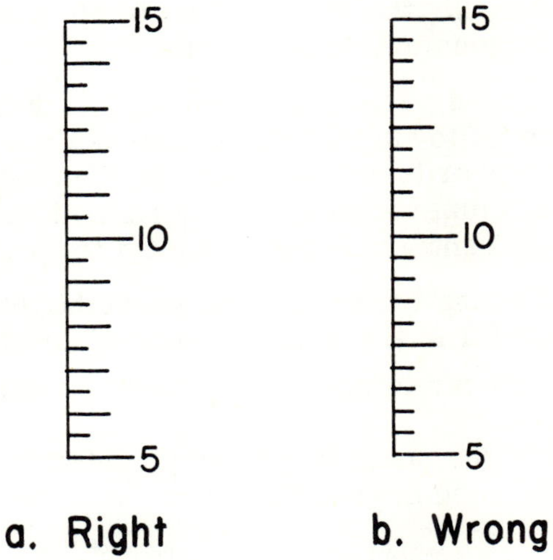

FIG. 1:15 DIVISIONS FOR SCALES

3. Select the subdivisions of the scale to fit the natural subdivisions of the variable. Do not, for instance, divide a scale representing feet into tenths and twentieths, if measurements are made in feet and inches.
4. Do not emphasize the wrong subdivisions. A scale in which the principal subdivisions are 5, 10, 15 and so on, with graduations at every 0.5 should have alter-

nate long and short strokes to show the units and the half units (Fig. 1:15 a), not a long stroke in the middle with the rest of the strokes short (Fig. 1:15 b), as the positions 2.5, 7.5, etc., are not secondary subdivisions.

EXERCISES

1. Write defining equations for each of the following scales: (a) a speedometer scale, (b) a clock dial with the hour hand as indicator, (c) the D scale of a simple slide rule, (d) the T scale of a trigonometric slide rule.

2. Compare the values of m and l_o in the defining equations of each of the following thermometer scales: (a) Fahrenheit thermometer for ordinary weather, (b) Fahrenheit thermometer for oven temperatures, (c) Fahrenheit thermometer for clincial use, (d) centigrade thermometer for ordinary weather.

3. The following table lists values of the temperature base factor, F_{tb}, useful in the measurement of orifice flow,*

a) Using a linear temperature scale, construct a scale for F_{tb}.
b) Complete the scale by graphical interpolation to show a point for each degree F.

TEMPERATURE BASE FACTORS - F_{tb}

These factors are used when the base temperature of the co-efficient is 60 degrees F

Temperature base °F	Factor	Temperature base °F	Factor	Temperature base °F	Factor
45	0.9712	65	1.0096	85	1.0481
50	0.9808	70	1.0192	90	1.0577
55	0.9904	75	1.0288	95	1.0673
60	1.0000	80	1.0385	100	1.0769

* Orifice Meter Constants, Handbook E-2, p. 35; American Meter Company, Inc., (1939). Reproduced by permission of copyright owner.

4. Write the defining equation of (a) a logarithmic-hyperbolic scale and (b) an hyperbolic-logarithmic scale, so that each may be constructed on a linear reference frame.

5. Compare the absolute and relative errors of (a) linear, (b) logarithmic, (c) hyperbolic scales at the points $x = 0.1$, $x = 1.0$, and $x = 10$.

6. Construct a scale from a graph occurring in recent technical literature. The following references may be used.

1) Ketones, pp. 4-13. Union Carbide Corporation, Chemicals Division, 1964.
2) Flow Meter Engineering Handbook, 2nd Ed., pp. 10, 48; Minneapolis-Honeywell Regulator Co., 1946.
3) Engineering Data Book, 6th Ed., pp. 19, 38, 59; Natural Gasoline Supply Men's Association, 1951.

7. Devise means for geometrical construction of each of the following scales: (a) $s = \sqrt{x}$, (b) $s = x^2$, (c) $s = \sin x$.

8. Using the logarithmic scales of the Supplement, prepare logarithmic scales with moduli of 1.5, 3.2 and 7.8 inches. Use projection with parallel lines and projection through a point.

9. Using elementary physics laboratory equipment, assemble apparatus for optical modification of scale moduli. Use this apparatus and the scales of the Supplement to prepare an hyperbolic scale with $r = 6$, $m = 15$ inches.

10. Have one of the plates of the Supplement photographed on standard 4 x 6 inch photographic plate. From this prepare scales with other moduli by (a) optical projection and (b) photographic enlargement.

11. From the scales of the Supplement, prepare a logarithmic scale of modulus 7.5 inches by Photostat reproduction.

12. Prepare a set of standard logarithmic scales by ruling with stylus and carbon on a Multilith master plate.

13. Copy a few of the scales previously prepared (a) on exposed film or (b) on scratch board. Have prints made of these scales.

14. Using a piece of 2-cycle log paper, construct a scale for log n having a modulus of 6 inches. Increase the number of scale points on this scale by graphical interpolation.

chapter 2

CALCULATING WITH SCALES

The position developed in the preceding chapter is that of the scale as a graphical means of representing a function. In the present chapter there are descriptions of several means whereby relationships among functions may be represented by the relative positions of their scales with respect to one another. Such arrangements of scales are generally considered to be computation devices; actually, a more precise understanding of their nature may be obtained by regarding them simply as graphical representations of functional relationships. The proper approach to the problem of associating the scale arrangement with the functional relationship that is to be represented is through the defining equations of the scales involved; in this manner the problem is expressed in general rather than in specific terms.

ADJACENT SCALES

2.1 Construction from Formula

Two scales are constructed on opposite sides of the same base line. One of the scales is defined by $s = m_1 f_1(x) + l_{01}$; the other, by $s = m_2 f_2(y) + l_{02}$. At any point s, then,

$$m_1 f_1(x) + l_{01} = m_2 f_2(y) + l_{02} \tag{2.1}$$

This is the general "defining equation" for a system of adjacent scales. The representation of any given functional relationship by adjacent scales resolves itself simply into a matter of defining the moduli, zero points, and scale functions of the two scales.

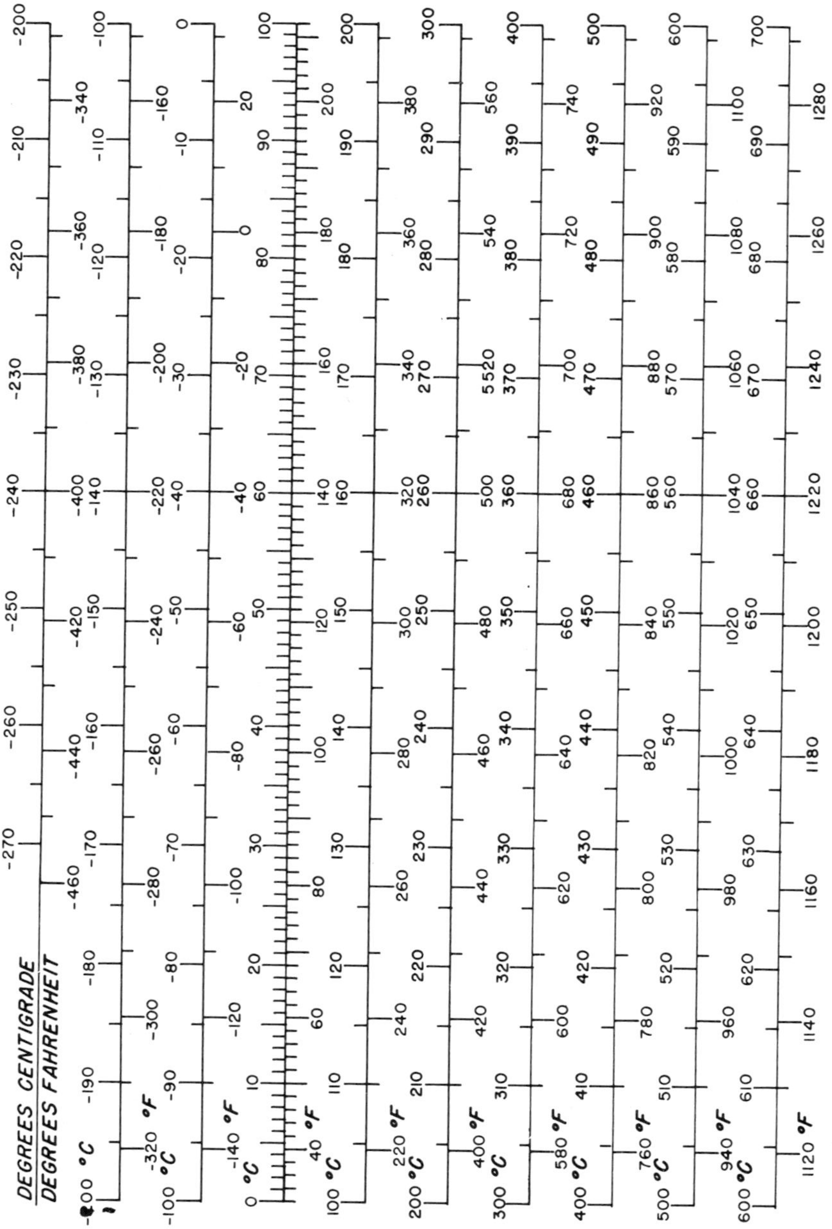

FIG. 2:1 ADJACENT SCALES FOR CENTIGRADE-FAHRENHEIT CONVERSION

Consider, for example, the relationship of the centigrade and Fahrenheit temperature scales, which may be expressed as

$$°F - 32 = 1.8°C.$$

Let the parameters of Equation 2.1 be defined as follows:

$m_1 = 0.1$ inch	$f_1(x) = °F$	$l_{01} = -3.2$ inch
$m_2 = 0.18$ inch	$f_2(y) = °C.$	$l_{02} = 0.$

Then $0.1\,°F - 3.2 = 0.18\,°C$, or $°F - 32 = 1.8°C$, is the relationship expressed by a pair of adjacent scales constructed on the basis of the above definitions of their parameters. The device is illustrated in Figure 2:1.

2.2 Construction from a Graph

In Sec. 1.7 a method is described for constructing a scale from a graph. The resulting scale, together with the reference scale on which it is constructed, forms a scale pair which is a device similar to that described in the preceding section. Had the reference scale been one for Fahrenheit temperature, and the scale constructed from a graph one for centigrade temperature, then the scale pair would have been a device identical with that shown in Figure 2:1.

This method of construction may be applied to functional relationships that are available as formulas, by first constructing a graph of the relationship, and then converting the graph into an adjacent scale arrangement. The principal application of this method, however, is to functions that are available only as graphs. The adjacent scale arrangement shows the functional relationship in exactly equivalent form; however, it is not only more compact, but also much easier to read and interpolate. Figure 2.2 shows a graph that may be readily converted into an adjacent scale arrangement; the device constructed from this graph is in Figure 2.3.

It is plain that neither the reference scale nor the scale constructed from the graph need be defined; the only requirement is that corresponding values of the two variables lie on the same point. This being the case, it is wise to select scales for the two variables that provide a satisfactory dis-

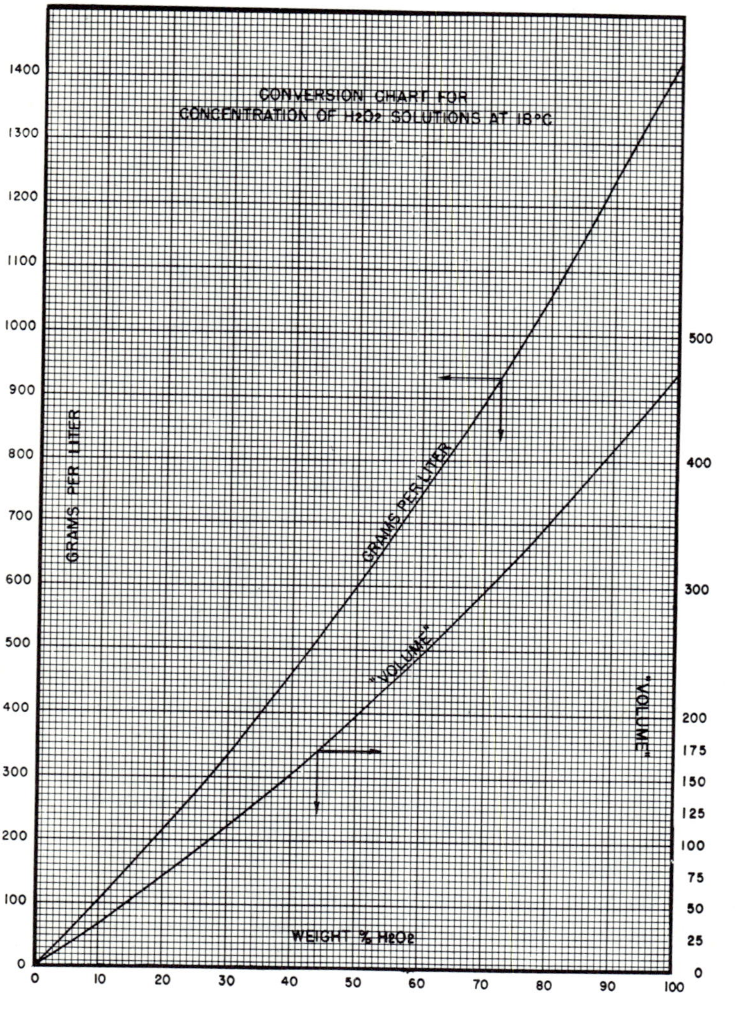

FIG. 2:2 GRAPH SHOWING CONCENTRATION (GRAMS/LITER) OF HYDROGEN PEROXIDE AND VOLUME, AS RELATED TO WEIGHT PERCENT
(From Hydrogen Peroxide, p. 32, Pennsylvania Salt Manufacturing Company, 1951)

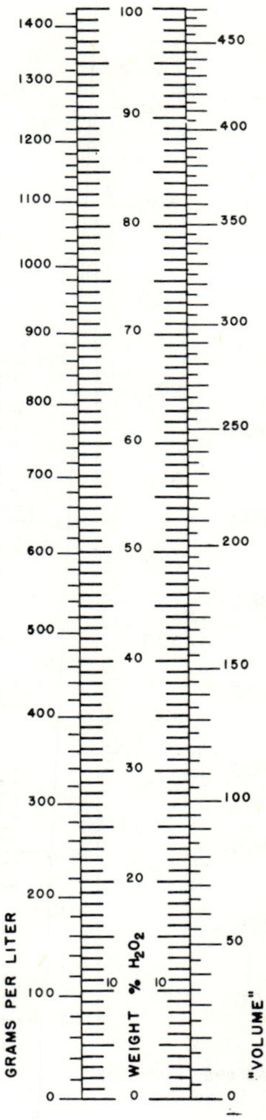

FIG. 2:3 DUAL ADJACENT SCALE SHOWING RELATIONSHIPS OF FIGURE 2:2

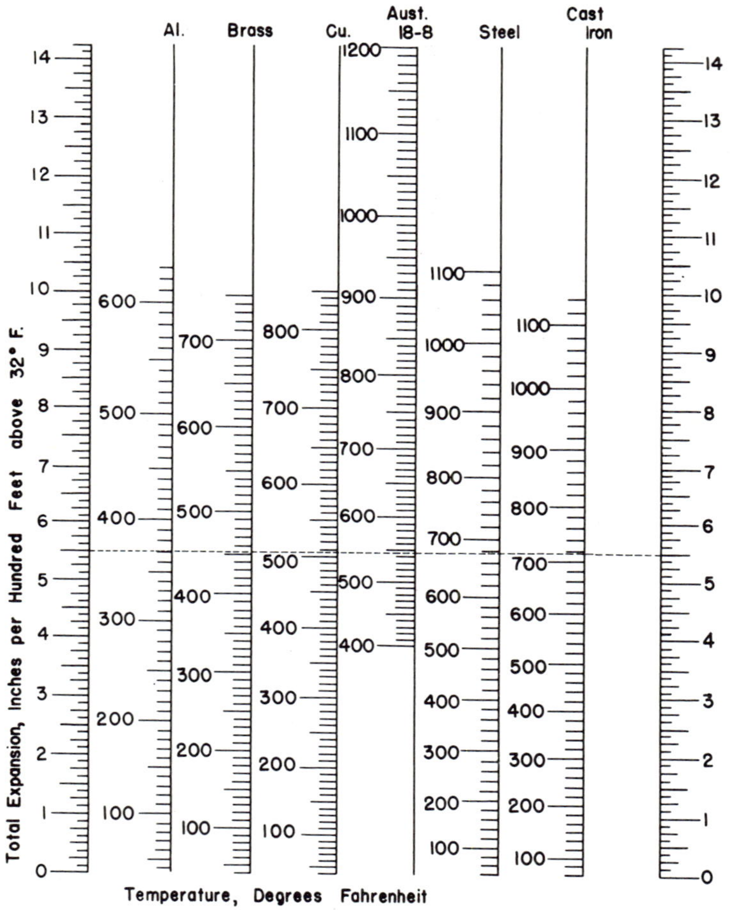

FIG. 2:4 MULTIPLE ADJACENT SCALE

tribution of points, easily legible, with no marked irregularities or discontinuities. Other factors being equal, a standard scale for one of the variables has the advantage of facilitating reading, because the trained eye more readily reads those scales with which it is familiar.

PARALLEL SCALES WITH INDEX LINE

2.3 Representation of Multiple Relationships.

There are numerous examples in the scientific literature of two or more variables that may be expressed as functions of a single variable. Such groups of related functions may be represented by constructing a separate pair of adjacent scales for each functional relationship. This leads, however, to excessive duplication of the scale for the common variable. One method of avoiding this duplication is the type of construction shown in Figure 2.4; the scale for the common variable is shown at each end of the group of scales, whereas scales of the other variables occupy intermediate positions. An index line (a straight line) drawn between equal values of the common variable on the extreme scales intersects each of the other scales in the corresponding values of the variables that they represent.

The reading of scales in this arrangement is greatly facilitated by the use of a T-square set to intersect equal values on the two extreme scales. It is clear that this arrangement is not intended as an aid to computation so much as a source of information. If that information happens to be data that are frequently used in the drafting room, the scales may be mounted on the drawing board, where they are not only readily available but also easily read. This feature of the arrangement is sometimes employed by manufacturers of equipment as a means of keeping both the names of their products and essential data concerning their equipment before the eyes of the engineer.

2.4 Radial Arrangements

Another arrangement with advertising appeal is that shown in Figure 2.5. The scales of the various functions are

a

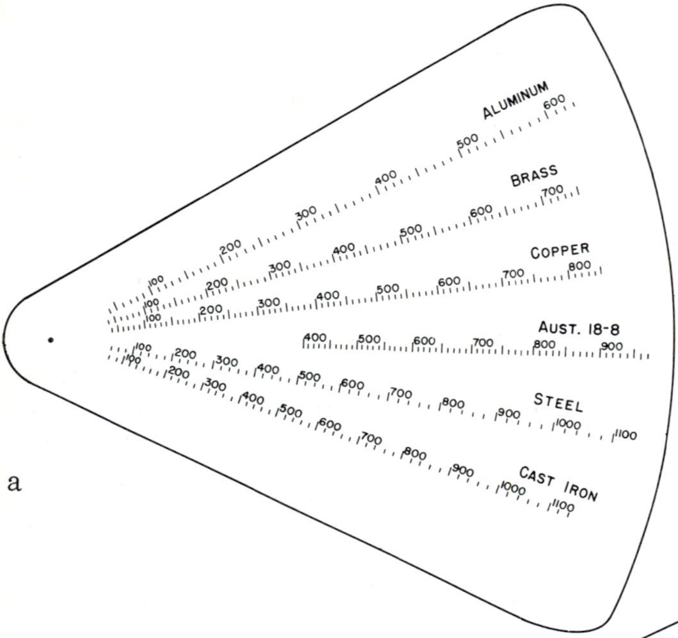

b

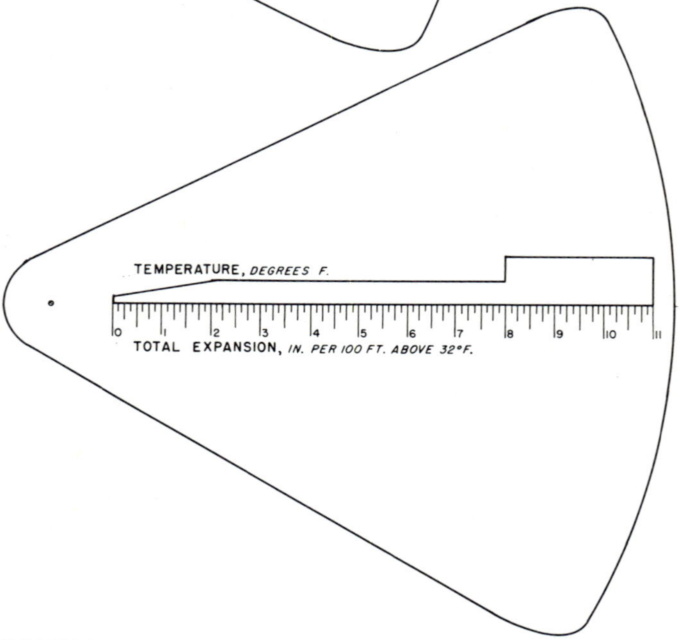

FIG. 2:5 RADIAL DEVICE FOR MULTIPLE ADJACENT SCALES
a) Inner Member; b) Outer Member

arranged on the radii of a disk which is then mounted concentrically under a second somewhat larger disk with a cut-out bearing the scale of the common variable. Rotation of the smaller disk brings each of its scales, in turn, into a position adjacent to the scale of the common variable, forming an adjacent scale pair. The advertising value of the device may be increased by additional cut-outs featuring the names of products, additional information concerning them, etc., as a guide to the use of the device.

STANDARD SLIDE RULES

2.5 Simple Logarithmic Slide Rules

A simple slide rule for multiplication, division, and proportion consists of two logarithmic scales of equal modulus, usually about ten inches, placed adjacent, but capable of being moved longitudinally with respect to each other. Generally, one of the scales, the D-scale, is considered to be stationary; the other, the C-scale, moves. The value of l_{0D} is taken as zero, the origin of the reference frame being the left index of the D-scale (the index of a log scale is a point having the value "1"). The value of l_{0C} is then equal to the displacement of the index of the C-scale from the index of the D-scale. For simplicity, l_{0C} is abbreviated to l in the equations that follow. The equations of the two scales are:

$$s_C = m \log n + l$$
$$s_D = m \log n \tag{2.2}$$

If the index of the C-scale is placed at a on the D-scale, the value of l is m log a. If the cursor is then placed at b on the C-scale, its position from the origin is

$$s_C = m \log b + m \log a = m \log (a \times b).$$

This position with respect to the origin is measured on the stationary D-scale.

In short, the product of a × b is obtained by placing the index of the C-scale on the value a on the D-scale, locating b

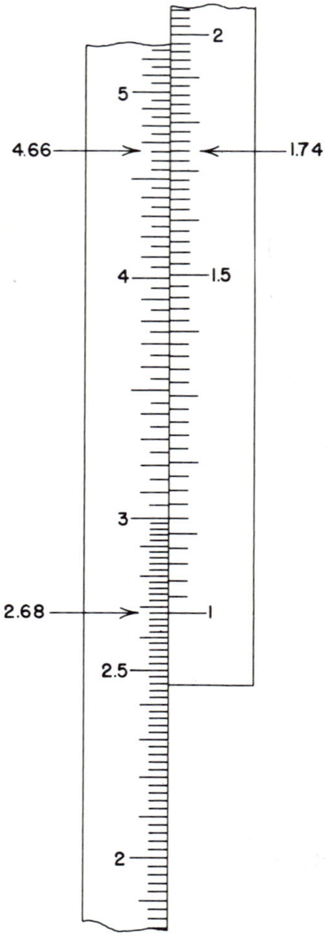

FIG. 2:6 A FORM OF MULTIPLICATION WITH STANDARD SLIDE RULE

on the C-scale, and reading the product at that point on the D-scale. In the example shown (Figure 2:6), the product of 1.74 × 2.68 is 4.66.

The slide rule does not designate the position of the decimal point; it only indicates the sequence of digits in the number.

All numbers having the same sequence of digits, regardless of position of the decimal point, lie at the same position on the C or D scale. Other products that would be obtained by exactly the same procedure as that outlined are .0174 × 268 = 4.66, .174 × 2680 = 466, and so on.

An important factor to remember is this: placing the index of the C-scale on any number on the D-scale produces simultaneously the products of that number multiplied by every number on the C-scale. In repetitive problems, it is frequently necessary to multiply a large group of different numbers by a single common number. A single setting of the C-scale accomplishes this; the actual products desired are obtained by setting the cursor on each of the different numbers (on the C-scale) in turn and reading the products on the D-scale.

Division may be effected by a procedure that is the reverse of that just described; in the example shown, 4.66 (on the D-scale) is divided by 2.68 (on the C-scale), the quotient 1.74 being read on the D-scale at the C index.

A more rapid method of division for repetitive operations is this: place the divisor (on the C-scale) opposite an index of the D-scale. This operation simultaneously divides every number on the C scale by the divisor. The quotients are read on the D scale opposite the dividends on the C scale. The cursor may be used to locate each quotient in turn. An example is the calculation of percentage distribution of items in a total such as the following:

No. Item	P = 56
No. Item	Q = 123
No. Item	R = 27
No. Item	S = 105
Total	= 311

The number 311 (C-scale) is placed at the D index, and the percentages are read (D-scale) opposite 56, 123, 27, and 105 (C-scale). With 311 set at the right D index, the Q, R, and S percentages fall on the D scale; with 311 at the left index, the

P percentage falls on the D scale. This shift is unnecessary, however, if the rule has a folded C-D scale, as the P percentage (DF scale) may be read opposite P (CF scale) with 311 at the right D index.

2.6 Other Logarithmic Slide Rule Scales

The slide rule in common use today includes scales whose designations and defining equations are as follows:

Designation	Defining Equation
A	$s_A = \frac{m}{2} \log n$
B	$s_B = \frac{m}{2} \log n + 1$
C	$s_C = m \log n + 1$
D	$s_D = m \log n$
K	$s_K = \frac{m}{3} \log n$
$\sqrt{\ }$	$s_{\sqrt{\ }} = 2m \log n$
$\sqrt[3]{\ }$	$s_{\sqrt[3]{\ }} = 3m \log n$
S	$s_{sin} = m \log \sin n + 1$
	$s_{cos} = m (1 - \log \cos n) + 1$
T	$s_T = m \log \tan n + 1$
L	$s_L = mn + 1$
CI	$s_{CI} = m (1 - \log n) + 1 = - m \log n + 1$*
DI	$s_{DI} = m (1 - \log n) = - m \log n$

* The factor +m is dropped, because it serves only to designate the left index of the scale as the origin. Actually, the function of the slide rule is identical, regardless of which index is taken as origin.

SLIDE RULE ROUTINES

2.7 Alternative Scale Operations

A number of the scales listed in the preceding section are specific in their application, in that no other scale on the slide rule may be easily substituted in a particular use. Examples are trigonometric scales and the square-root and cube-root scales. It is significant, however, that many of the scales on the slide rule may be used for multiplication and division; the C- and D-scales, the A- and B-scales, the CI- and DI-scales, and the folded scales. The C- and D-scales are the ones most commonly employed for isolated instances of multiplication and division; but in many repetitive slide rule operations it would be a gross waste of time to transfer a factor obtained in a previous operation to the D-scale for multiplication if the multiplication could be performed with the scale on which the factor falls. The following alternate methods for multiplication should be considered in devising routines for computation.

1. The folded scales have the same modulus as the C- and D-scales and are used for multiplication and division in exactly the same manner as the C and D-scales. As the index of the folded scale falls approximately in the middle of the rule, it is a very convenient scale to use in operations requiring that the cursor move past the index. Use of the C-scale in such operations requires an additional time-consuming step of shifting the right index of the C-scale into the position previously occupied by the left index, or vice versa. This step is avoided by use of the folded scale. If the final step in the routine (multiplication or division) is on the CF scale, the result appears on the DF scale; otherwise, it is found on the D-scale.

A further advantage in the use of folded scales lies in the value of the factor on which the scale is folded. Some folded scales are constructed so that the factor π on the DF scale is aligned with the index of the D-scale, and the index of the DF scale is aligned with $1/\pi$ on the D-scale. It follows from Equation (Sec. 2.5) 2.2 that any number on the D-scale may be divided by π simply by placing the cursor on that number and reading the position of the cursor on the DF scale. Multiplication by π is accomplished by placing the cursor on the number on the DF scale and reading the product on the D scale. This device is extensively used in trigonometric and other computations involving π, since introduction of the factor π into the computation is automatic. The slide rule routine used for a given computation must be so designed that, at the proper point in the routine, the computation will move from the D-scale to the DF-scale, or vice versa, in order that the desired division or multiplication by π will take place.

Another folded scale in use at present is folded on the factor 1/M, where $M = \log_{10} e$, and $1/M = \log_e 10$. This scale effects multiplication by 1/M in moving from the D-scale to the DF scale, and multiplication by M in moving in the opposite direction. Computations involving natural and common logarithms are frequently encountered in such fields as thermodynamics. These computations are greatly simplified by the adoption of routines based upon scales folded on 1/M.

2. The inverse scale, CI, may be used for multiplication and division in conjunction with the C- and D-scales; multiplication with the CI-scale is exactly like division with the C-scale, and division with the CI-scale is exactly like multiplication with the C scale.

3. The A- and B-scales may be used for multiplication and division in the same manner as the C- and D-scales; however, as they do not have the same modulus as the C and D scales, they may not be used interchangeably with those scales (see Sec. 2.9).

2.8 Efficiency of Slide Rule Routine

The efficiency of a routine may be measured in terms of the number of settings of the slide n_s and of the cursor n_c, and the number of operations n_o performed in the routine. Let this efficiency be defined as

$$\text{Efficiency} = \frac{n_o}{n_s + n_c}$$

Now consider two alternative methods of computing the product,

$$f = \frac{a \times b \times c}{d \times e}$$

which involves two multiplications and two divisions. The total number of operations is four.

In the first method the series of steps is as follows:

1. Set C-index on a (D-scale)
2. Set cursor on b (C-scale)
3. Set C-index on cursor
4. Set cursor on c (C-scale)
5. Set d (C-scale) on cursor
6. Set cursor on C-index
7. Set e (C-scale) on cursor, read P (D-scale) at C-index

The number of settings is 7; the efficiency is 4/7 or 57 per cent.

In the second method, the steps are:

1. Set cursor on a (D-scale)
2. Set d (C-scale) on cursor
3. Set cursor on b (C-scale)
4. Set e (C-scale) on cursor, read f (D-scale) at c (C-scale).

The number of settings is four; the efficiency is 4/4 or 100 per cent. Actually, the first setting of the cursor may be omitted, but it is included as an aid in setting d (C-scale) on a (D-scale).

Increases of 40 per cent or more in efficiency, may frequently be encountered, and they should be conscientiously sought. It is true that, in the example given, the first method is seldom used, even by beginners. Equal increases in efficiency, however, may often be realized in the methods employed by experienced persons for the solution of more advanced problems. An analysis of the problem, using the equations of the scales, may frequently be relied upon to reveal a saving of one or more steps in the solution.

The most common method of obtaining a multiple product, such as $f = a \times b \times c \times d \times e$, involves $2n_o - 1$ settings:

1. Set C-index on a (D-scale)
2. Set cursor on b (C-scale)
3. Set C-index on cursor, etc.

By use of the CI-scale, the number of settings may be reduced to n_o (if n_o is odd) or to $n_o + 1$ (if n_o is even). Thus, in the example given, the number of settings is four:

1. Set cursor on a (D-scale)
2. Set b (CI-scale) on cursor
3. Set cursor on a (C-scale)
4. Set d (CI-scale) on cursor, read f (D-scale) at e (C-scale).

Consider also the reciprocal of the product: $f = \frac{1}{(a \times b \times c \times \ldots)}$. If performed as a series of divisions, starting with the index of the D-scale and using the C-scale and C-index with each successive division, the efficiency is reduced by the large number of settings required. The same result may be obtained with an efficiency equaling or approaching 100 per cent with this procedure:

1. Obtain the product $a \times b \times c \times \ldots$ as previously described, using the C- and CI-scales alternately; but
2. Read the result on the DI-scale, rather than on the D-scale.

2.9 Choice of Slide Rule for a Routine

The various makes of slide rules do not have identical scales, nor are the scales arranged on the rule in the same manner. The nature and arrangement of scales on a rule has much to do with the adaptability of the rule to a particular routine. In some cases, the routine devised for a certain problem on one rule is as efficient as that devised for the same problem on another rule, although the two routines differ markedly. In other cases, the routine as worked out for one rule has great advantages over that for others, in which case the rule offering the greater advantage should certainly be chosen.

Consider, for example, the formula for obtaining an angle of an obtuse triangle when given the lengths of its three sides, a, b, and c. If s is the sum of these three sides, and A is the angle opposite the side a, then

$$\text{Sin}\,\frac{A}{2} = \sqrt{\frac{(s - b)(s - c)}{bc}}$$

The values of s, (s - b), and (s - c) are obtained before using the slide rule.

If the rule is equipped with the scales A, B, and S (S scale located on the slide), the routine generally followed is this:

1. Set cursor on (s - b) (A-scale)
2. Set b (B-scale) on cursor
3. Set cursor on (s - c) (A-scale)
4. Set c (B-scale) on cursor
5. Set cursor on B-index
6. Set B-index on A-index, read A/2 on S scale at cursor

Step 6 is necessary only because the S-scale lies on the slide; if it were stationary, A/2 would be read at the position of the cursor in step 5.

Step 5 may be eliminated by commencing the computation with a division (setting of the slide) rather than a multiplication (setting of the cursor), whereupon the cursor reaches its final position in step 4:

1. Set b (B-scale) at A-index
2. Set cursor on (s - b) (B scale)
3. Set c (B-scale) at cursor
4. Set cursor at (s - c) (B-scale)
5. Set B index on A-index, read A/2 on S scale

The number of operations is five (one multiplication, two divisions, one square root, one sine); hence the rule is used at 100 per cent efficiency in this routine.

The square root enters the picture at the point at which the routine leaves the AB scale (modulus 5 inch) and enters the S-scale (modulus 10 inch). The accuracy of the entire routine is limited to that of the A- and B- scales which, being of 5-inch modulus, have only half the accuracy of the C-, D-, and S-scales. In an effort to increase the accuracy of operations involving square roots, some slide rules have a $\sqrt{\ }$ scale, the modulus of which is twenty inches. The accuracy of the routine on this is limited by the C-, D-, and S-scales, rather than the $\sqrt{\ }$ scale. The increased accuracy is compensated for by a decrease in efficiency, as the steps in the routine for this problem are:

1. Set b (C-scale) at D-index
2. Set cursor on (s - b) (C-scale)
3. Set c (C-scale) at cursor
4. Set cursor at (s - c) (C-scale)
5. Read square root ($\sqrt{\ }$ scale)
6. Set cursor on square root (C-scale), read A/2 on S scale

The efficiency is 83 per cent.

It must not be construed that one rule is more efficient than another on all computations involving square roots; other problems may be cited in which the rule with the $\sqrt{\ }$ scale is the more efficient. The point is simply that, for a particular problem, one rule may have a higher efficiency than another, and this one should be chosen for repetitive solutions of that problem.

Another example of this difference in rules appears in solutions of the formula for obtaining the value of the work done in the isothermal expansion of an ideal gas.[1]

$$W = RT \ln \frac{P_A}{P_B},$$

where ln indicates the logarithm with base e (natural logarithm), or

[1] Perry, John H., Chemical Engineers' Handbook, 2nd Ed., p. 644. McGraw-Hill Book Company, New York, 1941.

$$W = 2.303\ RT \log \frac{P_A}{P_B}$$

where log indicates the logarithm with base 10 (common logarithm). P_A, P_B, T, and W are the variables. The constant R may be multiplied by the constant 2.303 to obtain a single constant, if the second formula is to be used in repetitive computations.

One rule has a stationary log-log scale (LL scale) and, on the same face a CF- and DF-scale folded so that the value 2.303 on this scale lies at the index of the C- and D-scales. The cursor set on a number on the LL-scale crosses the DF scale at the value of the natural logarithm of that number and the D scale at the value of its common logarithm. The moduli of the LL-, C-, D-, CF-, and DF-scales are all 10 inches.

A second rule has a stationary LL-scale with a modulus of 5 inches, so that the cursor indicates on the A-scale the natural logarithm of a number on the LL-scale.

A third rule has on its slide a logarithm scale (L-scale) with 10 inch modulus, which provides values of the common logarithms of numbers on the C-scale.

The routine for this problem, using the first rule, is:

1. Set P_B (C-scale) at D-index, read P_A/P_B (D-scale) at P_A (C-scale)
2. Set cursor at P_A/P_B (LL-scale)
3. Set R (CI-scale) at cursor, read W (DF-scale) at T (CF-scale), or
3. Set 2.303R (CI-scale) at cursor, read W (D-scale) at T (C-scale).

There are four operations; hence the efficiency is 4/3 or 133 per cent. The accuracy is that of a ten inch scale.

If the second rule is used, the routine is:

1,2. Same as preceding
3. Set index (B-scale) at cursor
4. Set cursor at R (B-scale)

5. Set index (B-scale) at cursor, read W (A-scale) at T (B-scale).

The efficiency is 4/5, or 80 per cent; a BI-scale would have saved one multiplication setting, increasing the efficiency to 100 per cent. The accuracy is that of a 5-inch rule.

With the third rule, the routine is:

1. Set P_B (C-scale) at D-index
2. Set cursor at P_A (C-scale)
3. Set C-index at D-index, read log (P_A/P_B) (L-scale) at cursor
4. Set cursor at log (P_A/P_B) (D-scale)
5. Set 2.303R (CI-scale) at cursor, read W (D-scale) at T (C-scale).

The efficiency of this routine is also 80 per cent, but the accuracy is that of a 10-inch rule.

2.10 Analysis of Routine

Thorough familiarity with the scale on a slide rule and experience in setting up routines are the requirements for devising rapidly the most efficient routine for a given problem. An analysis of the routine from the standpoint of the scale equations is strongly advised, however, not only as a check on the validity of the routine, but also as a means of eliminating excessive steps, or increasing the accuracy of the computation by a substitution of alternate scales.

Analysis of the foregoing problem as performed on the first rule is based on the equations of the scales involved, which are:

D-scale: $s_D = m \log n$ (a)

C-scale: $s_C = m \log n + 1$ (b)

CI-scale: $s_{CI} = -m \log n + 1$ (c)

LL-scale: $s_{LL} = m \log \log n.$ (d)

Step 1:

With P_B (C-scale) at D-index, $s_C = m$ and $n = P_B$; whence, by Equation b,

$$1 = -m \log P_B$$

The position of any other number n on the C-scale is now

$$s_C = m \log n - m \log P_B$$

The scale is read at $n = P_A$, giving

$$s_C = m \log P_A - m \log P_B$$

or

$$s_D = s_C = m \log P_A/P_B,$$

Step 2:

The cursor placed on P_A/P_B (LL-scale) gives

$$s_D = s_{LL} = m \log \log P_A/P_B.$$

Step 3:

With 2.303 R (CI scale) on cursor, $s_{CI} = m \log \log P_A/P_B$ and $n = 2.303R$,
whence

$$1 = m \log \log P_A/P_B + m \log(2.303R)$$

$$= m \log (2.303R \log P_A/P_B).$$

At the point which is read with the slide in this position, n = T

$$\therefore s_C = m \log T + m \log (2.303R \log P_A/P_B)$$

or

$$s_C = m \log (2.303RT \log P_A/P_B)$$

but

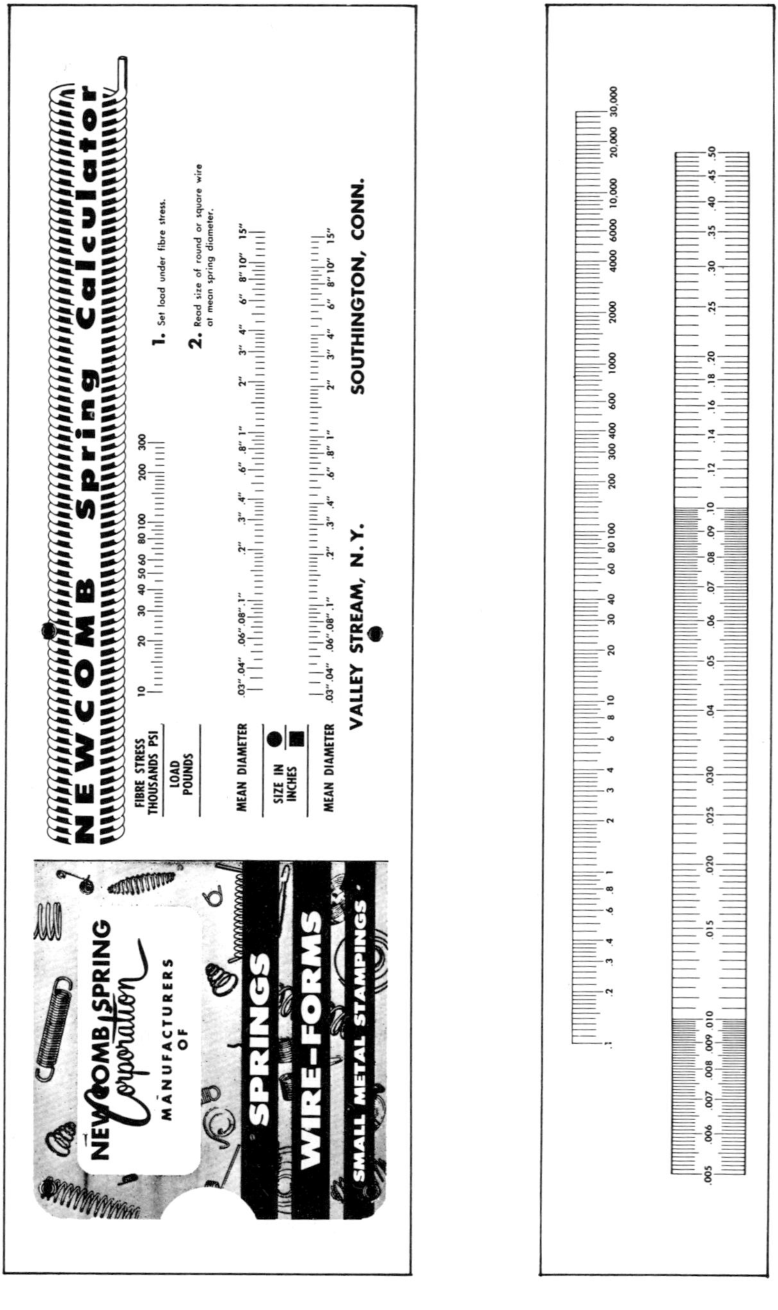

FIG. 2:7 A SIMPLE FORM OF SPECIAL SLIDE RULE
(Courtesy of Newcomb Spring Corporation)

$$s_D = s_C$$

$\therefore$ $W = 2.303RT \log P_A/P_B$ is the formula computed by this routine.

SPECIAL SLIDE RULES

2.11 Defining Equation

As its name implies, the "special" slide rule is designed to provide solutions of a single formula or group of formulas, not to serve as a general instrument of computation for a wide variety of formulas. All of the scales on the common logarithmic slide rule have wide applicability; the special slide rule has scales representing functions limited to one or a few formulas.

The simplest form of slide rule to prepare has a single slide with a scale on each edge operating against adjacent scales on the stationary part of the rule; see Figure 2:7. It is not necessary that the scales of either part have a common zero point, as is the case in the common slide rule; instead, each may be defined by the general scale equation.

Let the scales be designated as A, B, C, and D in the order in which they appear on the rule; A and D are on the stationary piece, B and C, on the slide. Considering the left end of the slide rule as the origin of the reference frame, the equations of the A and D scales are

$$s_A = m_1 f_1(A) + l_{01}, \tag{2.3}$$

and

$$s_D = m_4 f_4(D) + l_{04}. \tag{2.4}$$

The equations of the B and C scales, including the displacement L of the slide, are

$$s_B = m_2 f_2(B) + l_{02} + L \tag{2.5}$$

and

$$s_C = m_3 f_3(C) + l_{03} + L \tag{2.6}$$

For any setting of the slide, $s_A = s_B$ and $s_C = s_D$. Eliminating these terms and L from Eqs. 2.3 through 2.6 leaves

$$m_1 f_1(A) - m_2 f_2(B) + l_{01} - l_{02} = m_4 f_4(D) - m_3 f_3(C) + l_{04} - l_{03} \tag{2.7}$$

or

$$m_1 f_1(A) - m_2 f_2(B) + m_3 f_3(C) - m_4 f_4(D) + K = 0 \tag{2.8}$$

where

$$K = l_{01} - l_{02} + l_{03} - l_{04}.$$

Equation 2.8 is the defining equation of this type of slide rule.

2.12 Construction from a Formula

The defining equation of this slide rule (Eq. 2.8) may be adapted to numerous engineering formulas; an example is the formula for Supercompressibility Factor, f_{pv}, of gases.[2]

$$F_{pv} = \sqrt{\frac{1 + P_{fg} \times 9.16 \times 10^5 \times 10^{1.188G}}{T_f^{3.825}}}$$

where P_{fg} = Pressure, psi gauge

[2] Engineering Data Book, 6th ed., p. 16; Natural Gasoline Supply Men's Association, Mayo Hotel, Tulsa, Oklahoma, 1951.

G = Specific Gravity (Air = 1.00)

T_f = Temperature, °Abs F = 460 + °F

The formula may be rearranged by squaring both sides, substracting 1 from both sides, and taking logarithms. The resulting formula is

$$\log\left(F_{pv}^2 - 1\right) + \log T_f^{3.825} - \log P_{fg} - 1.188G - \log(9.16 \times 10^5) = 0.$$

There being, aside from the constant, two positive terms and two negative terms in this expression—just as in Equation 2.8—the two formulas may be matched up positive term for positive, negative term for negative, with the result that all four scales increase in value from left to right. Otherwise, one or more of the scales would have increased in value from right to left, a slightly less desirable form of scale. The constant, $-\log(9.16 \times 10^5)$ may be combined with any one of the other terms in the formula, divided among several terms, or incorporated in the K of Equation 2.8. In this instance it is combined with the T_f term. K for Equation 2.8 is set equal to zero.

The ranges of the variables and the functions appearing in the scale equations are these:

Scale	Variable	Range	Function	Range
Scale A	F_{pv}	1.060 to 1.005	$\log\left(F_{pv}^2 - 1\right)$	-0.91 to -2.00
Scale B	P	300 to 20	$\log P$	2.477 to 1.301
Scale C	T_f	150 + 460* to 50 + 460	$3.825 \log T_f - \log(9.16 \times 10^5)$	4.52 to 4.38
Scale D	G	1.00 to .75	$1.188\,G$	1.188 to 0.89

* The factor 460 is added to convert degrees Fahrenheit to Absolute. Graduations of the scale, however, will remain in degrees Fahrenheit.

The values of l_{01}, l_{02}, l_{03}, and l_{04} are selected to satisfy arbitrary demands as to the arrangement of scales on the rule. The values best satisfying these demands may be estimated from a sketch showing the positions of the ranges of the above functions on a common scale; see Figure 2.8. This

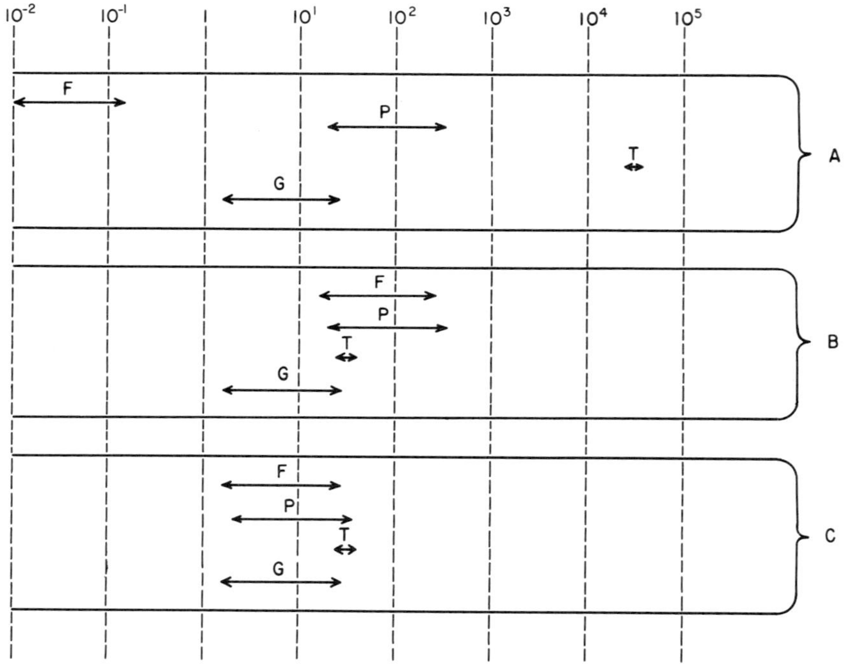

FIG. 2:8 SKETCHES LEADING TO ARRANGEMENT OF SCALES ON A SPECIAL SLIDE RULE

is equivalent to plotting the scale equations, taking $l_{01} = l_{02} = l_{03} = l_{04} = 0$. In this form, the slide rule lacks compactness; two shifts, each involving a pair of scales, may be effected to correct this. First, shift the F scale 3.2 m units to the right (M being the modulus of the rule) and the T scale 3.2 m units to the left; that is, let $l_{01} = 3.2$ m and $l_{03} = -3.2$m. The result is shown in Figure 2:8B. Second, shift the F and P scales m units to the left; that is, set $l_{01} = 2.2$ m, $l_{02} = -$m, $l_{03} = -3.2$m, and $l_{04} = 0$. These final values must be checked against the expression for $K = l_{01} - l_{02} + l_{03} - l_{04}$. In the present case, $K = 0 = 2.2 + 1.0 - 3.2 - 0$. The final positions of the scales are shown in Figure 2.8C; the equations of these scales are

$$\text{A scale: } s_A = m\left[\log (F_{pv}^2 - 1) + 2.2\right],$$

$$\text{B scale: } s_B = m(\log P - 1),$$

$$\text{C scale: } s_C = m\left[3.825 \log T_f - \log(9.16 \times 10^5) - 3.2\right],$$

$$\text{D scale: } s_D = m(1.188\ G).$$

The value selected for the modulus depends upon the degree of accuracy required and the length of the rule. In this example the value of m chosen is five inches. The rule thus constructed is shown in Figure 2:9; one of the sets of values satisfied by this position of the rule is G = .88, T = 70°F, P = 90, F = 1.017.

2.13 Construction of Slide Rule from Nomograph

Some formulas may not be readily manipulated to conform to the defining equation of the slide rule. If, however, a nomograph consisting of parallel scales can be constructed, by empirical or other means, to represent the relationship expressed in the formula, a slide rule may then be constructed by adaptation of the nomographic scales. To make the most of this, it is necessary to have a thorough understanding of nomographic methods, but a simple example serves to illustrate the relationship of the two devices.

A method is shown in Chapter 15 for the construction of a parallel line nomograph representing the formula for the Logarithmic Mean Temperature Difference, (MTD),

$$\text{MTD} = \frac{\Delta T_2 - \Delta T_1}{\ln \dfrac{\Delta T_2}{\Delta T_1}}$$

where ΔT_1 and ΔT_2 are the temperature changes experienced by two liquids as they pass through a heat exchanger. This

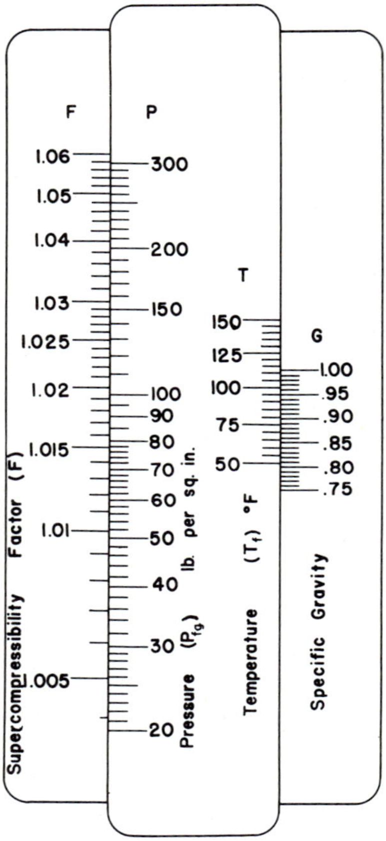

FIG. 2:9 SLIDE RULE FOR SUPERCOMPRESSIBILITY FACTOR OF GASES

formula may not be readily converted into the form of Equation 2.8; however, the nomograph constructed by empirical methods provides values of MTD corresponding to the two temperature differences. From the defining equation of nomographs of this type, one may write

$$2f_1\ (\mathrm{MTD}) - f_1\ (\Delta T_1) - f_1\ (\Delta T_2) = 0$$

in which the f_1's are identical, as indicated by the scales of the nomograph. This relationship may be adapted to the defining equation of the slide rule (Eq. 2.8) by defining the scales of the slide rule as follows:

A-scale, $s_A = mf_1 (\Delta T_1)$

B-scale, $s_B = 0$

C-scale, $s_C = mf_1 (\Delta T_2)$

D-scale, $s_D = 2mf_1$ (MTD).

The scales may be constructed by copying directly from the nomograph, the modulus of the MTD scale being twice that of the ΔT_1 and ΔT_2 scales. The resulting slide rule is shown in Figure 2:10.

METHODS OF REPRODUCTION

2.14 Slide Rule Blanks

The type of slide rule described in the preceding section is desirable in engineering practice because of the ease with which it may be reproduced, and because no cursor need be provided with the rule. Certain manufacturers of engineering supplies list slide rule blanks consisting of a stationary stock and a slide similar to that used in the examples given here. If such a blank is to be used, a Photostat of the blank is first prepared at 200 to 250 per cent size. Using this Photostat as

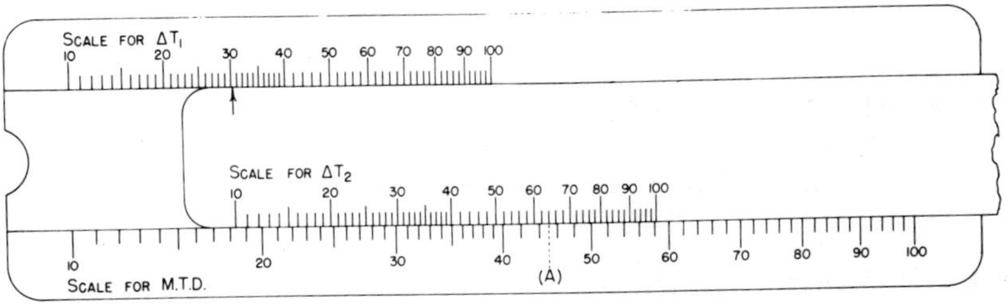

FIG. 2:10 SLIDE RULE FOR LOGARITHMIC MEAN TEMPERATURE DIFFERENCE

a guide to the positions and spacing of the scales, the scales are constructed in suitable form for photographing. The sizes of the graduations, the numbers, and the scale designations are selected so that, when reduced, they will be the size desired for the finished rule. Advertising material may be added to the rule, together with complete instructions for

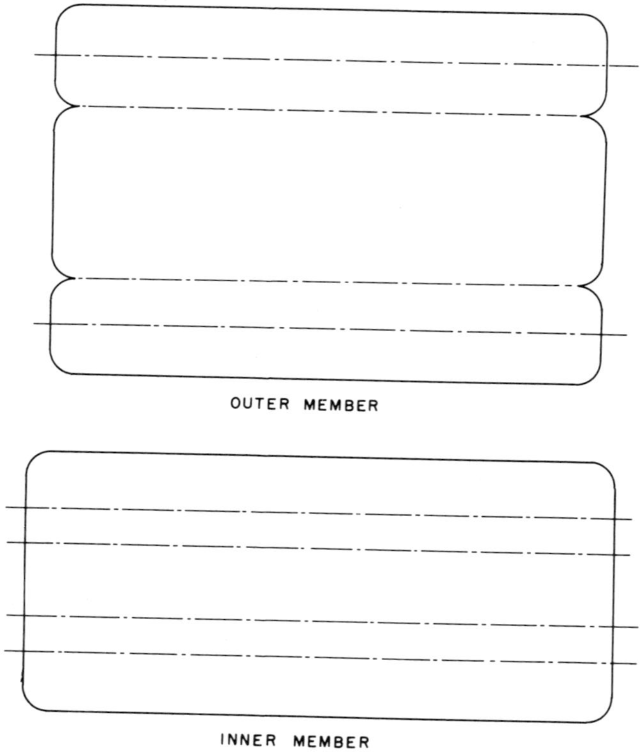

FIG. 2:11 MASTER FORM FOR A FOLDED PAPERBOARD RULE
(Dotted lines indicate positions of folds)

use of it. The finished master is then photographed and printed at a reduction that will exactly fit the blank.

If many copies are desired, a photo-offset method is recommended; however, the printing should be done with a light inking to avoid increasing the thickness of the lines. If only a few copies are required, the photographic reduction may be printed on photographic printing paper. Whatever the method of printing, the prints are then cut and cemented to the slide rule blank, the slide strip being cemented on first.

The number of manual operations involved in this method increases its cost considerably. The durability of the resulting rule depends largely upon the quality of the paper and the cement. It is possible to print directly onto the slide rule blank from the photo-offset plate, but the cost of printing is much higher than that on paper. Such factors as these must be given consideration in weighing the advantages of one method over another.

2.15 Folded Paperboard Rules

Figure 2:11 shows the master of a folded paperboard rule for the MTD formula (Sec. 2.12). This master is quite easily reproduced by photo-offset methods. The paper is folded mechanically, which not only reduces the cost, but also increases the accuracy with which the scales are matched. A method of folding is shown in Figure 2:12; and the finished rule is very easily handled. In large quantities this rule can be produced at very low cost. The possibilities of such rules as vehicles of advertising material have hardly been approached.

2.16 Machine Ruling

If a high degree of accuracy is desired, and expense is a negligible factor, the rule may be divided by mechanical scriber. The most satisfactory work of this kind is done by the manufacturers of slide rules, as they have access to the necessary materials, equipment, and trained personnel for superior craftsmanship in this art. Many special slide rules have been designed and reproduced in this fashion; e.g., chemical slide rules, which have a scale of gravimetric factors of the common chemical reagents, and statistical quality control slide rules with scales for parameters commonly used in that field.

Usually the rule carries, also, several of the common scales of the commercial slide rule, with special scales

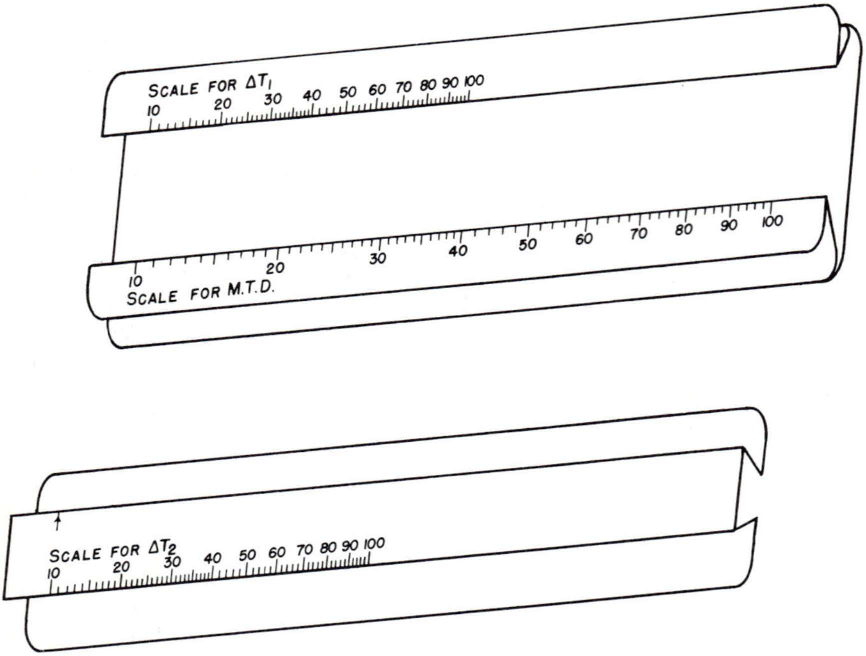

FIG. 2:12 METHOD OF FOLDING PAPERBOARD RULE AFTER SCALES ARE CONSTRUCTED

added or substituted for other scales. Such rules are too expensive to be distributed as advertisements. Their purpose is a more serious one. As the time that may be saved in computation with special slide rules is so great, and the rule is so readily adapted to engineering routines, it is highly recommended that this method be given primary consideration in the solution of every engineering problem to which it is applicable.

The manufacturer will furnish a list of the specifications required for special scales, such as

a) A table showing the position of each graduation on the scale, the length of each graduation, numbers to be printed on the graduations, etc.;
b) position of each special scale on the rule;
c) position and type of each common scale on the rule;

d) scale designations;
e) related matter, such as instructions, to be printed on the rule.

OTHER FORMS OF SPECIAL SLIDE RULES

2.17 Form for Multiple Products

The rule described in Section 2.10 ff. is adaptable to any product of the form, $P = A \times B \times C \times K$, where P, A, B, and C are variables or their functions, and K is a constant. If additional terms appear in the product, as in $P = A \times B \times C \times D \times E \times F \times K$, where D, E, and F are also variables or their functions, the simplest solution is the addition of a cursor to the rule, and scales to the slide for the additional variables.

Let the rule be of the form shown in Figure 2:13; the A-, B-, C-, and D-scales are located as before, and the E- and F-scales are located on the slide between the B- and C-scales. The mode of operation is to be this:

1. Set b on the B-scale opposite a on the A scale
2. Set the cursor at e on the E-scale
3. Set f (F-scale) on cursor, read d (D-scale) opposite c (C-scale)

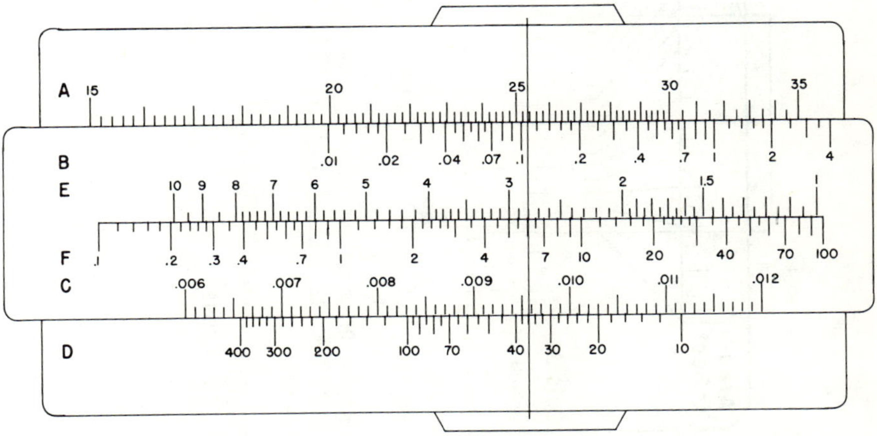

FIG. 2:13 SLIDE RULE FOR MULTIPLE PRODUCTS

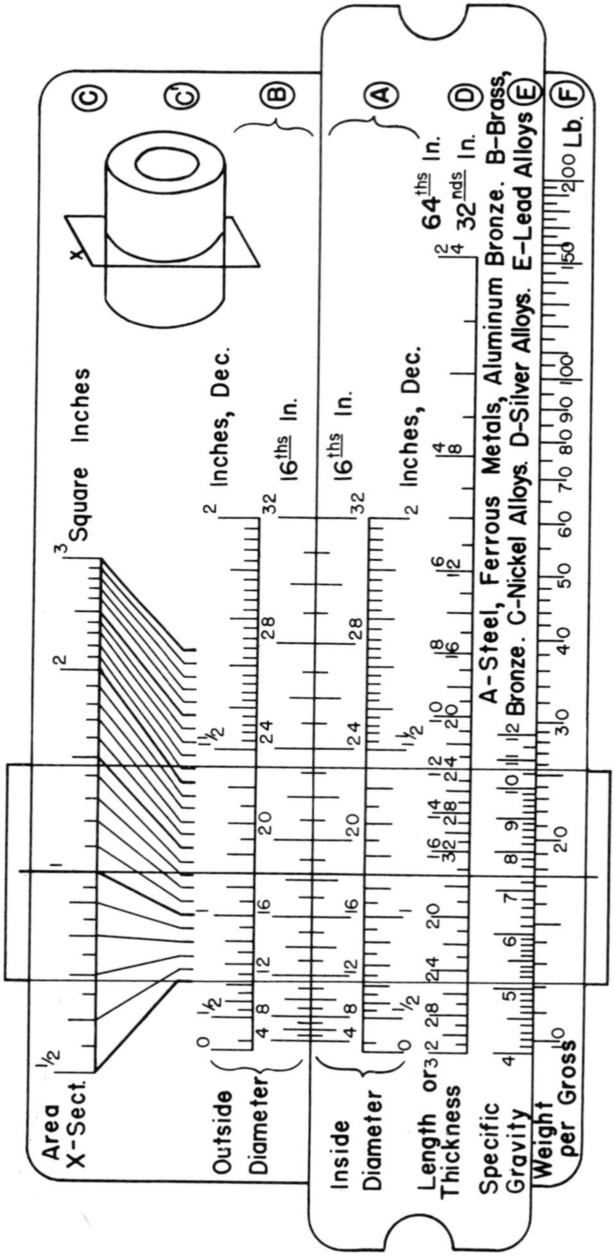

FIG. 2:14 SLIDE RULE FOR WEIGHT OF WASHERS AND BUSHINGS

From the scale equations of these six scales, the defining equation of this rule is

$$m_1f_1(A) - m_2f_2(B) + m_3f_3(C) - m_4f_4(D) + m_5f_5(E) - m_6f_6(F) + K = 0 \qquad (2.9)$$

where $K = l_{01} - l_{02} + l_{03} - l_{04} + l_{05} - l_{06}$.

The process of matching up terms in the formula with those in the defining equation and of selecting values for the zero points is identical with that discussed in Section 2.11.

If additional terms are present in the product, additional scales are added in pairs to the slide. If an odd number of terms, including the product, is to be represented, one of the scales will have the equation $s = 0$, as in the rule shown in Figure 2:10, or a constant value, $s = k$.

2.18 Transition Guide Lines

In engineering formulas there are frequent cases of variables that are present in two or more functional modifications, each requiring one scale in the defining equation of the slide rule. If the particular variable is a product obtained at an early stage of the computation, it is necessary to read its value on one scale, then commence the remainder of the computation with that value on another scale. The transfer of a number from one scale to another in this fashion is subject to error, and it increases the tedium of the computation. As an aid to this transition, guide lines may be placed from one scale to the other, connecting equal values on the two scales.

As an example, consider the formulas for computing the weight of washers and bushings, both of which may be considered as segments of hollow tubes. The area of the cross section is

$$\text{Area} = \frac{\pi}{4}(D^2 - d^2),$$

where D and d are outside and inside diameter, respectively. The volume of the segment is then

$$\text{Volume} = \text{area} \times \text{length},$$

and the weight is

$$\text{Weight} = \text{volume} \times \text{specific gravity}.$$

Conversion factors must be included, in order to utilize the units commonly used in making these measurements. The scales for outside diameter and inside diameter (Fig. 2:14) are defined by exponential (square) functions; consequently, they may not be applied directly to a multiplication step. The transition lines carry out the result obtained on the C'-scale to the C-scale, which is logarithmic. This scale then operates with the D and E scales to give a final product on the F-scale. The D-, E-, and F-scales are all logarithmic.

Outside diameter is located by the cursor on the B scale, and inside diameter is set on the cursor by moving the slide. The cursor is then moved to 0 on the A-scale, indicating a value on the C'-scale, which is then read out on the C-scale. The cursor is moved to this value on the C-scale, the slide is moved to place length on the D-scale under the cursor, and weight is read on the F-scale at the correct value of specific gravity on the E-scale.

DIFFERENTIAL GEAR RULES

2.19 Defining Equation

A differential gear consists of two bevel gears engaging a common pinion, which is mounted in a cutout in a large spur gear, as shown in Figure 2:15. Any two members of this system may be rotated through any given angle; the rotation of the third member is determined by the other two, the relationship being

$$R_s = \frac{R_1 + R_2}{2}$$

where R_1 and R_2 are the angles of rotation of the two bevel gears, and R_s is the angle of rotation of the spur gear.

In the gear rule, two such systems are connected by engaging their spur gears. As the rotations of the two spur gears are in opposite directions, if R_3 and R_4 are the angles of rotation of the bevel gears in the second differential gear system, the four rotations are related by the formula

$$R_1 + R_2 - R_3 - R_4 = 0 \tag{2.10}$$

If the nature of the scales to be used is such that positive rotations of all four bevel gears are preferred, the two spur

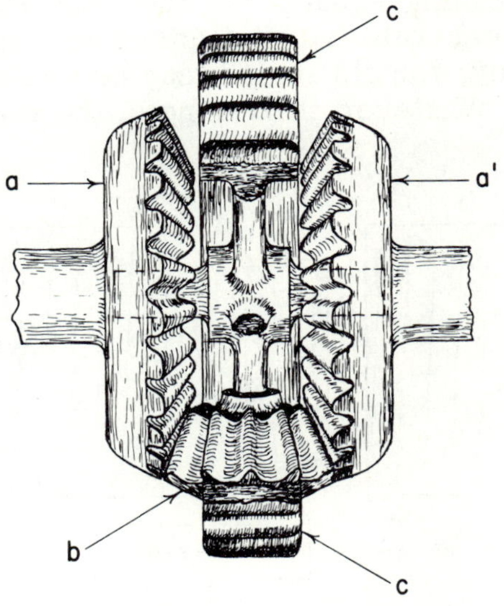

FIG. 2:15 DIFFERENTIAL GEAR
a, a', Drive Gears; b, Floating Bevel Gear Mounted in Spur; c, Spur Gear with Cut-away Showing Spur and Bevel Gear Mountings

gears may be connected to a common spur gear, or pinion; the relationship for such a system as this is then

$$R_1 + R_2 + R_3 + R_4 = 0 \tag{2.11}$$

Each of the bevel gears is mounted on the shaft of a separate drum bearing a slide rule scale. These scales may be defined as in Equations 2.3 through 2.6, substituting R_1 for S_A, R_2 for S_B, etc.; whence, the defining equation for the gear rule (from Eq. 2.10) is

$$m_1f_1\,(A) + m_2f_2\,(B) - m_3f_3\,(C) - m_4f_4\,(D) + K = 0 \qquad (2.12)$$

where $K = l_{01} + l_{02} - l_{03} - l_{04}$.

The gear rule accommodates the same formulas as the slide rule described in Section 2.10 and the formulas are fitted to the defining equation of the gear rule in the same manner. The gear rule has the advantage, however, of far greater accuracy, for its scales may be made of almost any desired length. Whatever their length, the scales are wound

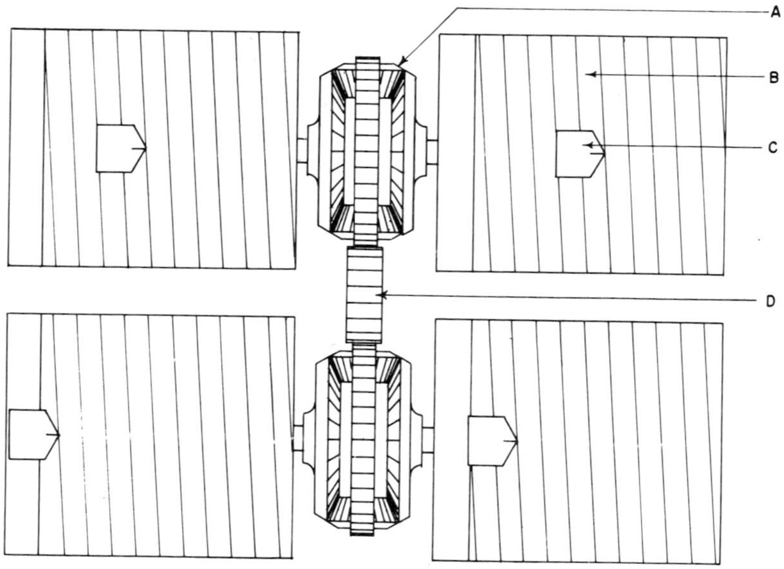

FIG. 2:16 ONE FORM OF GEAR RULE
A, Differential Gear; B, Drum with Spiral Scales; C, Traveling Indicator Following Spiral; D, Pinion Linking two Spurs of Differential Gears

on the drums in spiral fashion, and the drums are equipped with traverse indicators which follow the spiral to indicate the point at which the scale is read.

The prime precaution in the construction of the gear rule is the elimination of "play," a factor that does not enter into the construction of slide rules to any great extent. A loose connection in gears vitiates the entire purpose of the gear rule, which is the achievement of greater accuracy. The use of springs to increase the thrust that brings the bevel gears into contact with the pinions enables one to eliminate play effectively without the use of expensively machined mountings. A similar arrangement is recommended in the meshing of the two spur gears. Figure 2:16 shows a gear rule.

EXERCISES

1. Construct adjacent scale arrangements for the following: (a) The area of a circle in terms of its diameter. Use a linear scale for diameter, 0 to 10 inches. (b) The squares of numbers from 1 to 10 at intervals of 0.1. Make two charts, using a linear scale for N in one and a logarithmic scale for N in the other. (c) Specific gravity of liquids in terms of degrees A.P.I.*, for which the relationship is

$$\text{Degrees A.P.I.} = \frac{141.5}{\text{Specific Gravity } 60°\text{F}/60°\text{F}} - 131.5$$

Use a linear scale for degrees A.P.I., 0 to 150 degrees.

* Table of values is given in National Bureau of Standards Circular 154, and in various handbooks on liquid fuels.

2. Construct an adjacent scale pair for viscosity of water in terms of temperature, using the graph of Figure 1.6.

3. Using the chart*, prepare a set of parallel scales for surface tension of some paraffin hydrocarbons in terms of degrees F. Use a linear temperature scale at each end.

*CHART BY GEORGE GRANGER BROWN (Ex. 3.)
(University of Michigan)

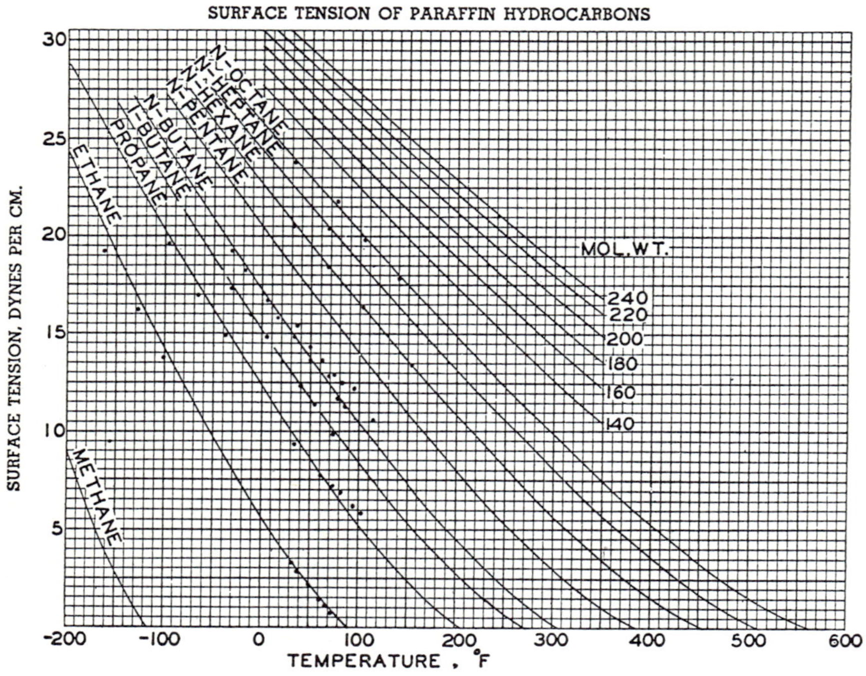

(From Engineering Data Book, 6th Ed., p 125; Natural Gasoline Supply Men's Association, Tulsa, Oklahoma, 1951)

4. Prepare a radial arrangement of the scales of Exercise 3.

5. The methods of Exercises 3 and 4 are applicable to many colligative properties of chemical compounds. Select several such properties that may be represented by similar charts.

6. If any number a on the C-scale of a common slide rule be placed at any number b on the D-scale, the ratio of a to b is equal to the ratio of any other number on the C-scale to the corresponding number on the D-scale. Show this to be true from the scale equations.

7. Devise slide rule routines for each of the following computations. In each case, determine the efficiency of the routine. (a) Dispersion of colors by a prism, $d = k \sin\frac{\alpha}{2}/\cos\phi$, where α is the angle of the prism, ϕ the angle of incidence, and K a constant. (b) Relative centrifugal force*, R.F.C. = $.00001118\ r\ N^2$, r = rotating radius, N = rotating speed. (c) The BCD brightness of a source of light, $B = 302\ F^{0.44}$,** where F is the brightness of the field.

*Chemical Rubber Company News, July-August 1951.
** Luckiesh, M, and Guth, S. K., Illuminating Engineering Vol. XLIV, No. 11, 1949.

8. Use the scale equations of the slide rule and analyze the routines devised in Exercise 7.

9. The factors that determine brightness in visual field at borderline between comfort and discomfort* are related by the formula,

$$M = \frac{B}{PF^{0.44}\left(Q^{-0.21} - 1.28\right)},$$

where M is an index of the sensation of visual comfort, B is brightness of source, P is position index, F is brightness of field, and Q is solid angle subtended by source. Ranges are $B\left|{}^{10{,}000}_{70}\right.$ footlamberts, $P\left|{}^{16}_{1}\right.$, $F\left|{}^{100}_{1}\right.$ footlamberts, $Q\left|{}^{0.13}_{0.001}\right.$ steradians. Design a slide rule for evaluating M. Mark the point M = 108 as the borderline between comfort and discomfort.

* Luckiesh and Guth (Problem 7).

10. Construct slide rules to perform the computations of nomographs shown in Figures 11.2, 11.3 and 11.4.

11. Construct a differential gear rule and equip it with four hundred-inch logarithmic scales. Such a rule may be used for computations having the accuracy of a five-place logarithm table.

chapter 3

GRAPH PAPERS

One of the most progressive steps taken by man in trying to deal with the mathematical problems associated with the material things of his environment was the discovery of analytic geometry. The geometry of the ancients regarded points, lines, areas, volumes and angles with respect to one another only through the association of such ideas as equality, similarity, congruence, perpendicularity, parallelism, etc. The result is an unwieldy set of theorems and propositions, which, though of unquestionable validity and exquisite beauty, prove to be poor tools for deductive reasoning. Indeed, it may be remembered that many a very obvious relationship of plane geometry remained unproved over a period of many years until someone arrived at the necessary association of theorems required for proving it.

The invention of the analytic geometry enabled man to represent the elements of space and his ideas concerning them by numbers and relationships of numbers (equations) through associating all points in space with a pair of axes. Equations and numbers are excellent tools for deductive reasoning, because the rules governing their conduct are few and simple. Consider, for example, how trifling is the effort involved in deriving the equations of the circle, ellipse, hyperbola and parabola from the general equation of the conic in analytic geometry. Compare with this the effort required to prove by the theorems of synthetic (non-analytic) geometry the family relationship of the conic sections.

The practical aspects of a coordinate system are increased by the readiness with which numerical relationships may be represented visually—a feature that has led to commercial applications far greater than any which could have been anticipated by René Descartes, the seventeenth century mathematician who is generally credited with the invention of analytic geometry. From the graphical representation of commercial or scientific data, the mathematician may in some cases derive an equation through which he is enabled to trace relationships to colligative variable factors.

As tools for the representation of data and its analysis, the graph papers are the most convenient forms in which the cartesian coordinates may be applied; their grids serve to associate every point in the plane immediately with a point on each axis. To accommodate his diverse needs man has devised graph papers with a wide variety of grids, a few of which are discussed in the following sections.

COMMON TYPES OF GRAPH PAPERS

3.1 Nomenclature and Defining Equation

The grid of a graph paper is determined by the nature of the scales employed on the axes. These scales may be linear, logarithmic, reciprocal, square root, etc., and may be used in any combinations on the two coordinates. The name of the graph paper indicates the nature of these two scales, but in commercial practice abbreviated names are commonly used. The order in which the two scales are named is of no consequence, because the relative positions of the two axes may generally be reversed by rotation of the paper.

The papers that are employed in this and subsequent chapters are listed in Table 3:1.

Taking the intersection of the X and Y axes as the origin, s_x is the distance from the origin to the point having the value x on the X axis, and s_y is the distance from the origin to the point having the value y on the Y axis. The moduli of the two axes are m_x and m_y, respectively. In the defining equations of the X axis for the normal distribution papers, z is the ordinate and t the abscissa of the normal distribution curve.[1]

[1] Peters, Charles C., and Van Voorhis, Walter R., Statistical Procedures and their Mathematical Bases, Ch. 10; McGraw-Hill Book Company, New York, 1940.

TABLE 3:1

SOME COMMON GRAPH PAPERS

Type of Graph Paper	Defining Equations of Axes
Linear-linear	$s_x = m_x x,\ s_y = m_y y$
Linear-log	$s_x = m_x x,\ s_y = m_y \log y$
Log-log	$s_x = m_x \log x,\ s_y = m_y \log y$
Linear-hyperbolic	$s_x = m_x (x - 1)/x,\ s_y = m_y y$
Linear-loglog	$s_x = m_x x,\ s_y = m_y \log(\log y)$
Linear-normal distribution	$s_x = m_x x,\ s_y = m_y \int_{-\infty}^{t = y} z\,dt$
Log-normal distribution	$s_x = m_x \log x,\ s_y = m_y \int_{-\infty}^{t = y} z\,dt$

A particular paper may be selected for the representation of certain data because it helps to clarify the representation, facilitates analysis of the data, or adapts the data readily to certain computations. Indeed, the same data may for one purpose be plotted on one type of paper and, for another purpose, on another type of paper. Examples of data plotted on the various types of graph papers included in Table 3:1 are shown in Figures 3:1 through 3:7. Figure 3:8 shows a set of data plotted on linear-linear graph paper; Figure 3:9 is a plot of the same data in logarithmic-reciprocal form.

3.2 Rectification

In Figure 3:9 the data fall on a straight line, and the graph is said to be rectified. The process of defining the axes of a graph paper on which a plot of the data will fall on a straight line is called rectification. One of the purposes of rectification is to analyze the data; that is, to find an algebraic expression that contains the same relationship among its variables

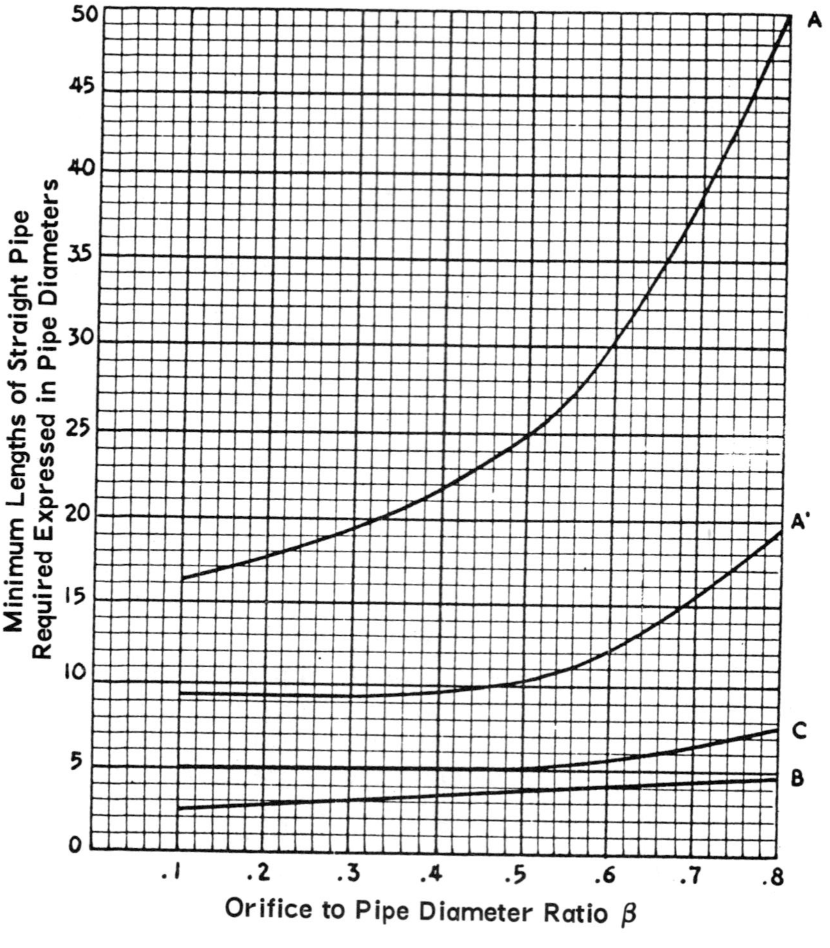

FIG. 3:1 LINEAR-LINEAR GRAPH SHOWING INSTALLATION DATA FOR VALVES
From Gas Measurement, Committee Report No. 2, p. 11, American Gas Association, New York

Flange F_c

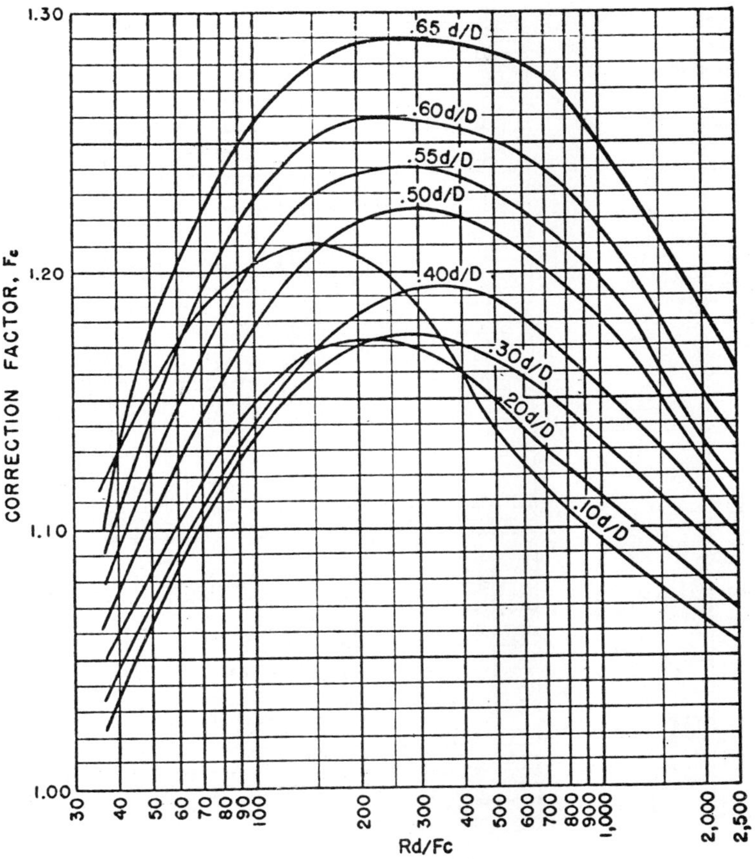

FIG. 3:2 LINEAR-LOGARITHMIC GRAPH SHOWING DENSITIES OF ETHANE AT VARIOUS TEMPERATURES AND PRESSURES

From Spink, L. K., Principles and Practice of Flow Meter Engineering, 7th ed., p. 118; The Foxboro Company, Foxboro, Mass., 1949

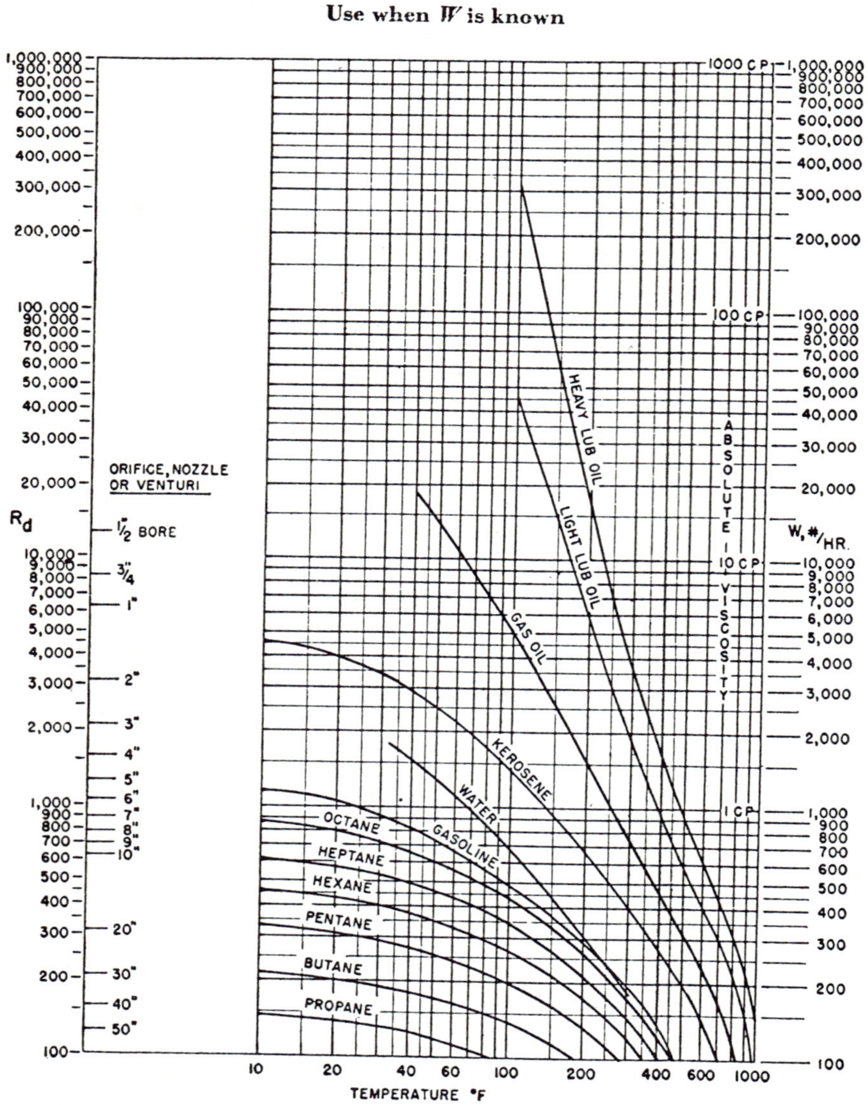

FIG. 3:3 LOG-LOG GRAPH SHOWING VISCOSITIES OF SOME LIQUIDS AT VARIOUS TEMPERATURES
From Spink, p. 109; See Fig. 3:2

as that shown by the data. When the data can be made to fall on a straight line, the equation is very readily derived from the analytic geometrical equation of the straight line together with the defining equations of the axes of the graph paper. This process is discussed in somewhat greater detail in Section 5.16.

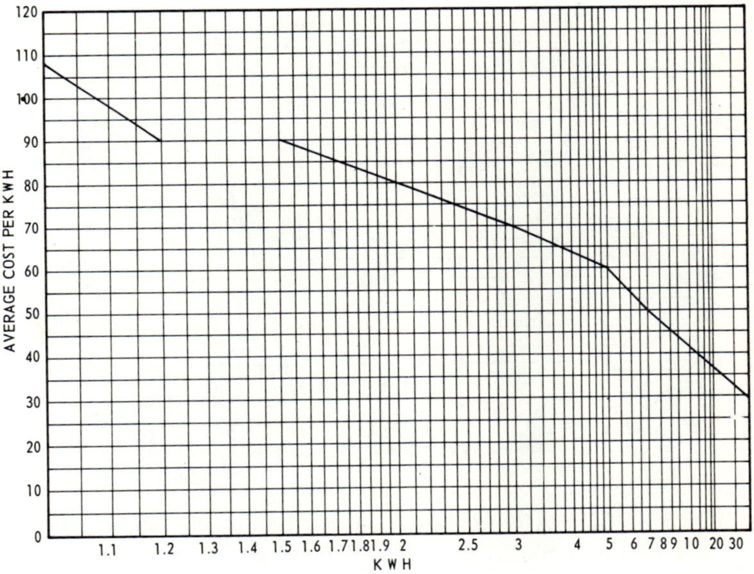

FIG. 3:4 LINEAR-HYPERBOLIC GRAPH SHOWING POWER RATES
From Graph Papers, Keuffel & Esser, Hoboken, New Jersey

Another purpose of rectification is the construction of a nomograph to represent the data. In this case a family of curves is involved showing the relationship of three variables, and it is desired to find a graph paper that will provide a family of straight lines as the plot of the data. It is not always possible to achieve rectification with standard graph papers, and it becomes necessary to prepare a special graph paper which will rectify the data.

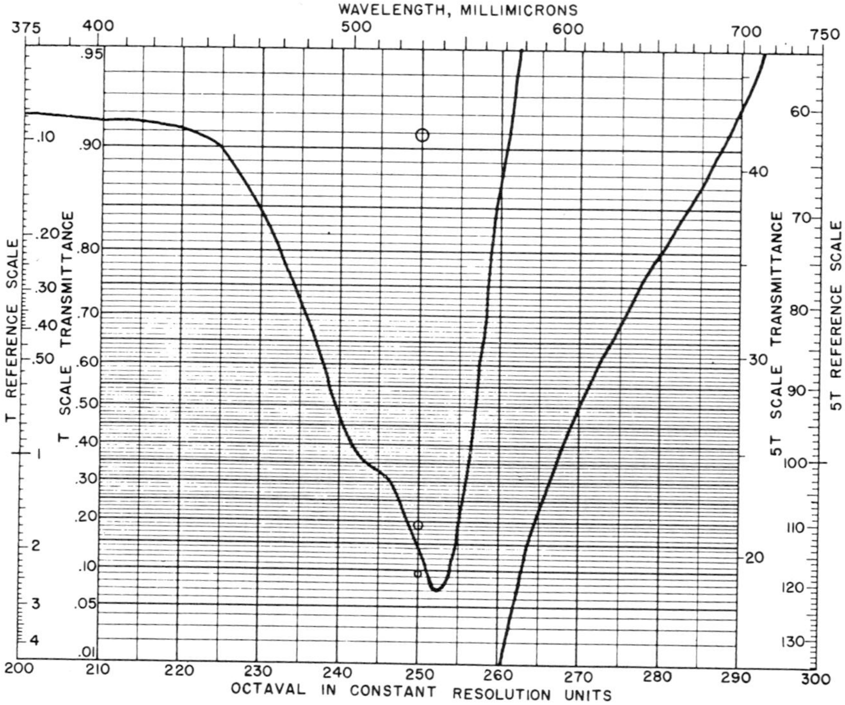

FIG. 3:5 LINEAR-LOG-LOG CHART USED FOR PLOTTING TRANSMITTANCE OF A DYE SOLUTION vs WAVE LENGTH OF LIGHT
From Peacock, W. H., The Application Properties of the Certified Coal Tar Colorants, Fig. 7; American Cyanamid Company, Bound Brook, N. J.

The method is quite simple. A plot is first made on any convenient type of graph paper, and a scale is prepared from the graph using the method described in Section 1.7 and illustrated in Figure 1:5. The scale thus produced is substituted for the original abscissa scale; the original ordinate scale is retained. The grid built on these two coordinate scales rectifies the data. This method is illustrated in Figure 3.9 for the vapor pressures of water. The curve here is a single one, not a family; however, it will be found that

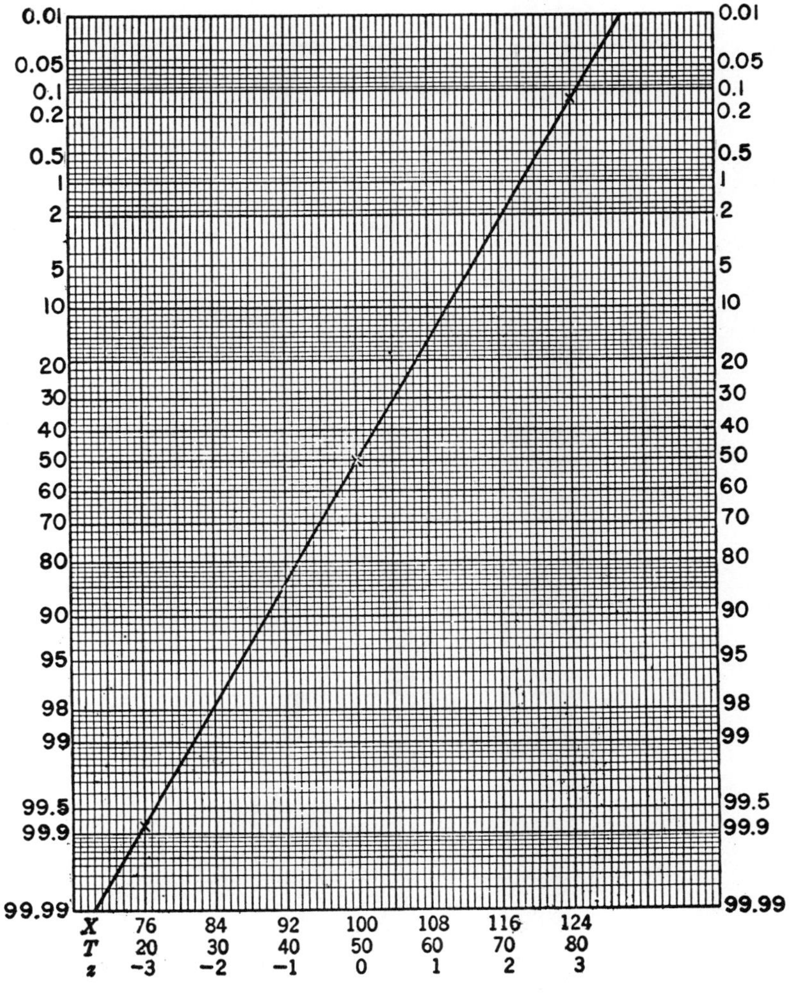

FIG. 3:6 LINEAR-NORMAL DISTRIBUTION CHARTS
From Dixon, Wilfred J., and Massey, Frank J., Jr., Introduction to Statistical Analysis, 2nd. ed., p. 56; McGraw-Hill Book Company, Inc., New York, 1957.

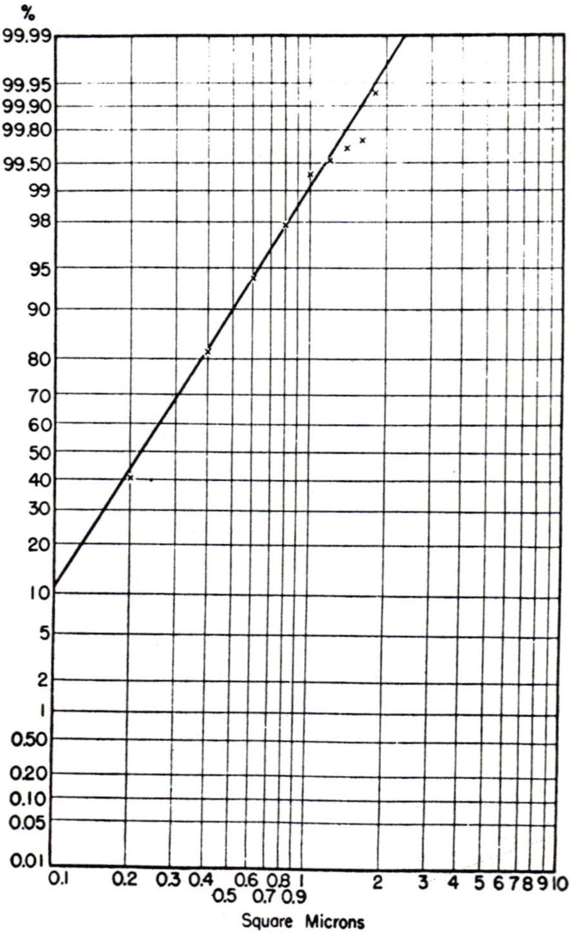

FIG. 3:7 LOGARITHMIC-NORMAL DISTRIBUTION CHART
From Kottler, F., J. Franklin Inst., 250-5:420, 1950

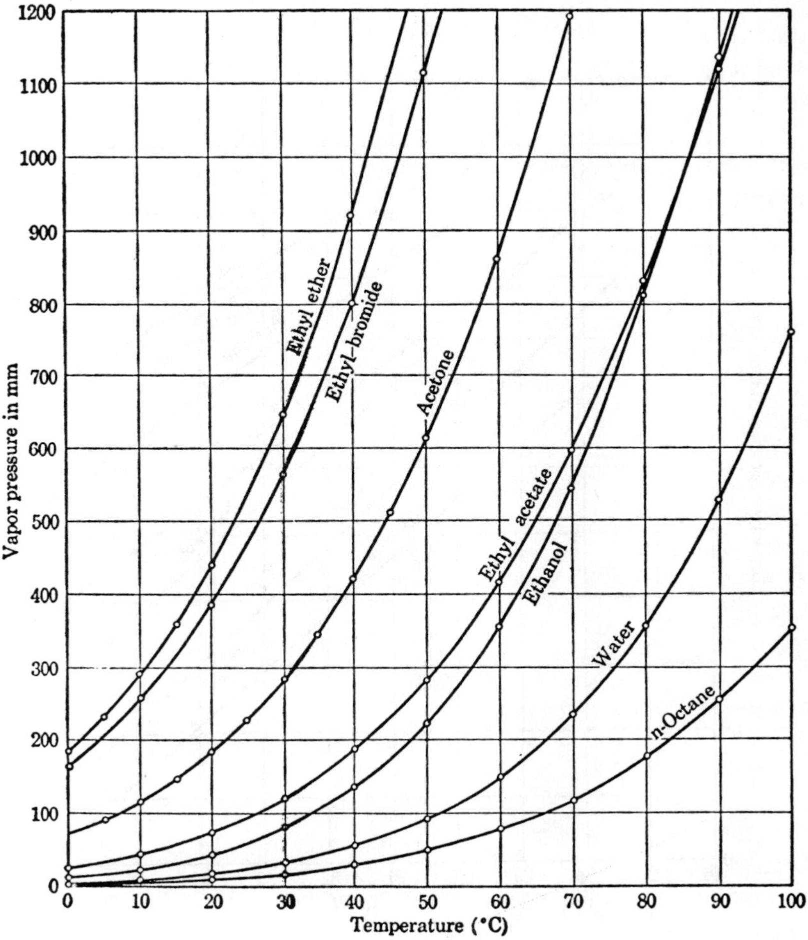

FIG. 3:8 VAPOR PRESSURE-TEMPERATURE CURVES FOR SOME COMMON LIQUIDS, PLOTTED AS A LINEAR-LINEAR GRAPH

From Daniels, F., Outlines of Physical Chemistry, p. 170, John Wiley & Sons, Inc., New York, 1948

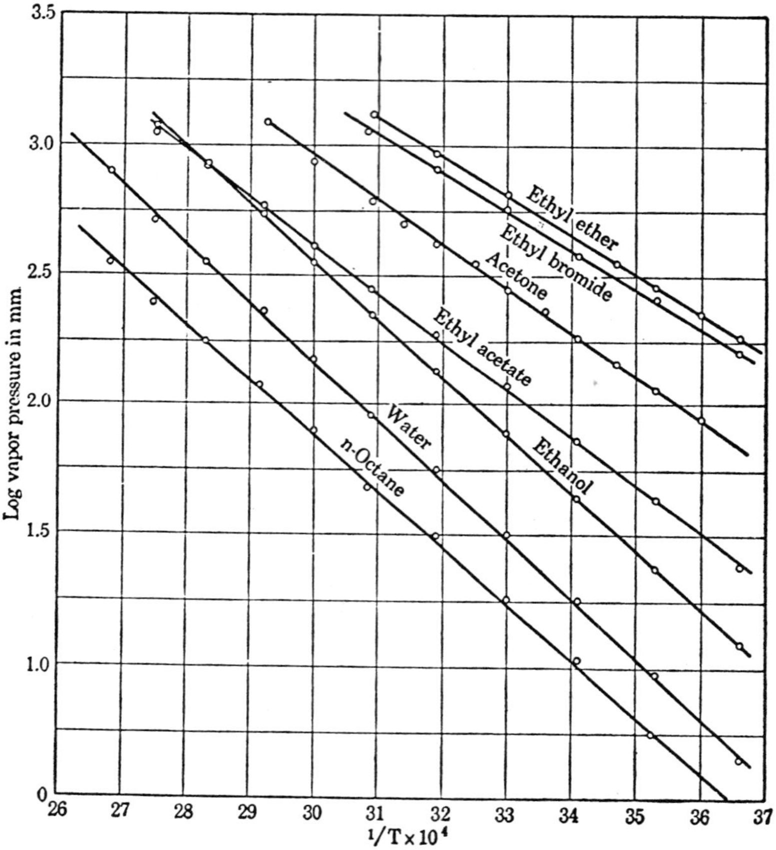

FIG. 3:9 LOG VAPOR PRESSURE vs 1/T CURVES FOR THE LIQUIDS SHOWN IN FIGURE 3:8
From Daniels, Ibid., p. 171

plots of the vapor pressures of many other liquids plotted on this grid also fall on straight lines.

The rectification achieved by a special paper of this sort is no help in the analysis of the curve, for no more is known of the nature of the abscissa scale than was known of the original unrectified curve. Such rectification may be used only for clarity of representation or for the construction of a nomograph.

CONSTRUCTION OF GRAPH PAPERS

Despite the large variety of graph papers obtainable commercially, it is still necessary to prepare custom-made papers for particular problems of analysis or representation. The method chosen for preparing the original master must take into account:

1) The method of reproduction to be employed:
2) Whether the graphs plotted on the paper are to be reproduced, as, for instance, illustrations for a scientific article;
3) Size requirements of the graph paper; and
4) Any special requirements of a particular laboratory.

Primary consideration must be given to the method of reproduction to be employed; the methods in most common use are (a) Ozalid, (b) Photostat, and (c) Offset printing.

Regardless of the method of reproduction, if the graph paper is to be printed in one color, the only way to emphasize the principal lines of the grid is to make them thicker. If the principal lines are too thick, however, there is a loss of accuracy in plotting points that lie between lines. In addition to this, a heavy grid tends to obscure lines plotted upon it. Consequently, it is necessary, if one is to avoid very heavy principal lines, to make subordinate lines as light as possible. Not over three different thicknesses should be employed.

3.3 Preparation of Master for Ozalid Reproduction

As there is no significant reduction in size during reproduction by Ozalid, the lines made on the master must be of the thickness expected on the final copy. If a lettering pen is used, sufficient emphasis may be achieved without reaching too great a thickness by using a pen in sizes #000, #0 and #1 (Leroy). There is not enough difference between the #000 and # 00, or between the #00 and #0 to provide the emphasis required in a grid.

If the ruling pen is used, it may be adjusted to approximately the same thicknesses of lines. With care the lines may be made somewhat finer. The danger of increasing the thickness of the lines through encrustation of the pen is far more serious in ruling a grid than in ruling a scale, because erasure and repair are far more difficult.

Regardless of the type of pen used, the paper or other materials on which the rulings are made must be properly surfaced. Thorough rubbing with a chalk or pumice bag or a gum eraser is essential. The surface thus prepared must be protected from contact with the flesh or other contaminating materials by a protective overlay.

Errors may be effectively corrected by scraping off excess ink with a knife, scalpel, or razor blade. Remaining ink is then removed with a fiberglass eraser, followed by pumice, or the fiberglass eraser may be dipped in pumice.

For this and other processes in which a transparent material may be used for the master, it is highly recommended that vertical lines of the grid be put in on one side of the material and horizontal lines on the other side. An error in the second set of lines may then be corrected without damage to the first set.

If finer lines are required than are obtainable by the pens suggested here, the #0 pen may be used for principal lines and the #00 pen for secondary lines. For the finest lines the # 000 pen is used, and further contrast between this and the # 00 pen can be achieved by using the same # 000 pen and diluting the ink. Results obtained with dilute inks differ in various laboratories, depending upon the operation of the Ozalid machine and the nature of the materials used. Hence, there should be some experimentation before actual use to determine the optimum dilution. Once this has been determined, a bottle of diluted ink should be prepared and retained for this purpose.

A finer line than that obtainable with ink may be produced

by the use of a dotted wheel scriber, such as is used with stencils, and carbon paper. The carbon paper may be placed face down on the upper surface of the tracing paper, and the scriber rolled directly on the back surface of the carbon. With this arrangement, the carbon obscures both the impression on the tracing paper and the mark of the stylus on the carbon sheet, making it difficult to line up the ends of the strokes perfectly. It is advisable to place a strip of tracing paper between the carbon and the impression sheet at each end of the grid to delineate clearly the ends of the strokes.

The carbon may also be placed under the tracing paper, face up, and the scriber operated in contact with the tracing paper. The results are more easily observed as the work progresses; however, the strips at the ends of the grids are still advised as an aid in lining up the margins of the grid.

Carbon lines may be alternated with lines in ink to produce the necessary graduations in thickness, or carbon lines may be used altogether. If all the lines of the grid are to be made with carbon, the dotted wheel scriber should be used for only the finest lines, a thin solid stylus for the secondary lines (a dulled small-blade scalpel is ideal) or a heavier wheel scriber, and a somewhat thicker stylus for the principal lines. As with the inked grid, it is advisable to place horizontal lines on one side of the paper and vertical lines on the other side.

Unless a fixative is used, the carbon lines are likely to be smudged as the master passes through the Ozalid machine. An aerosol-type fixative is available which may be sprayed directly from the container onto the master to avoid smudging the carbon impression with a brush.

The type of Ozalid printing employed depends upon the use to which the graph paper is to be put. If graphs plotted on this paper are to be reproduced as photographs, the Ozalid prints must be black line on white, and all lines in the master should be thick enough to allow for any reduction in size that may occur during the photographic process. If the Ozalid prints of the master are to be used merely for recording or plotting data, with no further reproduction anticipated, a blue-line print is preferable, because the points and lines of the graph stand out in greater relief.

3.4 Preparation of Master for Photostatic Reproduction

If the master is reproduced by Photostat, the print is black line on white paper; consequently, unless fine lines are used for the grid, the graph plotted thereon is likely to appear as if absorbed in the background, rather than in the desired relief. This disadvantage is overcome in part by the fact that the print may be prepared from the master by reduction, with proportional diminution in the thickness of the lines. A reduction of at least fifty percent should be planned for a graph paper 8-1/2 by 11 inches or larger. If the graph paper is to be smaller, a greater reduction is in order.

The use of carbon paper and styli (dotted-wheel scriber for the finest lines) is especially recommended for this method. No fixative need be used, as the master may be discarded after the first exposure. The negative is used for making all prints that are to be used as graph paper.

If ink is used, the combinations of pens may be the same as those suggested for Ozalid reproduction. It is not necessary to increase the width of the lines, even though the print is made at fifty percent reduction. The finest lines may be made with diluted ink for this process also, although if the ink is diluted too much it produces a very faint line. A well-sharpened HB or 2H pencil may be substituted for diluted ink in ruling the finest lines, but it is easier to produce a line of uniform thickness with ink.

Vertical and horizontal lines of the grid may be placed on opposite sides of a high quality tracing paper and reproduced very satisfactorily by Photostat, if the master is placed on a glossy white background for Photostating.

3.5 Preparation of Master for Offset Reproduction

Masters intended for reproduction by photo-offset are prepared in the same manner as those intended for Photostats. The same considerations apply; the only difference is that copies may be printed in color, if desired, instead of black. The use of color printing is, indeed, highly recommended for all graph papers; for this reason, offset methods of preparing special graph papers are favored over Ozalid and Photostat. If the graph papers are to serve as originals for further reproduction by photographic methods, orange or red may be used instead of black, as those colors photograph virtually the same as black. If the photographic process is intended to include the graph but not the grid, the grid may be printed in light green or blue, as these colors do not photograph well.

The master for the offset method may be ruled directly on an offset master plate. The only recommended procedure for this method is ruling with stylus on carbon paper. The carbon paper is placed face down on the master plate, strips of tracing paper are inserted to delineate the ends of the strokes, and the lines of the grid are ruled, using a dotted wheel scriber for the lightest lines and solid styli of different thicknesses for the other lines. Both horizontal and vertical lines of the grid are ruled on the same surface. For this reason, the correcting of an error is especially difficult. Furthermore, as all trace of the carbon must be removed in correcting an error, damage to other lines or to the surface of the plate of the grid is almost unavoidable. The only satisfactory method of correcting an error is to re-rule the entire grid on another plate.

In view of these various limitations, the preparation of an offset plate by direct ruling is often more expensive than the preparation of a photo-offset plate. Furthermore, the lines cannot be made so fine by direct ruling. It appears, then, that of all the methods available for the production of graph papers, the photo-offset is the most desirable.

A further advantage of either offset method over Ozalid and Photostat reproduction is the possibility of printing graph papers on tracing paper, so that the graphs plotted thereon may be reproduced by the relatively inexpensive Ozalid prints or blue-prints rather than by photographic methods. This is commonly done in many commercial graph papers, and may be done as well with special graph papers.

EXERCISES

1. Using the scales of some common graph papers or the scales prepared in Chapter I, prepare the following types of

graph paper: (a) Linear-reciprocal, (b) hyperbolic-hyperbolic, (c) logarithmic-hyperbolic. Prepare masters that may be reproduced by Ozalid, Xerox, Photostat, and photo-offset methods.

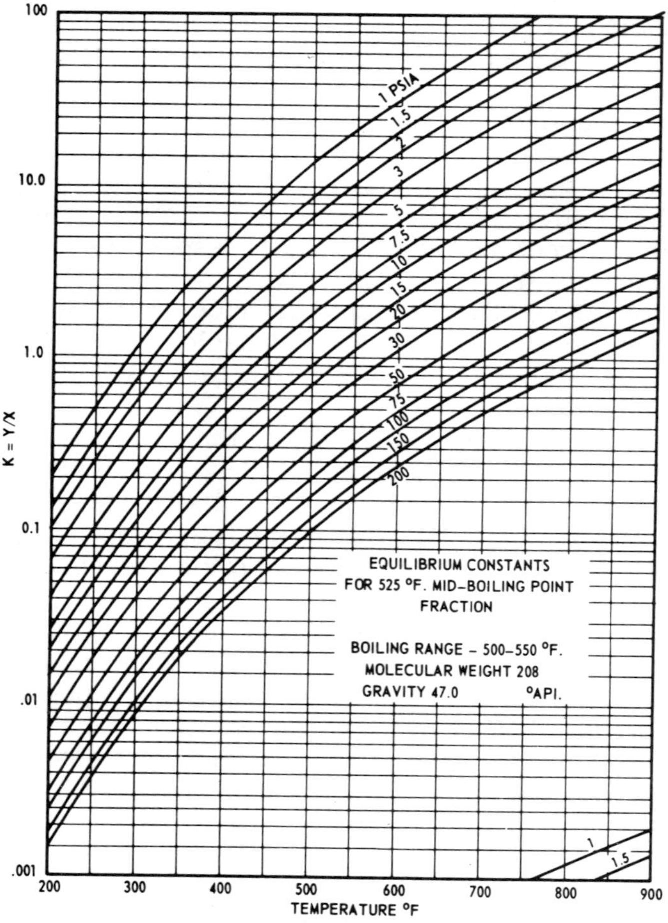

FIG. 3:10 EQUILIBRIUM CONSTANTS FOR HYDROCARBONS IN THE BOILING RANGE 500-550°F (From NGSMA Data Book, 6th ed., p. 118; Natural Gasoline Supply Men's Assoc., Tulsa, Okla., 1951)

2. Prepare a graph paper on which a plot of the viscosity of water as a function of temperature is a straight line. See Problem 6 in Chapter 1.

3. Figure 3:10 shows a graph on semi-logarithmic paper of the equilibrium constants for hydrocarbons in the boiling range of 500 to 550°F, at pressures ranging from 1 to 200 psi a. Without altering the vertical scale, modify the horizontal scale to make one of the curves (e.g., 10 psi a.) a straight line. Replot the remaining curves on this modified graph paper, and note to what extent these curves are rendered straight. There is, obviously, a family relationship here.

4. Prepare graph papers on which plots of the following functions appear as straight lines: (a) $y = \sin(n\pi x)$, where $n = 1/2, 1, 3/2, 2$, etc., (b) $y = e^{nx}$, where n = any integer.

chapter 4

CALCULATING WITH GRAPH PAPERS*

As a device for repetitive computation the common graph papers are often overlooked. Yet their grids are made up of scales whose defining equations may be correlated in many ways to conform to a graphical representation of certain engineering formulas.

The method of approach here does not, as a general rule, involve the plotting of a graph; although in some cases lines or families of lines are added to the grid to facilitate computations. The scales of the graph paper are used like nomograph scales, with solutions of a particular formula indicated by the values of the scale points at their intersections with an index line, usually a straight line.

Graph papers present greater diversity of geometrical arrangements than does the slide rule; hence, it is possible to perform certain computations which are impossible on the slide rule. This effect is greatly heightened by the wide variety of scale combinations available in commercial graph papers. In addition to this, graph papers have the further advantage that the position of the decimal point is usually indicated, even in logarithmic computations.

4.1 Defining Equations

Most of the computational procedures described in this chapter are based upon the use of a straight line as index line, whose points of contact with the graph paper scales correspond to the values of the variables represented by those scales. The defining equations of these procedures are derived from the defining equations of the graph paper scales and one of the equations of the straight line of analytic geometry.

*From material published in American Journal of Physics, 17:114 (1949).

The three common forms of the equation of the straight line of analytic geometry, as applied in nomography, are

a) The two-point form,

$$y_1(x_2 - x_3) - y_2(x_1 - x_3) + y_3(x_1 - x_2) = 0 \tag{4.1}$$

in which (x_1,y_1), (x_2,y_2), and (x_3,y_3) represent three points located on the straight line. The coordinates of any one of these points may be considered as variables, the other two points being sufficient to determine the line;

b) the intercept form,

$$\frac{x}{a} + \frac{y}{b} = 1, \tag{4.2}$$

in which x and y are coordinates of a point intersected by the index line, and a and b are the X and Y intercepts;

c) the point-slope form,

$$y = Mx + b, \tag{4.3}$$

in which x, y and b have the same meanings as in Equation 4.2, and M is the slope of the index line. The meanings of these forms are indicated in Figure 4:1.

HYPERBOLIC PAPER

4.2 Defining Equation

The vertical axis of hyperbolic paper is subdivided linearly, $y = mx + 1_o$, and the horizontal axis is subdivided by an hyperbolic scale, $x = m(p - 1)/p$. For most purposes m, which is the length of the scale, may be taken as unity; whence $x = (p - 1)/p$. On this scale x is the distance from the origin to the point marked p; the divisions of the horizontal axis are marked in values of p. p has finite values except at the terminus of the scale ($p = \infty$), at which point $x = 1$.

Several useful applications of this paper are indicated by considering the two-point form of the equation of the straight line (Eq. 4.1). In the event one of the points on the line lies on the right-hand scale ($p = \infty$), the values of x for the three

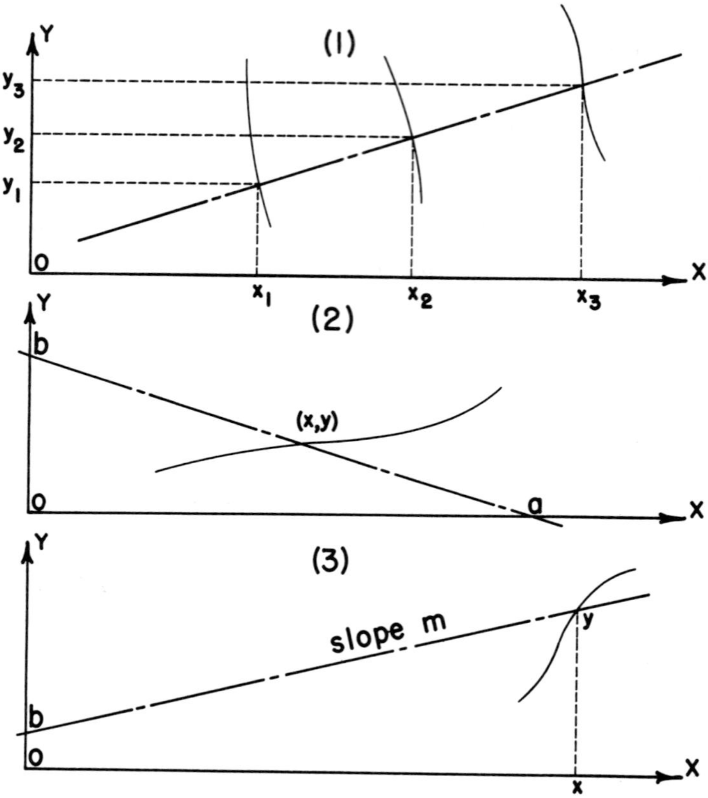

FIG. 4:1 EQUATION OF THE STRAIGHT LINE
a) Two-point form; b) Intercept form;
c) Point-slope form

points of Equation 4.1 are:

$$x_1 = \frac{(p_1 - 1)}{p_1}$$

$$x_2 = \frac{(p_2 - 1)}{p_2}$$

and $$x_3 = 1.$$

Retaining the values of y and substituting these values of x leads to the expression,

$$y_1p_1 - y_2p_2 + y_3(p_2 - p_1) = 0 \tag{4.4}$$

which may be considered as the two-point form of the equation of the straight line for hyperbolic paper.

4.3 Averages

An application of this form that is immediately apparent is the computation of the weighted average,

$$A = \frac{w_1N_1 + w_2N_2}{w_1 + w_2}$$

or $$A(w_1 + w_2) - w_1N_1 - w_2N_2 = 0$$

where N_1 and N_2 are two numbers which are averaged with the weighting w_1 and w_2, respectively, and A is the average. This equation is identical with Equation 4.4 with the substitutions $y_1 = N_1$, $y_2 = A$, $y_3 = N_2$, $p_1 = w_1$, and $p_2 = w_1 + w_2$.

The procedure for solving for the weighted average is indicated by these steps:

a) Locate the number N_1 on the vertical scale passing through w_1 (X axis).
b) Locate N_2 on the vertical scale passing through ∞ (X axis).
c) Pass the index line through these points and read A on the vertical scale passing through $w_1 + w_2$.

This procedure is illustrated in Figure 4:2.

Other substitutions may be used to realize solutions of this formula, but there is another purpose to be considered. The accumulation of averages of more than two numbers may be accomplished by a simple extension of the procedure outlined, taking the average of the first two terms as the

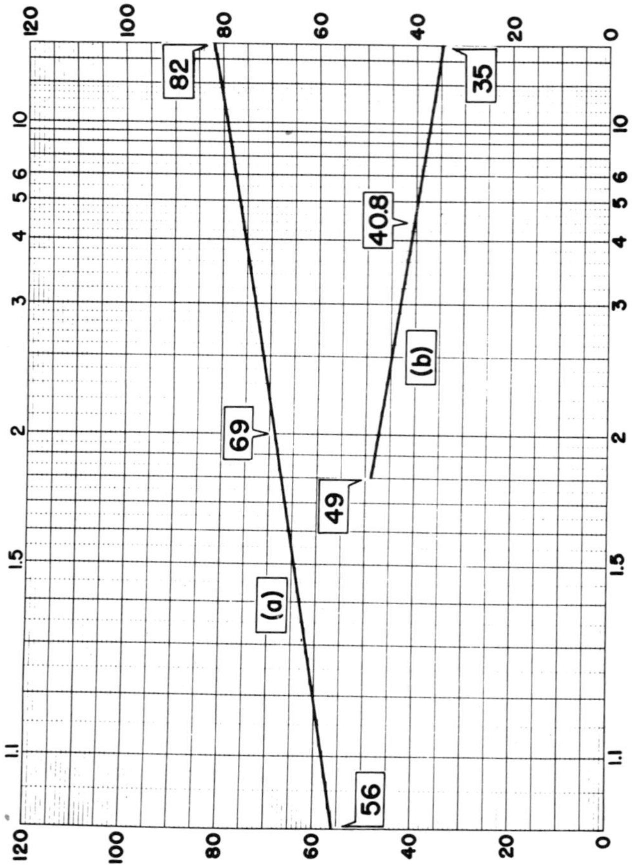

FIG. 4:2 METHOD FOR WEIGHTED AVERAGES

a) Simple Average, $\frac{(56 + 82)}{2} = 69$; b) Weighted Average, $\frac{(1.8 \times 49) + (2.6 \times 35)}{(1.8 + 2.6)} = 40.8$

first term in the second average, the second average as the first term in the third, etc.

Such an average may be written

$$A = \frac{w_1N_1 + w_2N_2 + w_3N_3 + w_4N_4 + \ldots}{w_1 + w_2 + w_3 + w_4 + \ldots}.$$

This expression may be reduced to separate simple averages:

$$A_1 = \frac{w_1N_1 + w_2N_2}{w_1 + w_2},$$

$$A_2 = \frac{(w_1 + w_2)A_1 + w_3N_3}{w_1 + w_2 + w_3},$$

$$A_3 = \frac{(w_1 + w_2 + w_3)A_2 + w_4N_4}{w_1 + w_2 + w_3 + w_4}$$

and so on. The procedure indicated by these expressions is:

1. Locate the weighted average of the first two numbers as indicated above. From that point pivot the index line to N_3 on the $p = \infty$ scale.
2. Locate the average of the first three numbers at the vertical scale passing through $w_1 + w_2 + w_3$. From that point pivot the index line to N_4 on the $p = \infty$ scale, and so on, for subsequent terms.

In repetitive operations, it is not uncommon that many similar averages must be computed involving the same sequence of weightings. The computation is greatly facilitated by marking off vertical lines at the points w_1, $w_1 + w_2$, $w_1 + w_2 + w_3$, etc. A device for computing averages in which the weightings are 1.5, 0.5, 1.2, and 0.8 is shown in Figure 4:3 with the solution of a typical problem.

The arrangement of the horizontal scale on this paper is such that the straight edge used for the index line and the point used for a pivot are more easily manipulated by a left-handed person than by a right-handed one. But all those

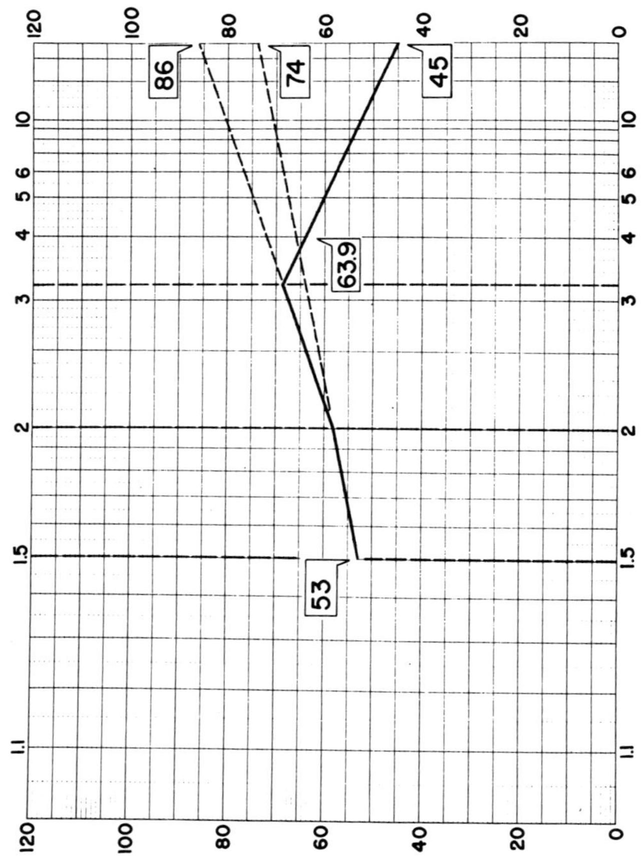

FIG. 4:3 COMPUTATION OF AVERAGE OF FOUR NUMBERS WITH SUCCESSIVE WEIGHTINGS OF 1.5, 0.5, 1.2, and 0.8

persons who are right-handed may facilitate the procedure further by inverting the paper before writing in the values of the scale points.

4.4 Weighted Sums

The weighted sum is

$$S = w_1N_1 + w_2N_2 + w_3N_3 + \dots$$
$$= (w_1 + w_2 + w_3 + \dots)A.$$

If, in Equation 4.4, p_1 is given the value 1, and y_3, the value 0, then

$$y_1 = y_2p_2 \tag{4.5}$$

But y_2 is A, and p_2 is $w_1 + w_2 + w_3 + \dots$; hence, y_1 is the weighted sum, S. Which is to say, a straight line passing through the point $p = \infty$ on the X axis and the point that indicates the average intersects the Y axis at the value of the weighted sum.

A drawback to this method of determining sums is the limited range of the Y axis. This defect is overcome by projecting the upper part of the Y axis onto a horizontal line at the top of the graph paper, using the point $p = \infty$ on the X axis as a center of projection. There is an obvious relationship between this projected scale and the abscissa scale which enables one to determine immediately the value of any point on the horizontal scale; (see Figure 4.4).

4.5 Quadratic Equations

The quadratic equation $y^2 + Ay + B = 0$ may be solved by applying following substitutions to Equation 4.4:

$$y_1 = -(A + B) \qquad y_2 = y \qquad y_3 = -A \qquad p_1 = 1$$

and

$$p_2 = y.$$

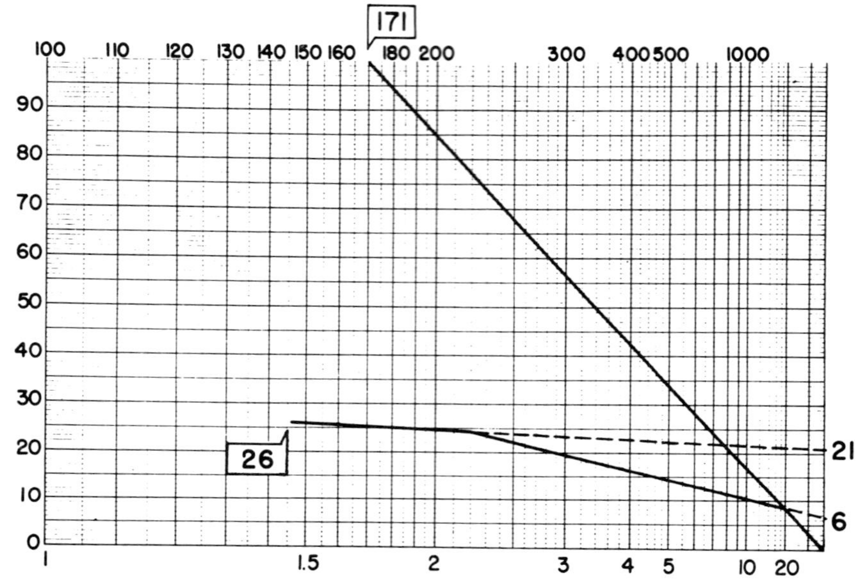

FIG. 4:4 METHOD FOR OBTAINING WEIGHTED SUMS
1.4 (26) + 0.7 (21) + 20 (6) = 171

The procedure indicated by these substitutions is:

> With suitable vertical scales on the extreme sides of the graph paper, locate on the left-hand scale the value of $-(A + B)$, and on the right-hand scale, the value of $-A$. Place the index line through these two points, and move along this index line until a value of p is found which is equal to the value of y at that point. This value is a root of the equation.

The roots of the equation $y^2 - 5y + 6 = 0$ are shown in Figure 4:5.

If only one root appears, the other is a negative root, or if no root appears by the first substitution, both roots are negative. As may be readily determined, negative roots may be found by making the substitutions $y_1 = -(A - B)$ and $p_2 = -y$, together with the substitutions in Equation 4.4 previously employed. To solve for negative roots, the procedure is:

> Lay the straight edge between -(A - B) on the left-hand scale and -A on the right-hand scale. Locate the abscissa at which the value of p is equal to -y. This is a negative root of the equation (Figure 4:6).

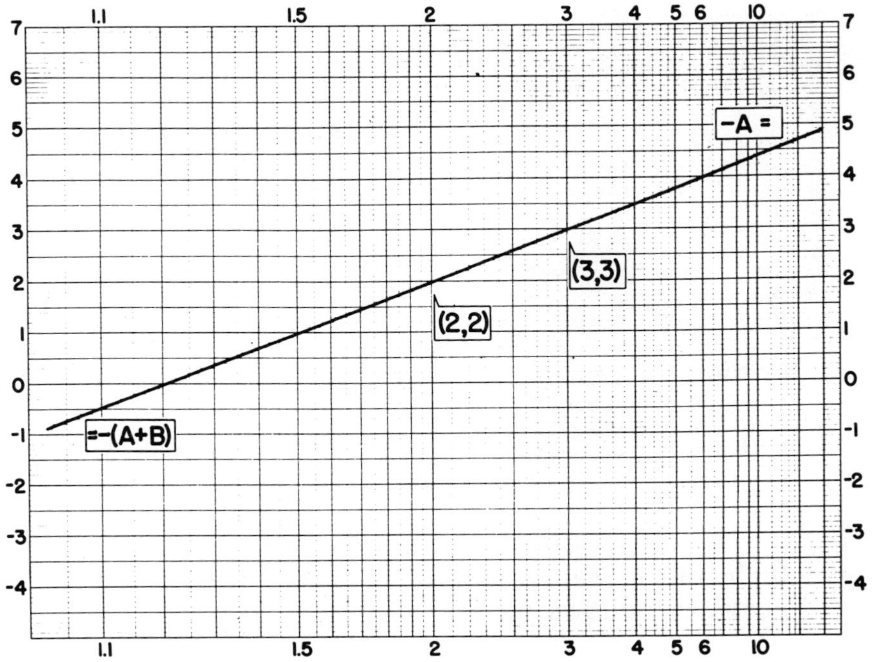

FIG. 4:5 SOLUTION OF THE QUADRATIC EQUATION, $y^2 - 5y + 6 = 0$ Roots are $y = +2$ and $y = +3$

The substitutions given here for the positive and negative roots of quadratics are satisfactory only for equations whose roots have absolute values within a range of 1 to 10. Below this range the roots lie off the hyperbolic scale; above this range accuracy is lost because of the crowding of scale

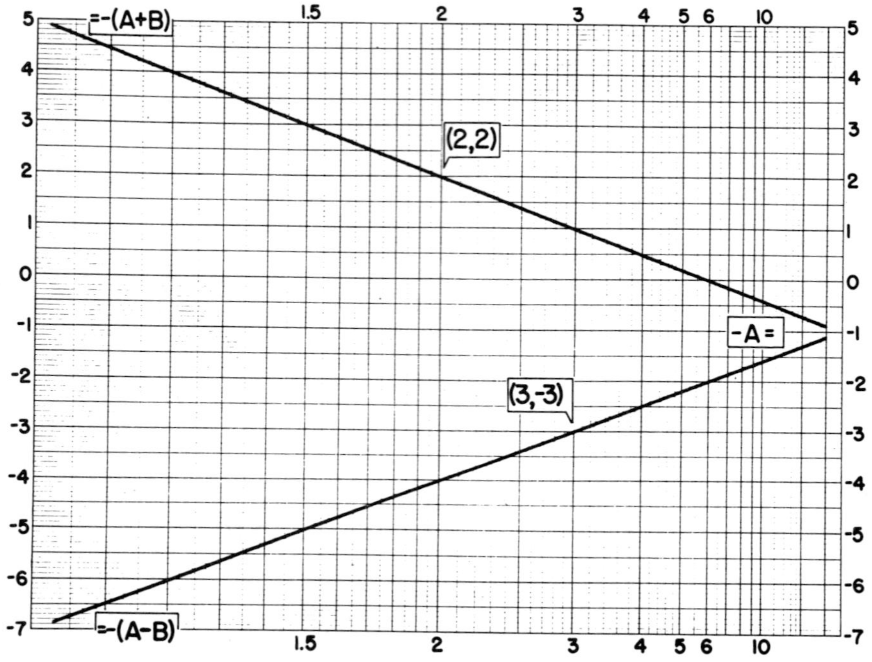

FIG. 4:6 SOLUTION OF QUADRATIC EQUATION WITH A NEGATIVE ROOT

points in the higher values of p. This difficulty can be circumvented as follows:

1. Multiply the scale points of the hyperbolic scale by some factor r, so that the range of the roots lies near the middle or lower portion of the scale.
2. Divide B in the equation of the quadratic by r to obtain a factor used in place of B in the above substitutions mentioned.
3. Proceed as before to determine the roots of the equation; see Figure 4:7. It is usually sufficient to take as the value of r some factor, such as $r = 10^{\pm p}$, where p is a digit—1, 2, 3, . . .

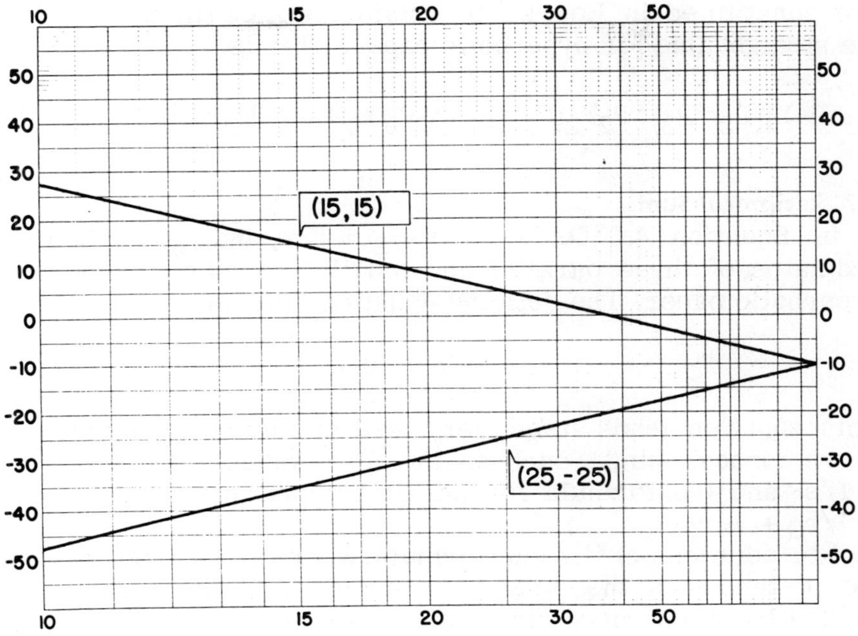

FIG. 4:7 SOLUTION OF QUADRATIC EQUATION:
$y^2 + 10y - 375 = 0$

4.6 Defining Equation, Finite Values of p

If all points on the straight line have finite hyperbolic abscissas,

$$x_1 = \frac{(p_1 - 1)}{p_1},$$

$$x_2 = \frac{(p_2 - 1)}{p_2},$$

and

$$x_3 = \frac{(p_3 - 1)}{p_3}$$

are substituted in Eq. 4.1 to obtain the general equation of the straight line for hyperbolic paper. This is

$$p_1 y_1(p_2 - p_3) - p_2 y_2(p_1 - p_3) + p_3 y_3(p_1 - p_2) = 0. \qquad (4.5)$$

4.7 Reciprocal Sums

In Equation 4.5 let $y_1 = 0$, $y_2 = 1/2$, and $y_3 = 1$, be the ordinates of three parallel and equally spaced scales on the hyperbolic paper. The defining equation then becomes

$$1/p_1 + 1/p_3 = 2/p_2 \qquad (4.6)$$

This equation lends itself very well to solving the formulas of reciprocal sums, such as the lens formula, $1/s_i + 1/s_o = 1/s_f$ and the formula for parallel resistance, $1/R = 1/r_1 + 1/r_2 + 1/r_3 + \dots$.

For solutions of the lens formula, where s_i, s_o, and s_f are the distances of the image, the object, and the focal point, respectively, one may substitute $p_1 = s_i$, $p_2 = 2s_f$, and $p_3 = s_o$. For convenience, the middle scale is renumbered to show values of s_f rather than $2s_f$; see Figure 4:8.

LINEAR PAPER

A number of the calculations that are possible with hyperbolic graph paper may also be performed with linear paper. Linear paper has the advantage of being much more familiar because of its widespread use. Furthermore, as its scales are linear, mathematical relationships are expressible in the simple terms of Equations 4.1, 4.2, and 4.3.

4.8 Simple Averages

Simple averages and certain weighted averages are easily calculated on linear graph papers employing the principles already considered. More involved averages and multiple averages are more difficult to calculate on this paper than on hyperbolic paper.

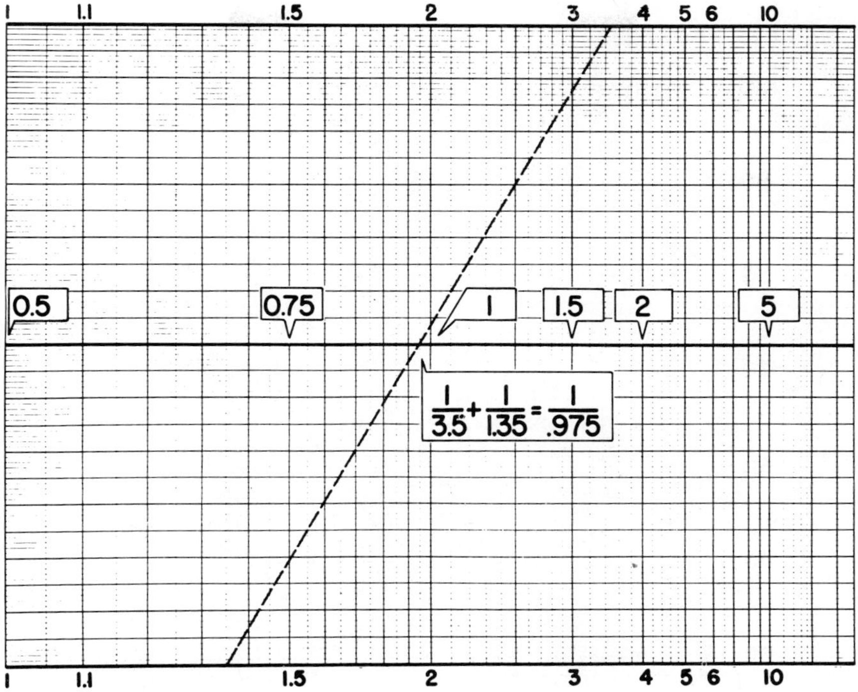

FIG. 4:8 A METHOD FOR RECIPROCAL SUMS
Solution of lens formula, $s_1 = 1.35$, $s_o = 3.5$, $s_f = .975$

4.9 Proportionality

There are two common methods for the representing of proportionality on linear graph paper. In the first method, two vertical axes of the paper are scaled over the ranges to be employed, one reading up, the other, down. The zero points of the two scales are then connected by a straight line, as shown in Figure 4:9. To solve for d in the equation $a/b = c/d$, a is located on one scale, b on the other, and a straightedge is passed through both points. The intersection of the straightedge with the diagonal is marked, and the straightedge is pivoted on this point until it intersects c on the first scale. In this position the straightedge lies at d on the second scale.

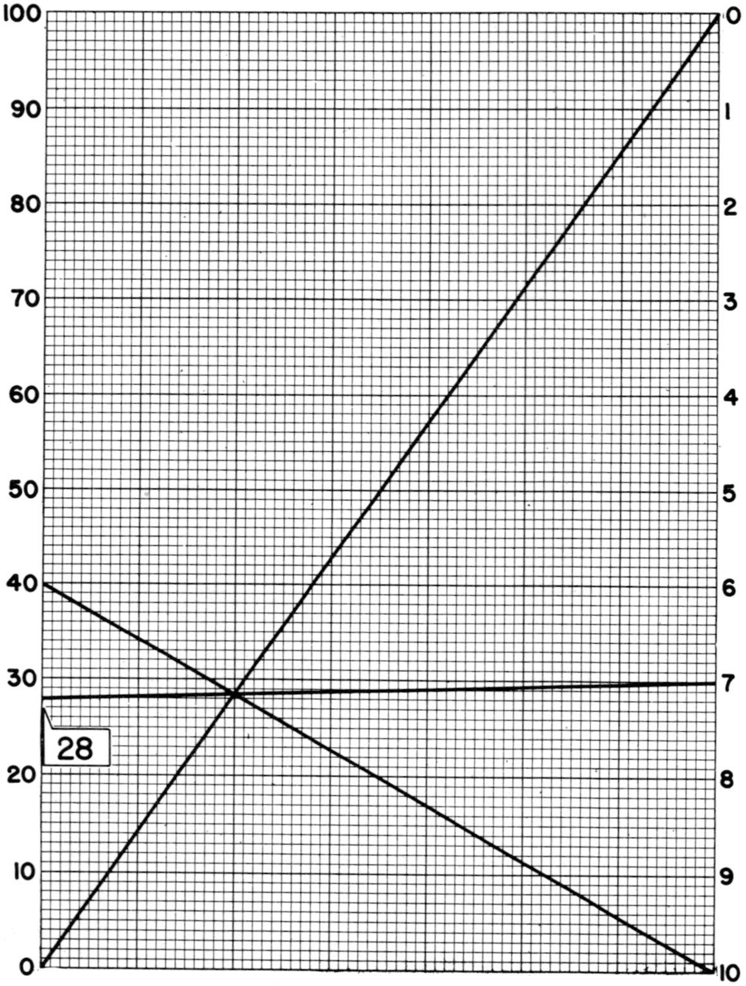

FIG. 4:9 A METHOD FOR THE PROPORTIONALITY, a/b = c/d

If the diagonal be taken as the X axis, the left-hand scale as the Y axis, and the length of the diagonal as unity, the points of intersection of the straightedge with the three lines in its first position are (0,a), (x,0), and (1,-b). Substituted in Equation 4.1, these values lead to the expression,

$$a/b = x/(1 - x).$$

In the same manner, in its second position the straightedge satisfies the relationship,

$$c/d = x/(1 - x).$$

As x is the same in both positions of the straightedge, it follows that $a/b = c/d$.

A second method of representing proportionality is that shown in Figure 4:10. This form is useful in converting grades from some odd basis to a basis of 100. Suppose the total of points that may be earned on a quiz is 72. This number as ordinate is plotted against 100 as abscissa, and a straight line is drawn through this point and the origin. All numerical grades (actual points earned) on this quiz are then easily converted to percentage grades by reference to this line and the abscissa and ordinate scales.

Interpolation is an application of proportionality in cases wherein deviations from linearity of the interpolated function may be disregarded. This is true in the case of logarithms, if the interpolation is made between very close values. If, for example, a five-place table is used to obtain a value for the logarithm of 1.9227, one may read from the table log 1.922 = 0.28375 and log 1.923 = 0.28398. Interpolation involves only the last two digits of the numbers and their logarithms. A cross-plot, as shown in Figure 4:11 establishes the position of a pivot point which is then used to determine interpolated values; in the present case, the value shown on the scale is log 1.9227 = 0.28391. Logarithms are not linear functions, but over such a narrow range non-linearity may be neglected (the value interpolated from six-place tables is 0.2839112).

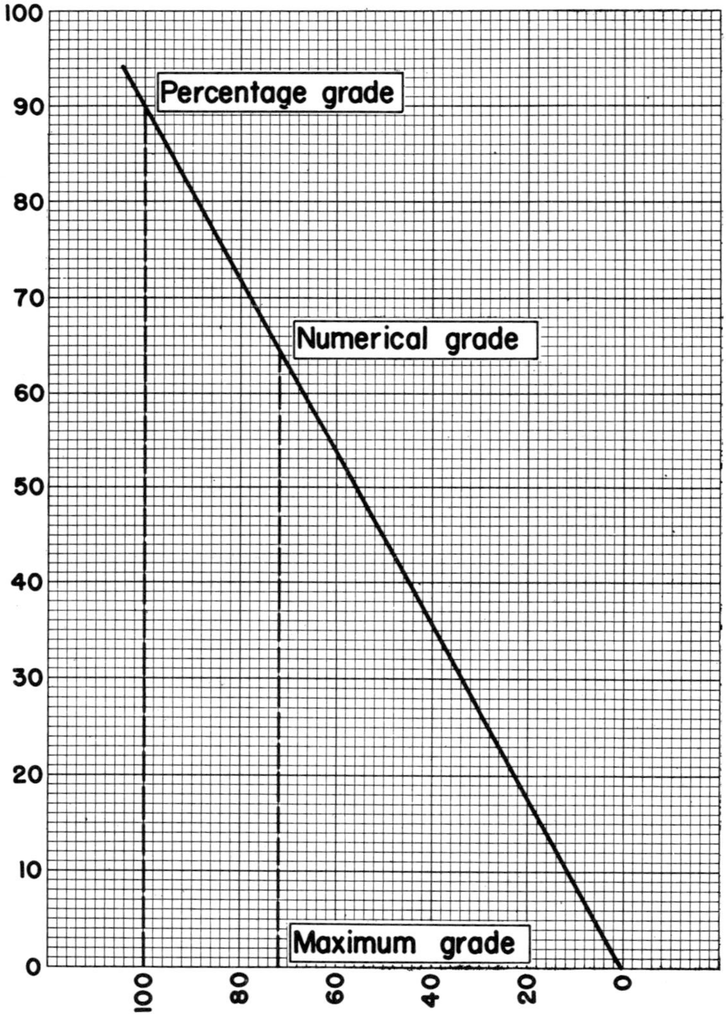

FIG. 4:10 CONVERSION OF GRADES TO PERCENTAGE BASIS

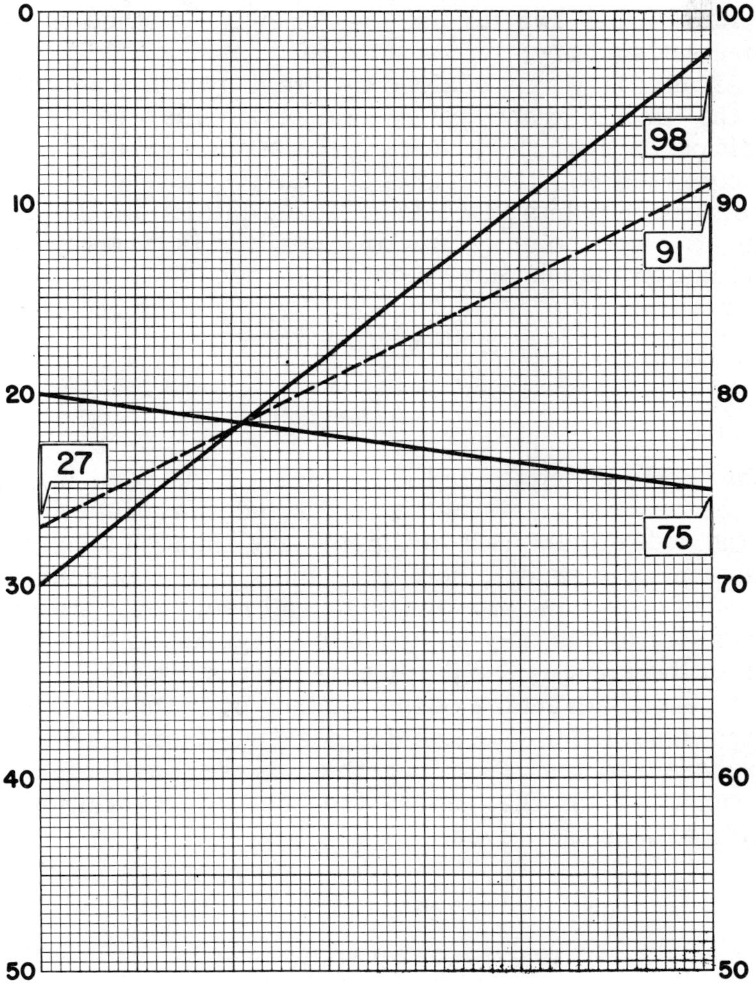

FIG. 4:11 INTERPOLATION BY PROPORTIONALITY

4.10 Reciprocal Sums

In Section 4.7 a method is described for solving reciprocal sums, such as those occurring in the lens formula. For sums of more than two terms, the following method is possibly more satisfactory.

An auxiliary axis of slope M = 1 is drawn through the origin of the linear graph paper (the origin may be taken at any position on the paper). Regardless of the moduli of the X and Y axes, the equation of this auxiliary axis is y = x. An index line through points on the X and Y axes intersects the auxiliary axis at some point (x,y) whose position is given by Equation 4.2. But on the auxiliary axis, x = y; hence,

$$y/a + y/b = 1;$$

or

$$1/y = 1/a + 1/b.$$

That is, the ordinate (or the abscissa) of the point intersected on the auxiliary axis is the reciprocal sum of the intercepts of the index line. This ordinate may then be taken as the y intercept of a second index line, the x intercept being a third term of the sum. In this manner, reciprocal sums of any number of terms may be evaluated. Figure 4:12 shows a solution of the formula for parallel resistors.

A second advantage of this method over that of Section 4.7 is the inclusion of positive and negative terms. The axes may be chosen so that their negative, as well as their positive, portions lie on the paper; the method of solution is then identical with that given.

SEMILOGARITHMIC PAPER

4.11 Defining Equations

One axis of semilogarithmic paper is linear, conforming to the equation $s = mx + 1_o$; and the other is logarithmic, the defining equation being $s = m \log x + 1_o$. Either axis may be taken as the X axis, the other as the Y axis. If the Y axis is taken as logarithmic, the defining equation of a straight index

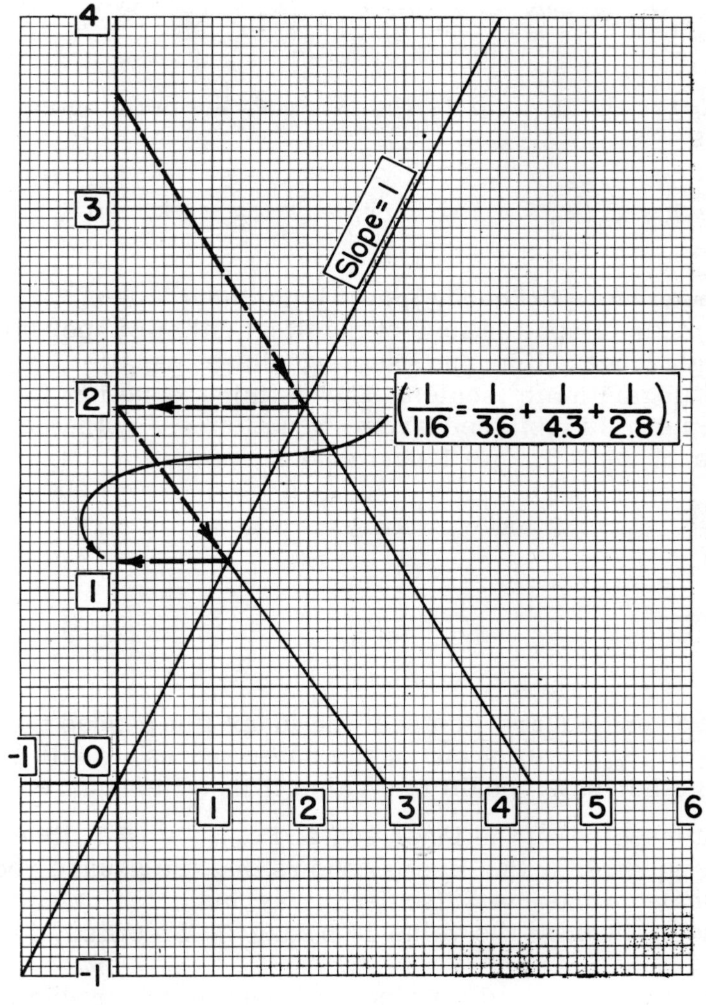

FIG. 4:12 METHOD FOR RECIPROCAL SUMS;
PARALLEL RESISTORS

line, from Equation 4.1, is

$$(x_2 - x_3)\log y_1 - (x_1 - x_3)\log y_2 + (x_1 - x_2)\log y_3 = 0. \tag{4.7}$$

If the logarithmic axis is the X axis, the defining equation becomes

$$y_1 \log x_2/x_3 - y_2 \log x_1/x_3 + y_3 \log x_1/x_2 = 0 \tag{4.8}$$

4.12 Operations with Equally Spaced Logarithmic Scales

A number of applications of semilogarithmic paper involve a straight index line intersecting three parallel, equally spaced, logarithmic scales. The defining equation for this arrangement is obtained from Equation 4.7 with the substitution of $x_1 = 0$, $x_2 = 1$, and $x_3 = 2$:

$$2 \log y_2 = \log y_1 + \log y_3,$$

or

$$y_2^2 = y_1 y_3 \tag{4.9}$$

Squares and Square Roots: If the index line passes through the origin (log 1 = 0) of one of the outer scales, it will intersect the other outer scale at a value that is the square of the value intersected on the inner scale. For, if $y_1 = 1$, then $y_2^2 = y_3$. In Figure 4:13 the line (a) shows that the square root of 2000 (scale A) is 44.7 (scale Y), as indicated by an index line passing through the value 1 on scale B. Likewise, $44.7^2 = 2000$.

Simple Products and Quotients: To obtain the product a × b = c, pass an index line through a (scale A) and b (scale B). Place a pivot at the point of intersection with the center scale (scale Y) and pass the index line through this point and the origin of one of the outer scales; say, scale A. The intersection on scale B is the product desired. This relationship is readily deduced from Equation 4.9 for the two positions of the index line and their common point on the middle scale:

$$a \times b = y^2 = c \times 1 \tag{4.10}$$

This device is illustrated by lines (a) and (b) in Figure 4.13, which show the product 8 × 250 = 2000.

Combined Multiplication and Division: If the second position of the index line passes through some point other than 1, such as d, on scale A the product obtained is (a × b)/d = c. This is demonstrated by substitutions in Equation 4.9 similar to those above. In Figure 4:13 lines (b) and (c) show that the

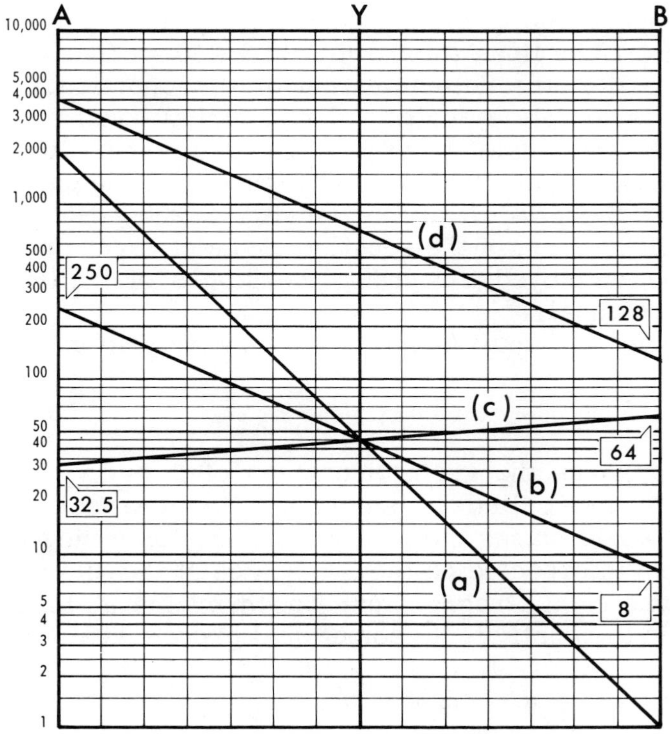

FIG. 4:13 SEMILOGARITHMIC PAPER
For Squares, Roots, Products, Quotients, Geometrical Means, and Progressions

value of the product (8 x 250)/64 is 32.5. This device may be extended to multiple products, such as are found in calculations involving ideal gases. For example, to obtain the volume V_2 from the expression $V_2 = V_1T_2P_1/T_1P_2$, first the product $V' = V_1T_2/T_2$ is obtained. This point on scale B is the starting point for the second product, $V_2 = V'P_1/P_2$.

Proportionality: In more general terms, the defining equation of the device consisting of two index lines crossing at a common point on the middle scale is

$$y_3/y_1 = y_2^2 = y'_3/y'_1 \tag{4.11}$$

The proportionality involving y_1, y_3, y'_1, and y'_3 is illustrated Figure 4:13 by lines (b) and (c): 32.5/250 = 8/64.

In reality, the determination of a proportionality in this device is simply a matter of measuring equal distances on scale A and scale B. These equal distances may also be measured by parallel lines. In Figure 4:13 the parallel lines (b) and (d) show the proportionality 8/250 = 128/4000. The point-slope form of the equation of the straight line (Eq. 4.3) may be employed for proof; the slopes of the two parallel lines are equal.

Geometric Mean: Equation 4.9 may be written

$$y_1/y_2 = y_2/y_3, \tag{4.12}$$

showing y_2 as the geometric mean of y_1 and y_3. Obviously, a single index line is employed; its intersection with the middle scale marks the position of the geometric mean of the numbers intersected on the two scales at the outer extremes.

Geometrical Progression: The numbers a, b, c, d, ... are said to constitute a geometrical progression if

$$a/b = b/c = c/d = \ldots = k;$$

that is, if the ratio of each number to its successor is the same. An example is the series, 1/4, 1/2, 1, 2, 4, 8, ...; the value of k in this series is 1/2. Given any member of the series and the value of k, the rest of the members of the series are readily determined by successive multiplication of that member by k or its reciprocal.

Many practical applications of the geometrical progression, however, require one to determine the value of each member of the progression when given the first and last members and the number of members, but not the value of k. The value of k may be determined from the expression,

$$l = sk^{n-1}$$

where s is the smallest member of the series, l is the largest, and n is the number of terms in the series.

The problem is solved without determining the value of k by the use of semilogarithmic graph paper. The smallest number is located on the extreme left-hand vertical logarithmic scale, and the largest number, on the vertical logarithmic scale n spaces to the right (a space is any convenient unit of length, such as one-half inch, the spacing on most semilogarithmic graph papers). A straight line is drawn between these two points, and the values of intermediate members of the series are read on the logarithmic scales 1 space, 2, spaces, 3 spaces, etc., to the right of the smallest member.

An example of the geometrical progression is the tempered musical scale used on the piano, organ, and other keyboard instruments. The frequency of a given tone is known, and it is desired to know from this the frequencies of the rest of the tones in the scale. The octave of the tone has a frequency twice that of the tone; and, including the tone and its octave, there are thirteen notes in the chromatic scale.

Figure 4:14 shows the solution of this problem for the scale beginning with note “A” below middle “C” that tone has a frequency of 220 sec^{-1}. The value 220 is located on the left-hand scale, and the value 440, twelve spaces to the right. The index line is drawn through these points, and the frequencies of the other notes of the scale are read at the intersections of the index line with intervening vertical scales, as indicated.

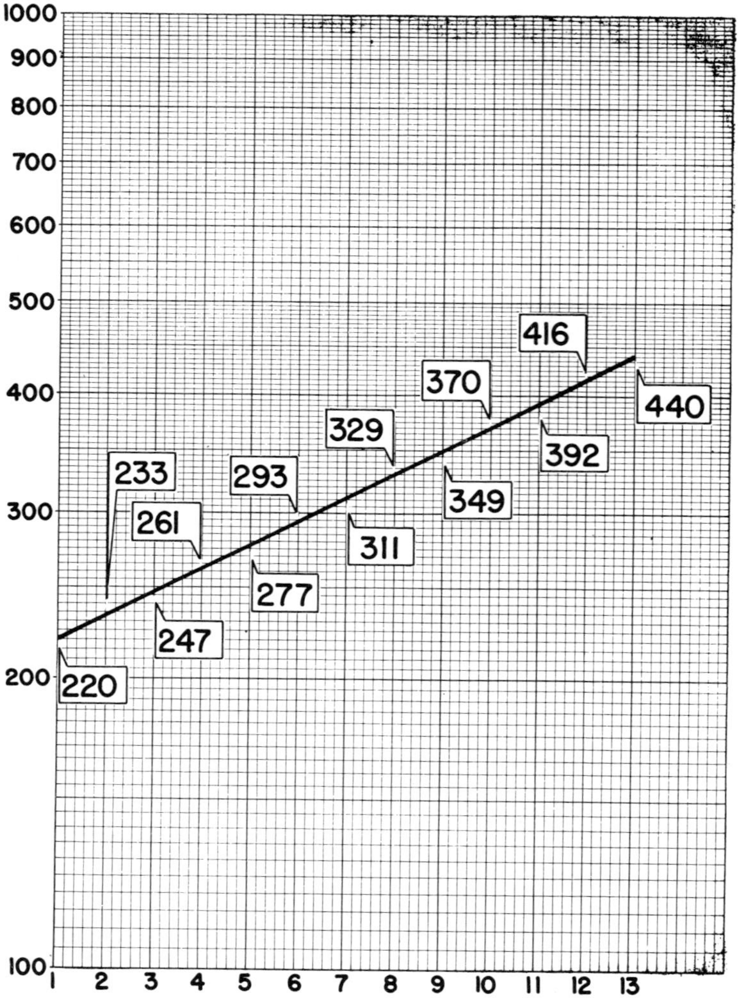

FIG. 4:14 A GEOMETRICAL PROGRESSION; FREQUENCIES OF TONES IN A CHROMATIC OCTAVE
A = 220 Vibration per second, A# = 233,
B = 247, C = 261, etc.

4.13 Fractional Powers

If the Y axis is logarithmic and the X axis is linear, the point-slope form of the equation of the straight line (Eq. 4.3) becomes log Y = Mx + log B. If the slope of the line is then measured in terms of the logarithmic vertical scale (that is, log M) the equation becomes

$$\log Y = x \log M + \log B,$$

or

$$Y = B \times M^{x} \tag{4.13}$$

The slope is determined by the point M on the vertical scale at x = 1; a line passing through this point and the origin has the slope log M. Figure 4:15 illustrates the method of obtaining the power $13.5^{0.73} = 6.67$ and the product $2.7 \times 13.5^{0.73} = 18.0$.

This method is more conveniently worked out with the graph paper mounted on a drawing board, to facilitate the drawing of parallel lines. For repetitive operations wherein the exponent is constant, a permanent vertical ruling corresponding to that exponent is an additional help.

4.14 Products of Fractional Powers

In Section 4.4, a method is described for obtaining the weighted sum, $S = w_1N_1 + w_2N_2 + w_3N_3 + \ldots$ in which the value of N_1, N_2, N_3, etc., are located on the linear axis of hyperbolic graph paper, and the values of w_1, $w_1 + w_2$, $w_1 + w_2 + w_3$, etc., on the hyperbolic axis.

If logarithmic scales are substituted for linear scales on this paper (a log-hyperbolic paper is not available commercially at present), the sum becomes

$$\log S = w_1 \log N_1 + w_2 \log N_2 + w_3 \log N_3 + \ldots$$

or

$$S = N_1^{w_1} N_2^{w_2} N_3^{w_3} \ldots \tag{4.14}$$

Operation of the device is identical with that for weighted sums; the final product is obtained by projection onto the left-hand scale.

In adapting this method to ordinary graph papers, one may cut two logarithmic scales from a sheet of semiloga-

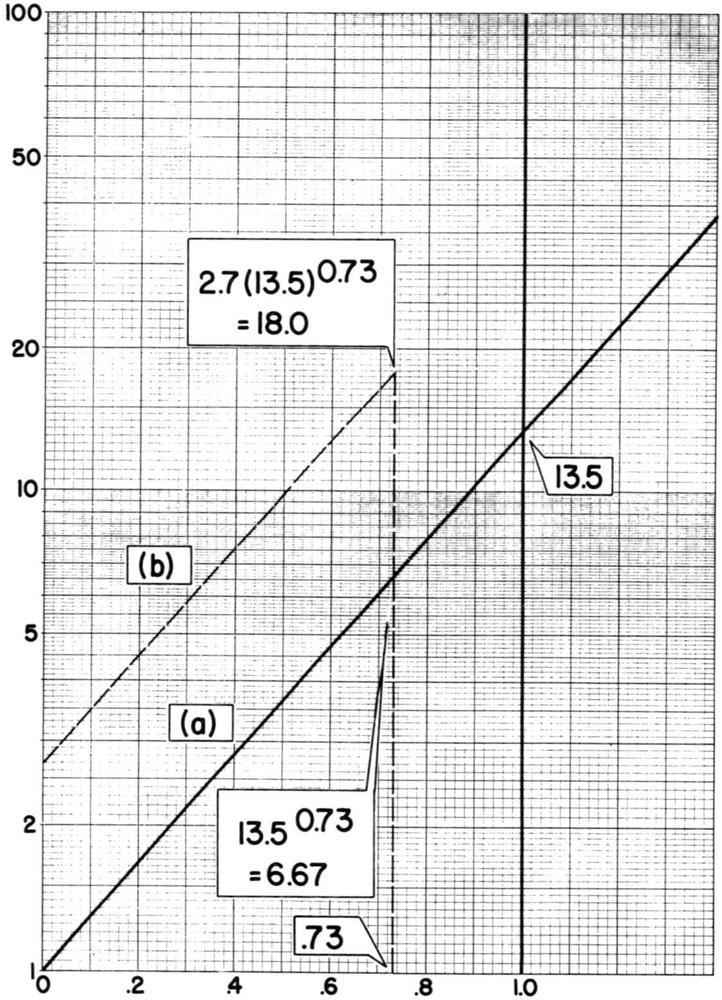

FIG. 4:15 METHOD FOR FRACTIONAL POWERS

rithmic graph paper and paste them along the extreme linear scales of a sheet of hyperbolic paper. As most applications of this device are for repetitive computations, however, it is generally more satisfactory to mark some vertical pivot axes on a sheet of semilogarithmic paper corresponding to the hyperbolic abscissas, $w_1, w_1 + w_2, w_1 + w_2 + w_3$, etc. The positions of these pivot axes may be determined by superimposing an hyperbolic reference scale on the graph paper, after the manner shown in Figure 4.16. This figure also shows the method of computing the product $1.57a^{0.61}b^{2.2}/c^{1.7}$. Nomographs for computing products of fractional powers based upon methods similar to this have been prepared by McMillen[1] and Burrows.[2]

4.15 Natural and Common Logarithms

The relationship between natural and common logarithms is

$$\ln \underline{a}/\ln 10 = \log \underline{a}/\log 10.$$

where <u>ln</u> represents <u>natural logarithm</u>; and <u>log</u>, <u>common logarithm</u>. However, ln 10 is 2.303 and log 10 is 1; hence,

$$\ln a = 2.303 \log a.$$

This expression is related to the point-slope form of the equation of the straight line (Eq. 4.3) by the substitutions, $y = \ln a$, $M = 2.303$, $x = \log a$, and $b = 0$. The graph paper is placed with its log scale as X axis and its linear scale as Y axis. The linear scale is divided into convenient units. A straight line A is drawn through the origin and the point (10, 2.303); see Figure 4:17. The natural logarithm of any number on the abscissa scale is located on the ordinate scale by projecting upward to the diagonal line, then across.

This device may also be employed to read logarithms to any base other than 10 or e. It is necessary to know only (1) the value of the modulus, $m_s = \log_s 10$ (where s is the base of

---

[1] McMillen, Ind. Eng Chem. 30:71 (1938)
[2] Burrows, Ind. Eng. Chem. 38:586 (1946)

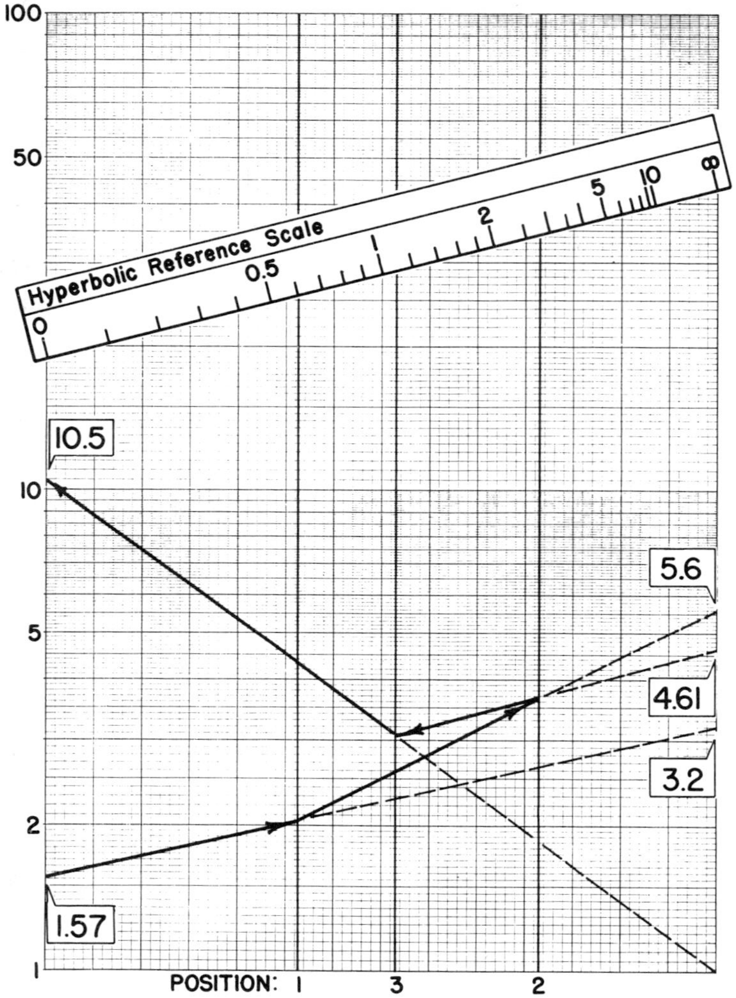

FIG. 4:16 PREPARATION OF SEMILOG PAPER FOR PRODUCTS CONTAINING FRACTIONAL POWERS

the logarithmic system) or (2) the $\log_s$ of any number a, from which the modulus may be determined by the expression, $m_s = \log_s \underline{a}/\log_{10} a$.

Obviously, common logarithms may also be reckoned by this device, the index line being drawn from the origin through the point (1, log 10). The linear scale then gives the values of logarithms of numbers on the log scale.

The index line may be made as long as desired to include a wide range of numbers. At the point of its contact with an edge of the graph, it is continued in the same direction from the opposite edge. This continuation is shown in part in Figure 4:17, line B. In using a continuation of the line, one must bear in mind the position of the decimal point of the logarithmic axis of the scale, and the scalar equivalent of the linear axis, as indicated on each continuation.

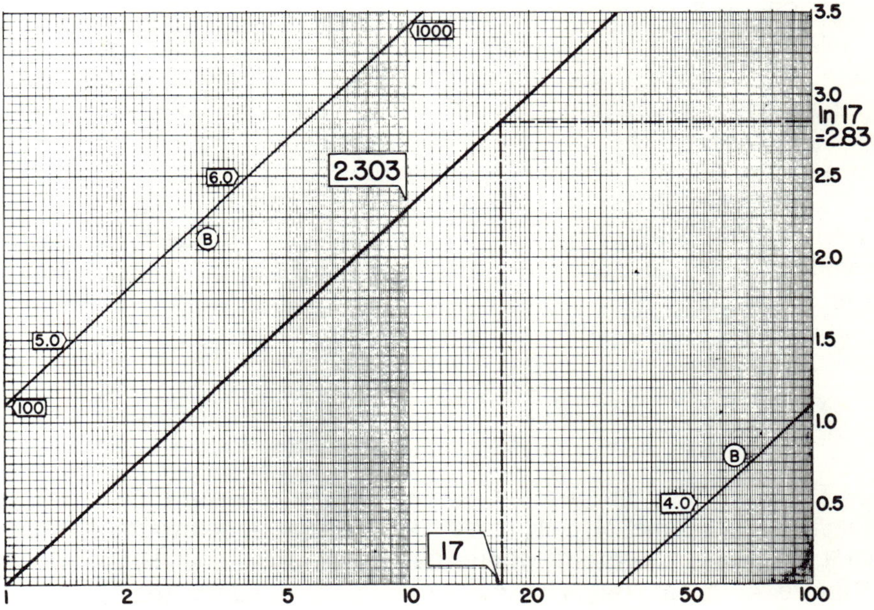

FIG. 4:17 RELATIONSHIP OF NATURAL AND COMMON LOGARITHMS

LOGARITHMIC PAPER

4.16 Defining Equation

The axes of logarithmic paper are defined by $x = m_x \log X$, and $y = m_y \log Y$. Most commercial forms of this paper are constructed with equal moduli on the two axes, $m_y = m_x = m$; without loss of generality, this modulus may be taken as unity. The defining equations thereby become $x = \log X$ and $y = \log Y$.

The point-slope form of the equation of the straight line (Eq. 4.3) is the one most commonly employed with this paper. With the above substitutions suggested for x and y, this equation becomes

$$\log Y = M \log X + \log B,$$

or

$$Y = BX^M \tag{4.15}$$

4.17 Simple Products and Quotients

If in Equation 4.15, $M = -1$, then $B = XY$, and if $M = +1$, $B = Y/X$. In Figure 4:18 at (a) is seen a sheet of logarithmic paper ruled with families of lines of slopes $M = -1$ and $M' = +1$, showing the evaluation of the product $23 \times 8.2 = 188.6$, and of the quotient $23/8.2 = 2.8$.

Extension of this principle to products of several terms is a simple matter of taking the first product as the Y of a new product. This is illustrated in Figure 4:18 at (b), which shows the operation

$$23 \times \frac{8.2}{0.52} = 98.1$$

Division may be performed, as an alternate method, using the lines of slope $M = -1$, by the device that amounts to finding that number X which, when multiplied by B, will give Y. In Figure 4:18, at (a), a reversal of the procedure followed for multiplication would realize the quotient $188.6/8.2 = 23$. This device, although it provides a simplification (by eliminating one set of diagonals) for the method in general, has the further

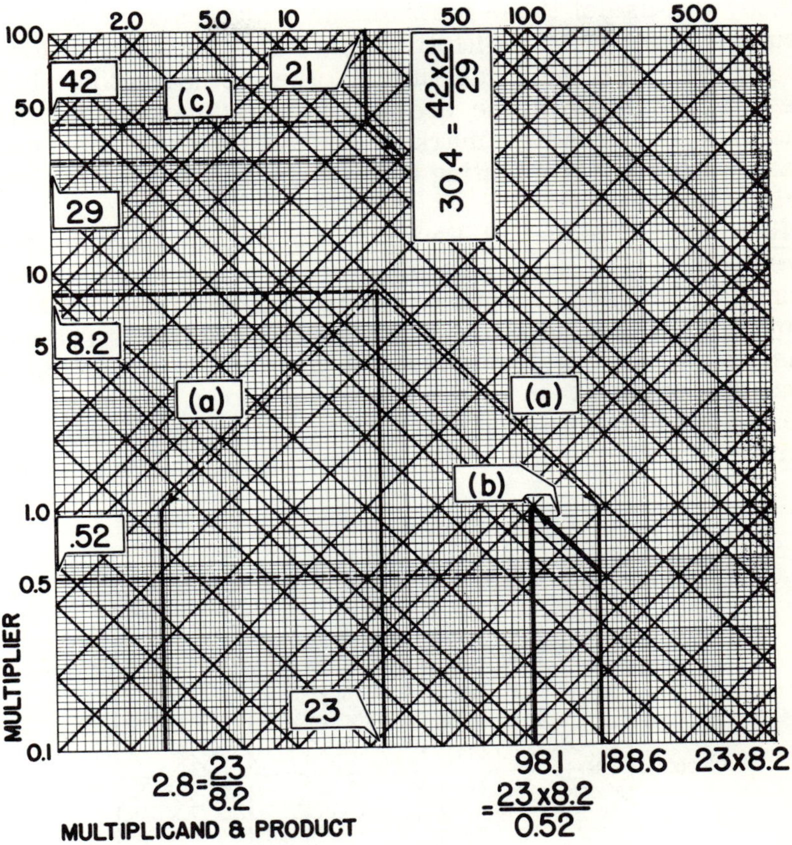

FIG. 4:18 METHODS FOR MULTIPLICATION AND DIVISION

advantage of offering a method for simultaneous multiplication and division. In evaluating the product B = XY/Z, the pivot is moved along the diagonal (M = -1) from the intersection of X and Y to the ordinate Z, rather than the ordinate 1. The abscissa of the point is the value of B. This scheme is illustrated in Figure 4.18 at (c), showing 42 × 21/29 = 30.4.

A 45° triangle (or a T-square set at an angle of 45° to the axes of the paper) greatly facilitates application of the above method and obviates the need of the diagonally ruled lines.

Indeed, the use of these instruments provides a rapid, fairly accurate method of multiplication, which has the following advantages over the slide rule: (1) the method automatically fixes the position of the decimal point; (2) commonly used constants are readily marked on the paper by horizontal lines, virtually eliminating the necessity of locating them during the calculation.

Reactance-Frequency Paper: A commercial graph paper, developed by the Bell Laboratories for solving quickly relationships between reactance, capacitance, inductance and frequency, makes use of the foregoing principles. The paper consists of two log-log grids, one of which is superimposed upon the other at an angle of 45° as shown in Figure 4:19. The lines 10^n on the diagonal grid intersect the Y axis at the points $2\pi \times 10^n$. Each grid is composed of two pencils of

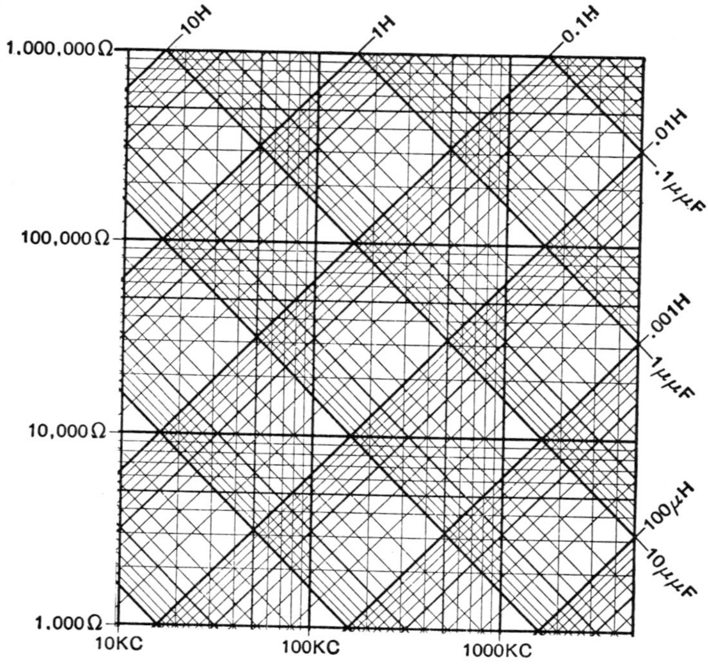

FIG. 4:19 SECTION OF REACTANCE-FREQUENCY PAPER

parallel lines at right angles to each other, the equations of the four pencils being:

a) The horizontal pencil, $y = \log X$ (where X is the reactance, measured in ohms),

b) The vertical pencil, $x = \log f$ (where f is the frequency, measured in sec.$^{-1}$),

c) The pencil with slope -1, $x + y = -\log 2\pi C$, (where C is the capacitance, measured in farads), and

d) The pencil with slope +1, $x + y = \log 2\pi L$, (where L is the inductance, measured in henrys).

At any point on the grid, $X_C = 1/2\pi fC$ (Eqs. a, b, c,) and $X_L = 2\pi fL$ (Eqs. a, b, d).

4.18 Products with Fractional Powers

If in Equation 4.15 M has values other than -1, then products with fractional or other powers result. For instance, by using slopes of -0.61, -2.2, and 1.7, the product

$$\frac{1.57 \times 3.2^{0.61} \times 5.6^{2.2}}{4.61^{1.7}} = 10.5$$

is evaluated in Figure 4:20. The primary difficulty in this method is not that of accumulating the product but of setting the slopes. There are three convenient methods of accomplishing this:

1. Use of an adjustable triangle (Lesh angle) on which the protractor is calibrated in slopes rather than in degrees (used in conjunction with the (T-square this device gives rapid readings easily);
2. Use of a protractor or ruled card, scaled in slopes, in conjunction with a T-square or right triangle;
3. Use of a chart of slopes to set the slope. The slope may be transferred to the graph paper by the T-square-triangle method of drawing parallel lines.

As these methods are obvious to those familiar with the use of drawing instruments, it is unnecessary to illustrate them here.

An elaboration of this method especially applicable to set formulas such as formulas with fixed exponents is shown in

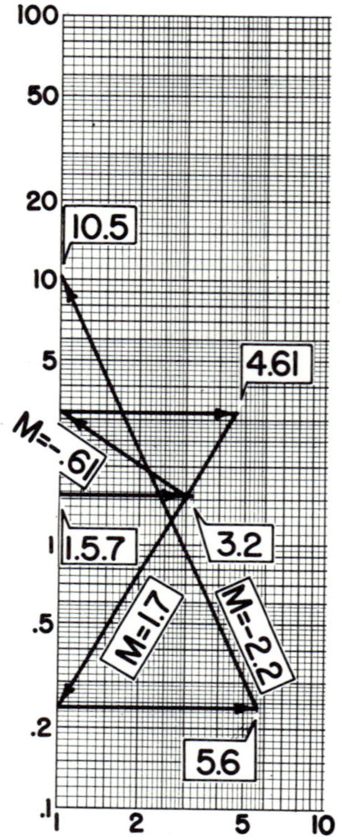

FIG. 4:20 PRODUCTS OF FRACTIONAL POWER TERMS

Figure 4:21. The product obtained in the lower section is projected upon the Y axis of the section just above it, where it serves as a term in the new product. This product from the second section is then projected upon the Y axes of the third section, and so on until the product is complete. This device is similar in principle to the "run-around" chart described in Section 6.2.

To adapt this method to logarithmic polycycle paper, it is desirable to transpose X and Y axes so that X axes are vertical and Y axes horizontal. This rearrangement must be taken into consideration in determining the slopes of the diagonal lines in each section. The method is illustrated in Figure

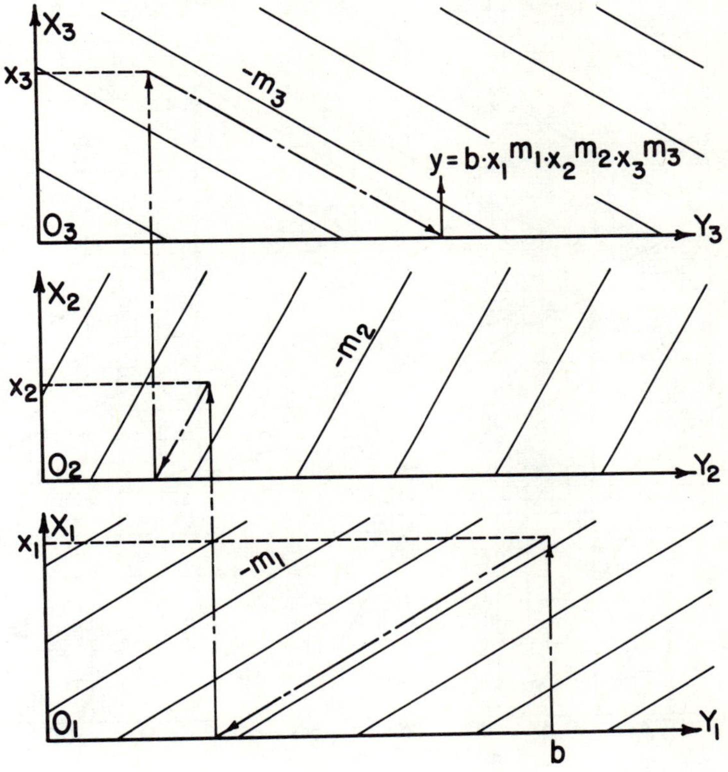

FIG. 4:21 METHOD FOR PRODUCTS IN FIXED MATHEMATICAL EXPRESSIONS

4:21. In Figure 4:22 application is made to the formula

$$x = \frac{1.57\, a^{0.61}\, b^{2.2}}{c^{1.7}}.$$

A third method of evaluating products with fractional powers is based on a transposed form of Equation 4.15: $Y = BX^M$. If $B = 1$, then $Y = X^M$. A family of lines of varying slopes is laid off on logarithmic paper. The value of any

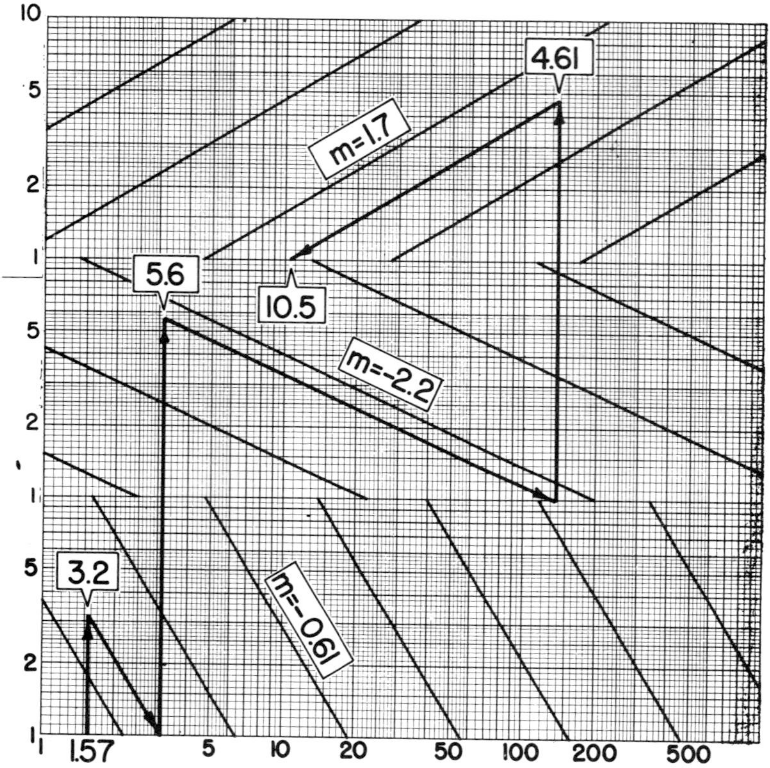

FIG. 4:22 CHART FOR THE EXPRESSION

$$x = -\frac{1.57\,a^{0.61}\,b^{2.2}}{c^{1.7}}$$

term, X^M is then the ordinate of the line with slope M at a point whose abscissa is X. A product of such terms may be accumulated with dividers or an equivalent instrument. Each term, X^{M_1}, $X_2^{M_2}$, etc., in the product is "lifted" with the dividers and transposed to the Y axis, where it joins the product of terms already there.

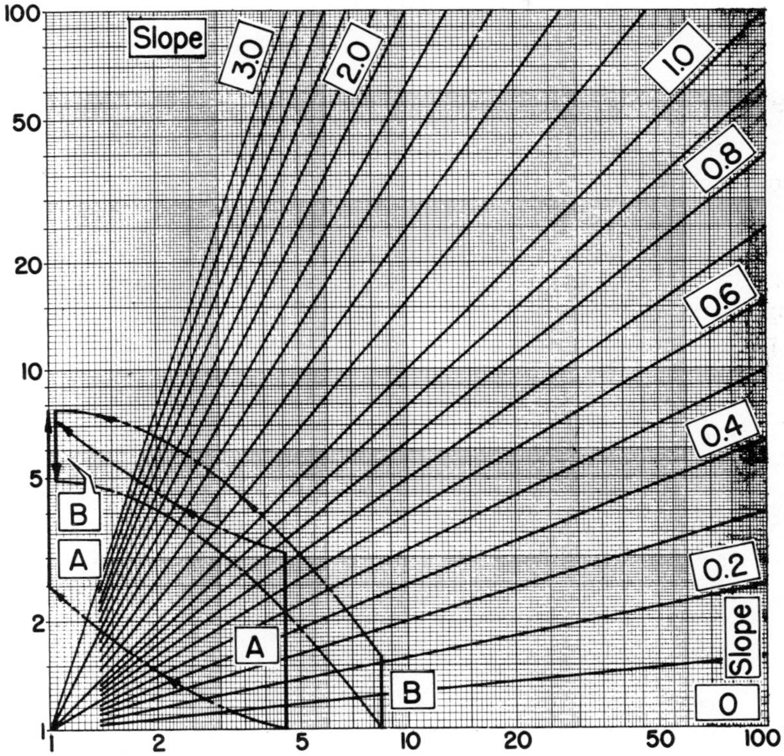

FIG. 4:23 PRODUCTS OF FRACTIONAL POWER TERMS

This method is illustrated in Figure 4:23, which shows the product

$$\frac{2.5 \times 4.5^{0.74}}{8.4^{0.88}}$$

Obviously, this is another application of the method of evaluating a product by adding the logarithms of its terms.

All graph papers used as illustrations in this chapter were supplied by Keuffel & Esser, Hoboken, New Jersey.

EXERCISES

Use graph papers to solve each of the following problems:

1. During one term a teacher gave eight quizzes occupying a total of 250 minutes, as follows: (1) 50 min.; (2) 25 min.; (3) 10 min.; (4) 20 min.; (5) 50 min.; (6) 20 min.; (7) 25 min.; (8) 50 min. Each quiz was graded on a basis of 100, and the quizzes for the term were averaged, each being weighted in proportion to the time it consumed. On a piece of hyperbolic graph paper mark off vertical lines for the sums, w_1, $w_1 + w_2$, etc., and obtain the weighted averages of grades for the following students.

Quiz No.:	1	2	3	4	5	6	7	8
Student								
"A"	85	75	78	64	88	60	72	77
"B"	56	64	70	63	78	72	58	75
"C"	90	92	98	82	80	77	85	95
"D"	92	72	70	83	63	74	98	92
"E"	74	79	85	88	66	62	72	80

2. Using the method for weighted averages and weighted sums, calculate (a) cost of a hundred pounds, and (b) the total cost of an alloy of the following composition: Tin, 146 lb at 26.00 dollars/cwt; Antimony, 74 lb at 21.50 dollars/cwt; and Lead, 1,780 lb at 6.50 dollars/cwt.

3. Find both roots of the following quadratic equations: (a) $x^2 + 7.6\,x - 5.3 = 0$, (b) $x^2 - 0.25\,x + 1.00 = 0$, (c) $x^2 - 1.78\,x - 4.20 = 0$.

4. Derive a method similar to the method for quadratics for solving the reduced cubic equations: (a) $x^3 + ax + b = 0$, and (b) $x^3 + ax^2 + b = 0$.

5. What is the focal distance of a lens through which an object at 25 inches casts an image at 12.5 inches? What is the

distance from lens to image of an optical system in which the object is placed 8 inches from a lens having a focal distance of 30 inches?

6. What is the resistance of a set of parallel resistors of 8, 15, 2.5, 4, and 10 ohms?

7. Evaluate the following: (a) $(17.3)^2$, (b) $\sqrt{46.4}$, (c) 12.6×27.2, (d) $68 \times 33.6/18.2$.

8. In the design of machine parts, that are identical except for size, it is best to manufacture them in sizes of which the distribution is approximately that of a geometrical progression. This is known commercially as a preferred number distribution. A certain manufacturer wishes to make a series of gears of a certain design in sizes from two inches to twenty inches in diameter. Determine what additional sizes would be included if the series were to contain (inclusive of the 2 and 20-inch sizes), 10 gears, 12 gears, 15 gears, or 20 gears. Round each size off to the nearest half-inch.

9. Evaluate the following: (a) $6.32^{2.5}$, (b) $1.96^{0.44}$, (c) $3.72 \times 7.6^{1.67}$, (d) $4.80 \times 1.88^{0.22}$, (e) $0.125 \times 7.6^{1.3} \times 0.26^{4.3}$,

(f) $\dfrac{0.0035 \times 73^{0.22} \times 6.6^{1.54}}{2.45^{0.8}}$ (g) ln 1.86, (h) ln 0.466.

10. An electrical circuit contains a reactance of 45,000 Ω (ohms) and a capacitance of 7.0 $\mu\mu$F (micromicrofarads). (a) What is the frequency of its oscillation? (b) What inductance would have to be matched with 45,000 ohms reactance to provide the same frequency?

chapter 5

GRAPHS

The graph is well known for its role in the presentation, storage, and correlation of information in industrial and scientific environs. In preceding portions of this book, the graph is employed in the construction of scales and scale pairs and in the graphical interpolation of scale points. In later chapters, the resolution of graphs into nomographs is presented, as well as various types of computations involving graphs. Clearly, in all fields of graphical method, it is essential to have a thorough knowledge of the nature and anatomy, as well as the powers and limitations, of the graph.

RELATIONSHIP OF GRAPH TO TABLE AND FUNCTION

5.1 Graph and Table

The graph is a series of points, each corresponding to an abscissa and an ordinate defined by the arguments and function values expressed in a table. A numerical table is the only feasible starting point in the construction of a graph; a graph that presents points corresponding only to the values given in the table is said to be uninterpolated.

The process of determining values of intermediate points in either the table or the graph is known as interpolation. Interpolation of values in a table is a computational procedure, generally made with the assumption that between adjacent values in the table a simple functional relationship, such as a linear relationship, may be applied; otherwise the computation required for interpolation is apt to be very laborious. In the graph this assumption is not necessary, as it is possible to approach more nearly the actual functional relationship by constructing a smooth curve that passes through all of the plotted points. Figure 5:1 shows (a) an uninterpolated graph, and (b) an interpolated graph of a set of data.

Under certain rather rigid conditions it is possible to extend the curve of a graph beyond the range of plotted points, a process known as extrapolation. Figure 5:1 (c) shows various forms of extrapolation which might be possible with the same data. It is obvious that extrapolation cannot be accurately executed on the sole premise that the extrapolated points lie on the same smooth curve as the plotted points; something must be known of the function beyond the region of the plotted points to serve as a basis for extrapolation.

5.2 Graph and Function

A graph may be constructed to represent a mathematical function by the indirect process of constructing a table first. Selected values of the independent variable, as arguments, and corresponding values of the dependent variable, as results, comprise the table, from which the uninterpolated graph is constructed. The extent to which the interpolated graph corresponds to the function is determined largely by the wisdom and labor put into drawing the curve. At best, the graph, like the table, represents the function over a range limited to the values of the argument of the table.

If the graph is constructed from a table of empirical data, means are available for finding a mathematical expression, or group of such expressions, whose functional relationships are approximately those of the graph. This process, known as analysis, makes use of the defining equations of the graph papers on which the data are plotted, and the equations of the straight line, conic sections, and other standard curves of analytical geometry. A mathematical expression thus derived from a graph may be reliably applied to computations involving values of its variables within the ranges represented in the graph. Application of the formula to values of a variable outside these ranges is a form of extrapolation.

GRAPHS OF REGULAR FUNCTIONS

5.3 Computation of Points

A graph of a function is prepared from a table of values computed from the function, one value of the function for each

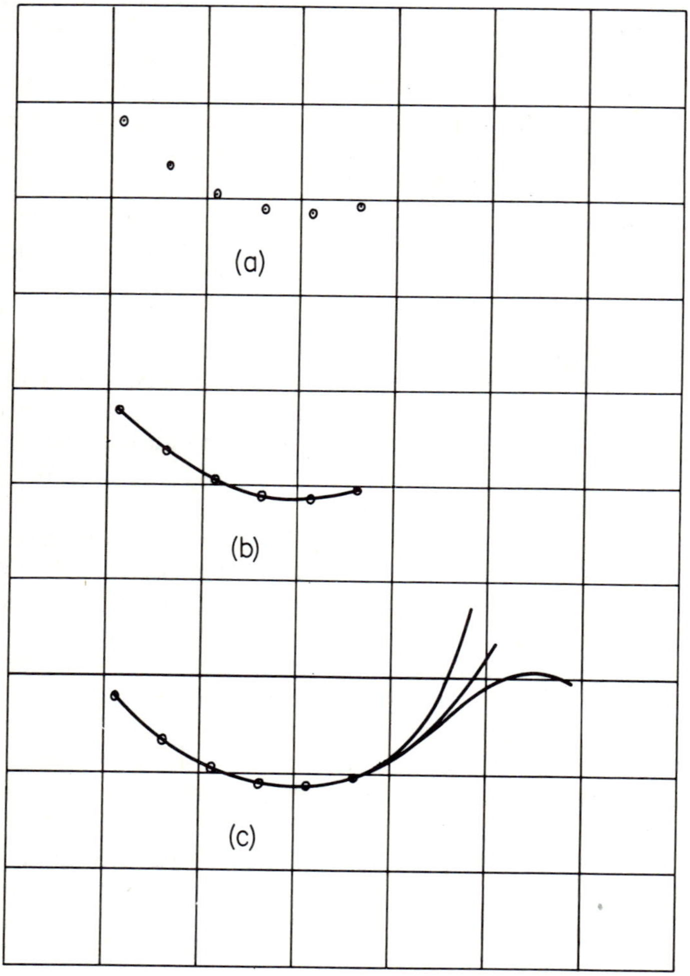

FIG. 5:1 GRAPHS
a) Uninterpolated; b) Interpolated; c) Various Curves Resulting from Extrapolation

value of the argument to be represented. If an interpolated graph is to be prepared, the accuracy of the curve will depend very largely upon the number and distribution of the computed values, and upon other information about the function.

If, for example, it is known that the function to be represented is a linear one, two values are sufficient; a third may be computed as a check. If it is known that the graph of the function is a circle, three values are sufficient, with a fourth for a check. Both the straight line and the circle are special cases of the general form known as the conic sections. If it is known only that the plotted curve is one of the conic sections, more than four points are required for the plot, for it is possible to construct either an hyperbola or an ellipse through the same four points.

In general, graphs encountered in representing engineering formulas are much less regular than any of the standard curves of analytical geometry, and a large number of points must be established if the interpolated graph is to be reasonably accurate. Nevertheless, full advantage should be taken of any additional information at hand concerning the nature of the curve.

Figure 5:2 (a) shows three points through which a straight line may be drawn, as indicated. However, it may be known that the function to be represented is a modified third degree function, which cannot be represented by a straight line. Additional points are plotted, (Fig. 5:2, b), showing that the three original points lie at a flex point of the curve. For a reliable curve to be drawn, even over the range of the original points, it is highly desirable to establish additional points on either side of this range as well as within the range itself.

The density of points required for accurate interpolation usually varies widely over the range of values represented. In Figure 5:2 (b), the density of plotted points must be considerably greater in the range of the flex point than in the extremities of the curve. To some extent this is accounted for by our additional knowledge of the nature of the curve; namely, a third degree curve has only one flex point, and approaches asymptotes in opposite directions in its extremes.

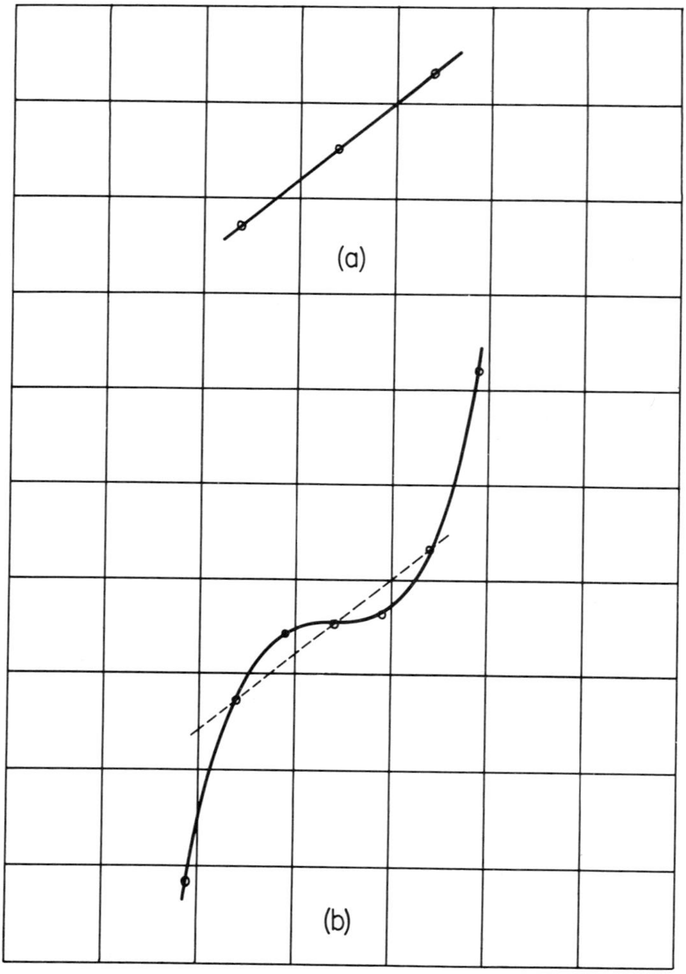

FIG. 5:2 POSSIBLE POSITIONS FOR THREE COLINEAR POINTS
a) On a Straight Line; b) On a Third-degree curve

Before any computations are made towards plotting the curve, careful thought should be given to the matter of choosing the simplest computations capable of giving the desired results. In cases such as the expression $y^3 + 6.2y + 0.135 = 0.15x^2$, it frequently happens that x is limited as to the values it may take (as, for instance, the statistical N, which is limited to integers). A table prepared to represent this formula would be virtually useless if it were constructed in any way other than with x as argument and corresponding values of y as results. The preparation of such a table is, nevertheless, laborious, for it requires the solution of a reduced cubic equation for every value of the argument.

If the table is prepared solely for the construction of a graph a great deal of labor and time may be saved by taking y as the argument and computing corresponding values of x. Insofar as the interpolated graph is concerned, the results are identical.

5.4 Properties that Aid in Plotting

There are various properties of the curve that are helpful in plotting it. In some cases these properties are more easily computed than are the positions of points; in others the computation of a single property, such as the position of a maximum of minimum, may obviate the computation required for the location of several points. Another computation, such as the location of a flex point, may be useful as a guide to the density of points required before the positions of any points are calculated.

Among the properties of the curve that facilitate the plotting of it are (a) the values of its x and y intercepts, and (b) the loci of its axes or points of symmetry, if any. These properties are determined from the equation of the curve; methods for determining them are given in most elementary texts in analytical geometry.

From the first derivative of the function, one may determine the slopes of the curve at various points and the positions of maxima and minima. The positions of flex points may be determined from the second derivative of the function. These properties and methods of obtaining them are treated in elementary texts in differential calculus.

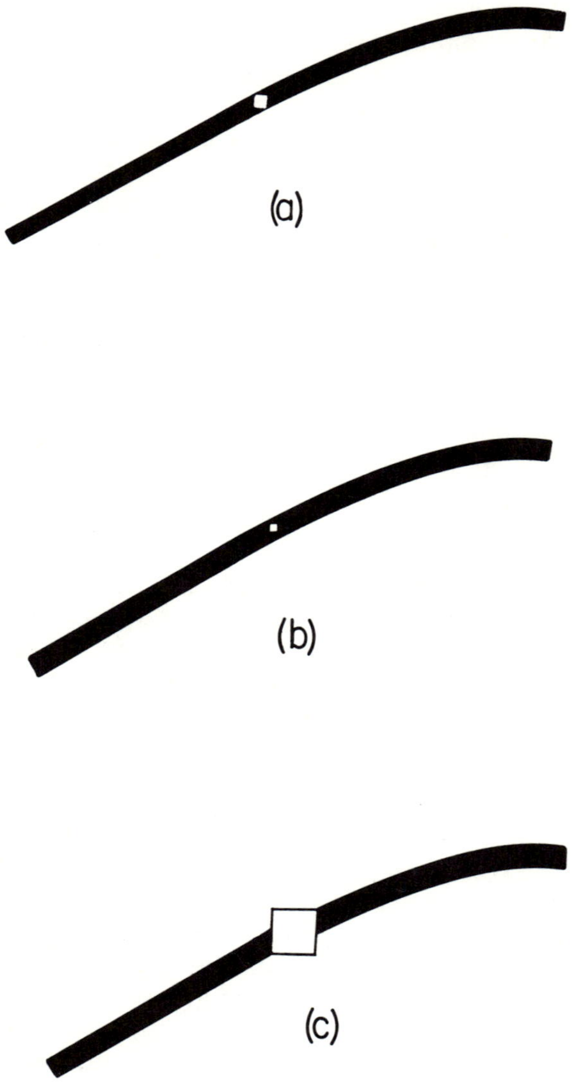

FIG. 5:3 SIZE OF POINT RELATIVE TO THICKNESS OF LINE

5.5 Accuracy of Computation

Every point in a graph, whether computed from a function or arrived at empirically, should indicate the degree of accuracy with which it was determined. In the case of a theoretical function, such as $y = \sin x$, $y = xe^{-x}$, etc., no limits of accuracy are implied; therefore, the accuracy of the computation should be determined by the moduli of the axes of the graph paper on which the plot is to be made and the thickness of the locus that is to be drawn.

If a locus of 0.01 inch thickness is to be drawn on a graph sheet having moduli of 0.1 inch on each axis, both x and y should be determined within ± 0.05 at most. A point whose abscissa is given as 0.23 and whose ordinate is given as 2.55 will then occupy itself within the area whose abscissas range from 0.18 to 0.28 and whose ordinates range from 2.50 to 2.60. This area, on the sheet mentioned, is a square 0.01 inch on a side; hence, the thickness of the point does not exceed the thickness of the locus. Figure 5:3 shows such a point in comparison with points of other dimensions.

A range slightly less than the thickness of the line might be desirable in some cases, but excessive accuracy in the computation of points is time-consuming and contributes nothing to the accuracy of the graph. In general, a safe rule is to carry the computation to a degree of accuracy such that

$$\frac{d}{4m} \leqslant \text{max. error} \leqslant \frac{d}{2m} \qquad (5.1)$$

where m is the modulus of the axis, and d, the thickness of the locus to be drawn.

The maximum error of Equation 5.1 is not necessarily the maximum error in the value of the parameter, but, rather, the maximum error in the value of the function plotted. The function that appears on the graph paper is determined by the equation of the locus and the defining equations of the axes of the paper. Thus, if the abscissa $p = x^2$ is plotted on hyperbolic graph paper for which the defining equation is $s = m(p-1)p$, the function appearing as a graph on the paper will have as its abscissas the values of $m(x^2 - 1)/x^2$. If the maximum error

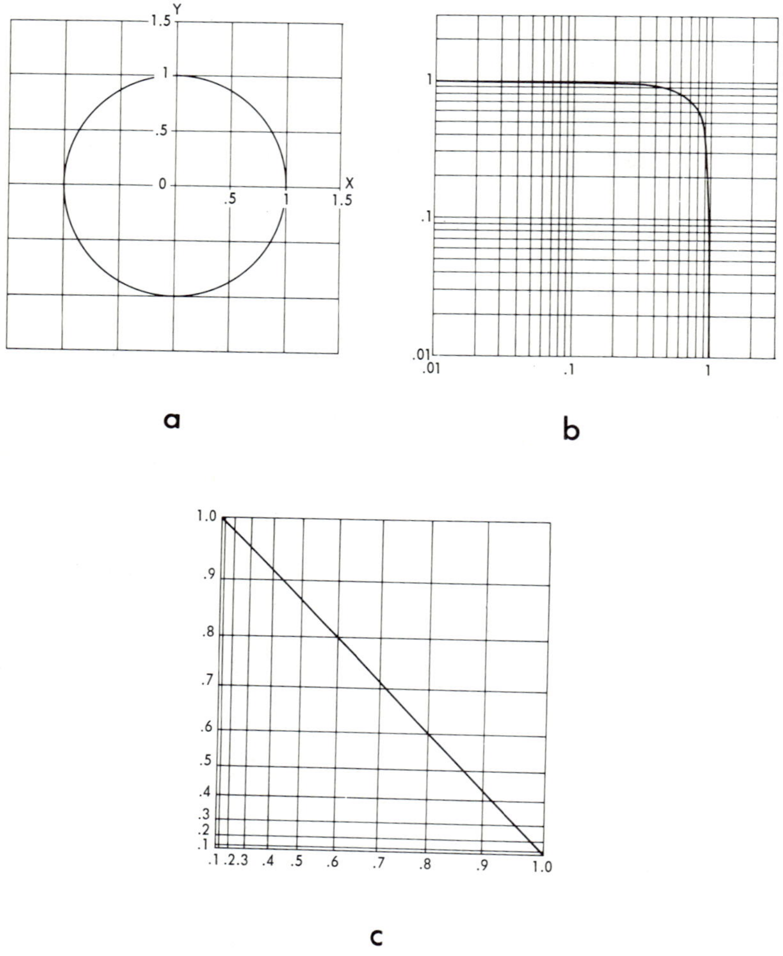

FIG. 5:4 GRAPHS OF THE EXPRESSION $x^2 + y^2 = a^2$
a) On Linear Graph Paper; b) On Log-log Graph Paper:
c) On a Special Graph Paper. Defining Equations of
the Axes in (c) are $s_x = mx^{1/2}$, $s_y = my^{1/2}$

of this function of x is d/2m, the maximum error of x may be given as Δx, then

$$\frac{(x+\Delta x)^2 - 1}{(x+\Delta x)^2} - \frac{x^2 - 1}{x^2} = \frac{d}{2m} \qquad (5.2)$$

The value of Δx in terms of x, d, and m may then be obtained by solution of this quadratic equation. Rigid treatment is not, however, in order in a case as complicated as this; a rough estimate of the variation of Δx with x is sufficient to insure the required degree of accuracy in computing abscissas of the graph without expending too much time on excessive computations.

5.6 Choice of Graph Paper

If the graph is prepared for the purpose of making further computations (see Chapter 6), the graph paper must be chosen to meet fully the conditions required in those computations. The primary requisites are (a) accuracy slightly in excess of that required in the computations, and (b) a form of graph readily adaptable to the intended computations.

The maximum allowable error in the values of the variables represented by the abscissas and ordinates of the graph are Δx and Δy, respectively. Depending upon the nature of the defining functions of the coordinates of the graph and the axes of the graph paper, an expression (of which Equation 5.2 is an example) may be set up for each coordinate. The values of x, Δx, and d thereby establish the value of the modulus m. If this value is dependent upon the value of x, its maximum value must be chosen; otherwise, for some value of x, Δx would exceed the allowable error of x. A similar procedure leads to the required value of the modulus of the Y axis. In this application, d is the minimum displacement of a point clearly discernible to the eye: about 0.025 inch.

The modulus of the axis multiplied by the range of the function measured on that axis gives the length of axis required for the plot; i.e., the size of graph paper to be chosen.

The following is a simple illustration of the selection of a modulus.

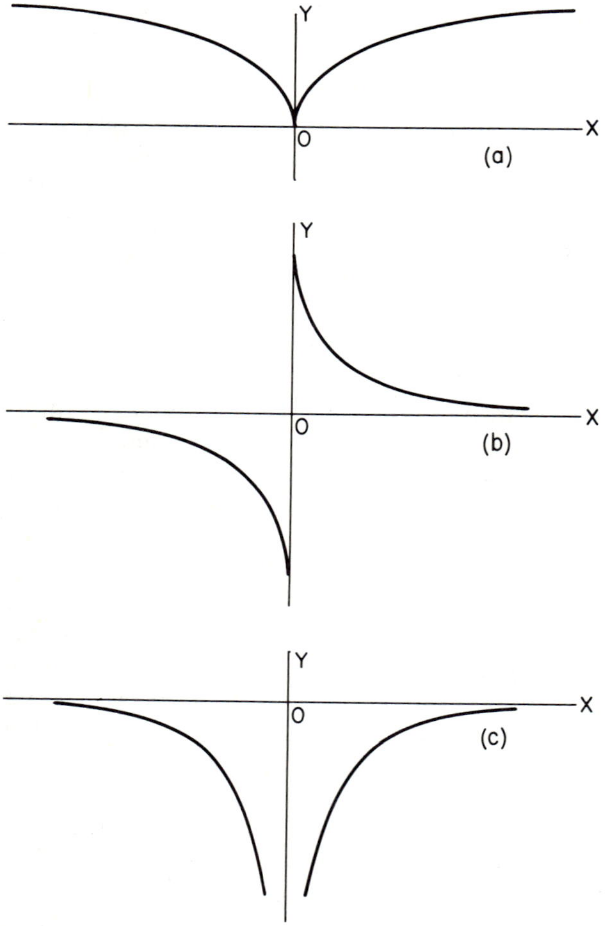

FIG. 5:5 TYPES OF CURVES
a) Curve Showing a Discontinuity; b) First Derivative, and
c) Second Derivative of Curve in Vicinity of Discontinuity.

In preparing the graph of freight rates for cotton (Sec. 6.4) it was known that, although the rates change at 5, 10, or 20 mile brackets, it would be necessary to read abscissas accurately to the nearest mile; i.e., $x = \pm 1/2$ mile. The minimum displacement of a point readily visible is about 1/40 inch; this may be taken as the value of d_x. From Equation 5.1,

$$d_x = 2\Delta s_x = 2\,m_x\; f\,(x + \Delta\, x) - f\,(x)$$

In this expression, $f(x) = x =$ no. mi.; hence,

$$d_x = 2m_x\;(\Delta x)$$

From the given values of $\Delta\, x$ and d_x, m_x has the value $\frac{1}{20}$ in./mi. The range is 0 to 500 mi.; the length of the X axis, therefore, $1/20 \times 500 = 25$ inches.

The following chapter indicates the form of graph most suited to the computation to be performed. This is a matter that is decided largely by the nature of the steps involved in the computation. The vast differences in the shape and form of the graph of a given function as plotted on different types of graph paper are shown in Figure 5:4; the equation of the locus in each case is

$$x^2 + y^2 = a^2,$$

a having values of 1, 2, 3, etc. This is generally known as the equation of the circle, after the shape of its locus on a linear graph paper that has X and Y axes of equal modulus. Obviously, the same equation describes an ellipse, hyperbola, and a straight line, as well as other curves, depending upon the nature of the graph paper upon which it is plotted. The wide variety of forms taken by a given curve on different papers is a very great aid in devising methods of computation that utilize graphs.

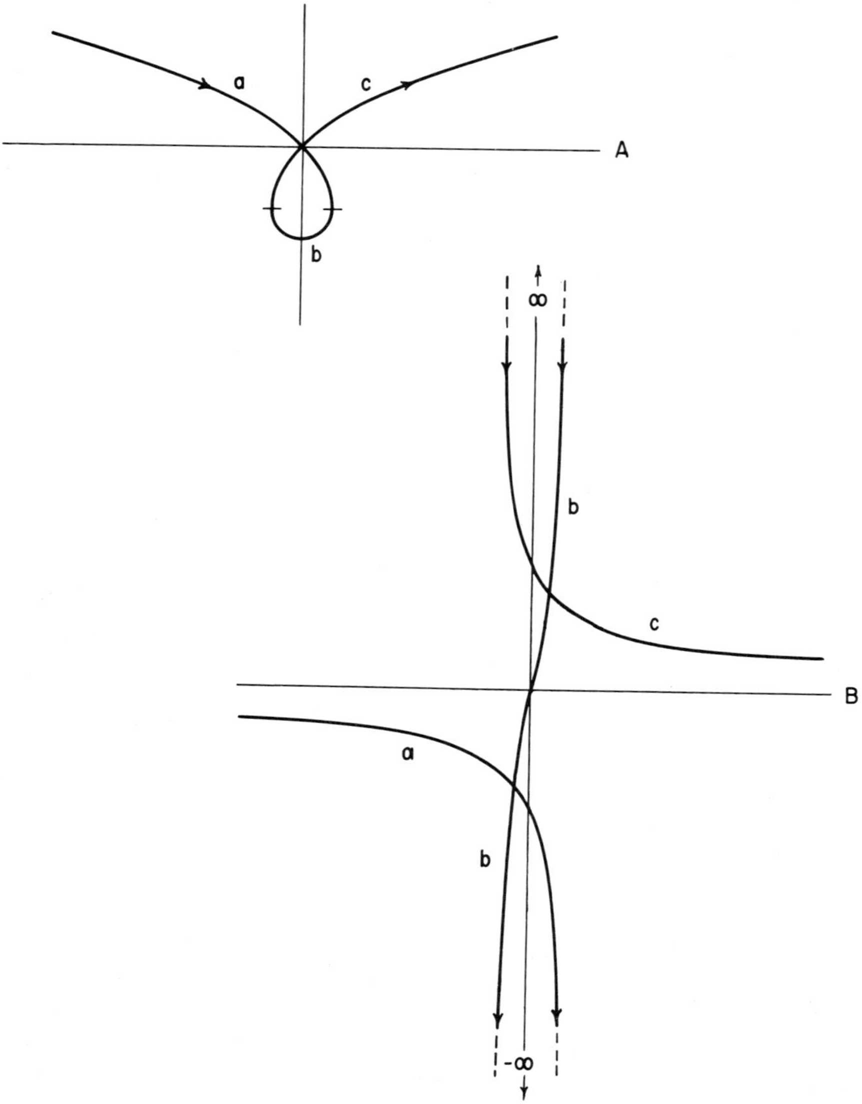

FIG. 5:6 TYPES OF CURVES
(A) Curve Showing a Loop; (B) First Derivative of Curve in Vicinity of the Loop. Sections a, b, and c of Derivative Correspond to Sections a, b, and c Respectively of Original Curve

INTERPOLATION AND EXTRAPOLATION

5.7 General Requirements for Interpolation

Various methods may be used for interpolating between the plotted points of a curve, but, whatever the method, there are certain principles that must be followed.

In the plotting of curves that represent regular functions each of the points on the plot is computed with an accuracy represented by the area of the point itself, possibly with even greater accuracy. There is no justification for drawing the locus otherwise than on the centers of these points. This should be taken as what it actually is, an error of judgment as to the true shape of the curve. The proper remedy is computation of additional points in the vicinity of the point in question, or of the slope of the curve at that point.

There should be no discontinuities or loops in the curve between plotted points. If the curve does contain such a characteristic, the position and nature of that characteristic are indicated by a plot of the first and second derivatives of the function in the vicinity (Fig. 5:5).

When drawn, the curve should be checked. Readings are taken at points between the plotted points, and the values of the abscissas and ordinates of those readings are inserted in the original mathematical expression. The computed values should check with the values from the graph within the limits of error previously decided upon; if they do not, additional points in the vicinity of the error should be computed to produce a curve whose points are within the allowable error.

5.8 Interpolation with Standard Curves

There are available various forms of small transparent curves, such as the French curve, which are readily adaptable to loci of small radius of curvature. In the construction of graphs and nomographs for purposes of computations, however, a large scale master is usually constructed and reduced photographically as a means of increasing the accuracy of the plots. In constructions of this type, the French curve is seldom applicable. Instead, a set of ships curves, including a wide variety of sizes and shapes of curves (exponential, second degree, third degree, etc.) are a virtual necessity.

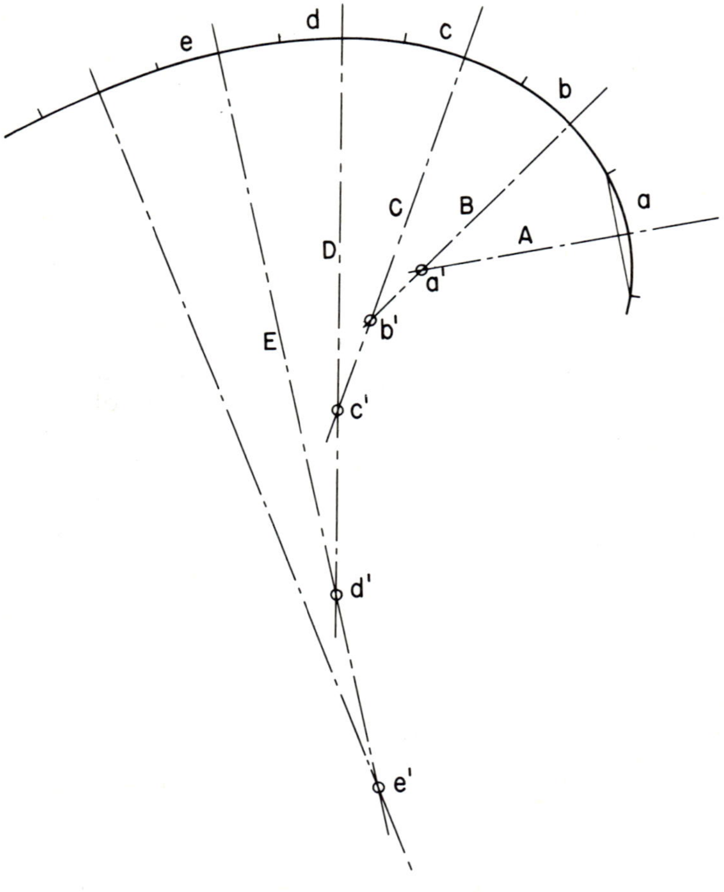

FIG. 5:7 CURVE DRAWN BY COMPASS
Points a' - e', Centers for Segments a-e; A-E, Perpendicular Bisectors of Chords Subtending Respective Segments

Whatever the type of curve used, the locus is usually sketched in lightly at first by freehand, until a smooth curve passing through all the plotted points is obtained. The standard curve is then matched to the sketch in sufficiently small segments so that, for a space beyond each end of the finished segment, the curve and the locus continue to coincide.

If one has available a very complete set of ships curves, it is sometimes possible to omit the freehand sketch, at least over portions of the curve. If this is done, it is advisable to draw the locus in small segments such that there are one or two plotted points at each end of the segment which continue to coincide with the curve.

In laying the standard curve on the sketch or the plotted points, the end of the curve having the greater radius of curvature must lie towards the end of the locus having the greater radius of curvature. It is advisable to draw the curve in pencil before inking it.

5.9 Interpolation with Flexible Curves.

In addition to the standard curves, there are various types of flexible curves, which are less expensive and less satisfactory to use. One of these, known as a spline, consists of a flexible steel or plastic strip held in place by steel points attached to lead weights, or "ducks." Of the various flexible curves, this is the one most easily shaped. It has the advantage of covering long portions of the locus in a continuous curve. The lead weights are a drawback, and one finds it difficult at times to reproduce a curve accurately, as is often done in drawing symmetrical loci with standard curves.

Other types of flexible curves require no weights, as they retain the shape to which they are bent. These range all the way from simple lead strips and lead strips covered with rubber or plastic, to elaborate perforated Vulcanite strips edged with clear plastic and a steel friction strip. They all have an overwhelming drawback; it is almost impossible to eradicate from the strip small humps which cause irregularities in the drawing of long smooth curves.

5.10 Interpolation with Compass

Some loci are composed of curves that may be subdivided into segments that deviate imperceptibly from circular arcs.

Successive segments differ only slightly in radius of curvature and position of the center with the result that the junction of two adjacent segments appears continuous. Such curves may be most easily interpolated with a compass.

If the plotted points are close enough, no preliminary sketch is required, but the curve should be struck in with pencil compass first. A trial-and-error method of locating the center of curvature and arriving at the true radius of curvature is about as satisfactory as any method, although a plot of the evolute (Sec. 5.11) is a more precise approach. The curve is drawn in successive short segments with various settings of the compass angle and different positions of the center point. The settings and center points must be carefully chosen to produce an arc that coincides with each plotted point in the segment. The position of the center point is marked and the arc is drawn in lightly in pencil. This procedure is then repeated on each succeeding segment. Very often, a definite pattern for the positions of the center point will quickly appear, thereby expediting the location of additional center points.

A typical example of a curve that lends itself readily to interpolation by compass is the upper range of the Cornu spiral (see Sec. 5.12). Figure 5:7 shows a portion of a spiral and positions of the center points for compass interpolation.

5.11 Interpolation with Pen and Cord; Involutes

Another device for interpolating applicable to spirals and other curves is a pen or pencil attached to a cord. If the cord is permitted to wind onto a center post as the pen makes its circuit around the post, the radius of curvature undergoes a constant reduction, with the result that the pen describes a spiral.

If the post is of uniform diameter, the spacing between successive turns of the spiral is constant. In the Cornu spiral of Figure 5:7, the spacing between successive turns decreases towards the center of the spiral. Such a spiral may be produced with pen and cord by using a post of non-uniform diameter. At the open end of the spiral, the size of the post becomes rather large.

There is a precise mathematical relationship between the locus of points of contact of the string with the post and the locus of points passed over by the pen; the locus of points on the post is the evolute of the curve described by the pen. Specifically, the evolute of a curve is the locus of its centers of curvature.

A very satisfactory method of interpolating curves of large radius of curvature is the use of the pen and cord with the evolute of the curve. This would appear to be simply substituting the drawing of one curve for the drawing of another. Such, however, is not the case, for the evolute need not be drawn with the high degree of precision required for the curve, itself (the involute). The evolute may be sketched in pencil and a row of pins pressed in along the curve of it, using a backing of fiberboard to serve as a guide for the cord, as shown in Figure 5:8 (a). Alternately, the evolute may be traced on paperboard and cut out; the cut edge of the paperboard serves to guide the cord as in Figure 5:8 (b). There should be a "dry run" of the pen as a check on the accuracy of the evolute before the involute is actually drawn in.

The equation of the locus to be interpolated may be written

$$y = f(x) \tag{5.3}$$

The first derivative of this curve is

$$\frac{dy}{dx} = f'(x)$$

and the second derivative is

$$\frac{d^2y}{dx^2} = f''(x)$$

At a point whose abscissa is a, the value of y is obtained by sutstituting a for x in the expression for y. In Equation 5.3 this may be written f(a). The value of the first derivative at the same point is obtained from the expression for dy/dx in the same manner and is designated f'(a). The value of the second derivative at that point is, similarly, f'' (a). If x' and y' are the coordinates of the point on the evolute corresponding to the

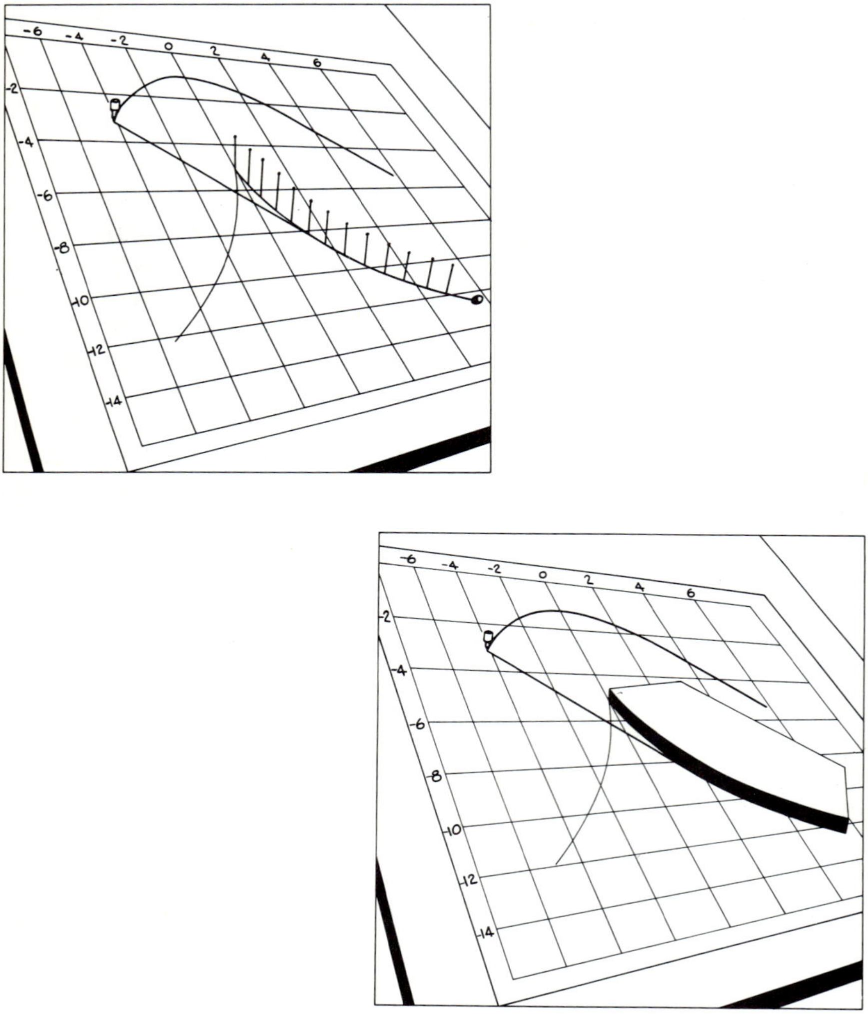

FIG. 5:8 CURVE DRAWN WITH CORD FROM EVOLUTE
(Upper) Evolute is Marked with Pins; (Lower) cut
from a Piece of Paperboard

point a—f(a) on the involute—it may be shown (see a text in advanced calculus) that

$$x' = a - \frac{1 + f'(a)^2}{f''(a)} f'(a)$$

and

$$y' = f(a) + \frac{1 + f'(a)^2}{f''(a)} \qquad (5.4)$$

These are the parametric equations defining the coordinates of the evolute. On the surface, it would appear that computing these coordinates would be very laborious; but actually, this is not so. One must derive the general expressions $f'(x)$ and $f''(x)$. At this stage it is usually desirable to insert actual values of the abscissas to obtain corresponding values of $f(a)$, $f'(a)$, and $f''(a)$ although for simpler cases a general expression for x'' and y'' may be derived before insertion of values of a. Note, also, that the expression $1 + f'(a)^2/f''(a)$ occurs in the equations for both x' and y'; once this factor has been computed, the remaining computations are greatly simplified.

One might, for example, choose this method of plotting the parabola $x^2 + 8y + 8 = 0$. Here $y = f(x) = -\frac{1}{8}(x^2 + 8)$. Hence $f'(x) = -x/4$, and $f''(x) = -\frac{1}{4}$. The values of these functions at the point $x = a$ are $f(a) = -\frac{1}{8}(a^2 + 8)$, $f'(a) = -a/4$, and $f''(a) = -\frac{1}{4}$. From these values, the corresponding point on the evolute is

$$x' = -a^3/16$$
$$y' = -\left[5 + 6(a/4)^2\right]$$

The evolute is symmetrical with respect to the Y axis, as is the parabola. It is necessary to construct only one wing of the evolute, as it can be used in one position for one wing of the parabola and reversed onto a mirror image of itself for the other wing. This evolute is shown in Figure 5:8 together with the construction of the parabola from it.

5.12 Extrapolation

In plotting regular functions there is seldom any justification for extrapolating the curve beyond the region of computed points. The computation of additional points for continuing

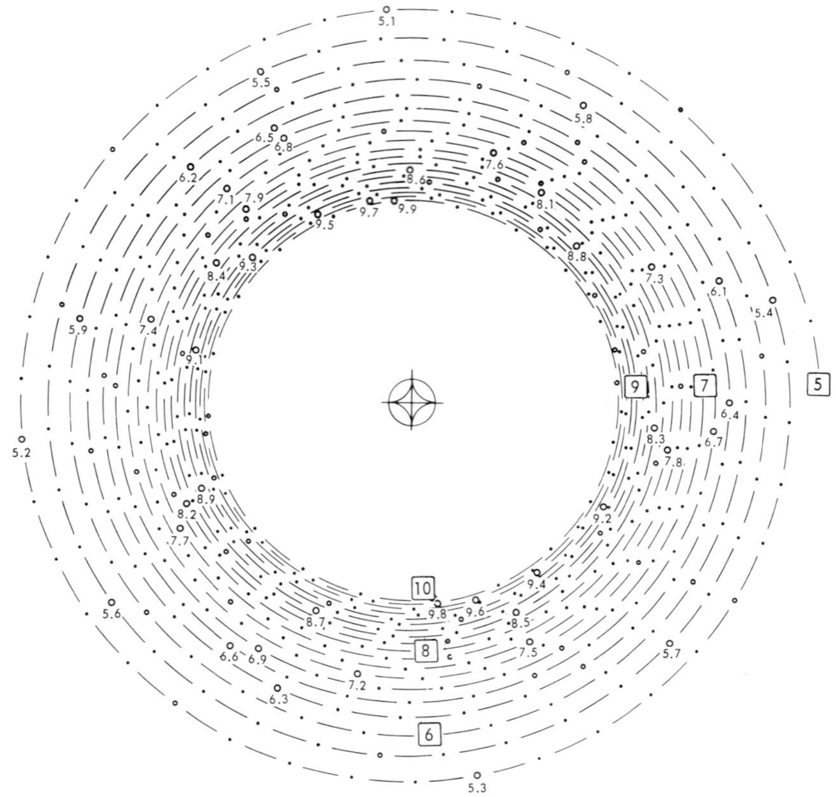

FIG. 5:9 EXTRAPOLATION OF CORNU SPIRAL
The Loci of Several Series of Numbers Provide a Safe Guide for Accurate Extrapolation

the plot is usually a simple matter, for which extrapolation is a poor substitute. There are, however, exceptions in which (a) the computation is extremely difficult or time-consuming and (b) enough information is available from the plotted locus to provide a high degree of accuracy in the extrapolated portion.

An example is the Cornu spiral, the parametric equations of which are

$$x = \int_0^v \cos \frac{\pi}{2} v^2 dv \text{ and } y = \int_0^v \sin \frac{\pi}{2} v^2 dv.$$

Values of x and y are given by Jahnke and Emde[1] for values of $0 \leqslant v \leqslant 5.5$. As computation of x and y from the expressions above is tedious and subject to cumulative errors, extrapolation is a desirable alternative.

Examination of the spiral through $v = 5.5$ indicates regularities which may be extrapolated with negligible error; the locus of points containing odd values of v, the locus of the points $v = .5, 1.5, 2.5, \ldots$, the locus of even values of v, etc., all of which spiral towards a center point $v = \infty$. This, together with the fact that the length of the arc of the spiral corresponding to unit increase in the value of v is constant, provides very close limits within which one may extrapolate the spiral to values of v up to 10 or 15. Any slight errors of extrapolation incurred within these limitations are non-cumulative, and a high degree of accuracy is attainable. Figure 5:9 shows the Cornu spiral[2] for values of $5.0 \lesseqgtr v \lesseqgtr 10.0$.

GRAPHS OF EMPIRICAL DATA

5.13 Plotting

Empirical data, unlike the values of parameters in a regular function, cannot always be determined to an unlimited degree of accuracy. If they could be, it would be possible to find one and only one curve that would pass through all of the plotted points.

---

1 Jahnke, E., and Emde, F., Tables of Functions, p. 35, Dover Publications, New York (1943)

2 Named for Marie-Alfred Cornu (1841-1902), professor of experimental physics at the École Polytechnique in Paris, who developed the graphical evaluation of field intensities.

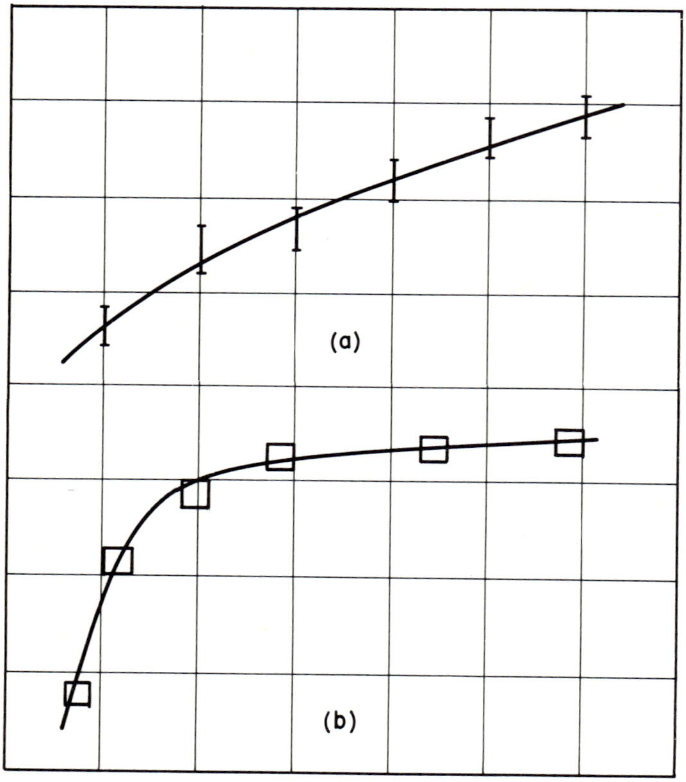

FIG. 5:10 EMPIRICAL DATA PLOTTED TO SHOW RANGES OF EXPERIMENTAL ERROR
a) In One Dimension; b) In Two Dimensions

Instead, experimental results are properly reported with an indication of their standard errors, as, for instance, t = 2.35 ± .07. This conveys the information that t is most probably found at the point 2.35, but will certainly (or with high probability) be found not lower than 2.28 or higher than 2.42.

In plotting empirical data it is highly desirable to indicate not a point, but a range of points that includes the error. In the example cited, the length of the t coordinate from 2.28 to 2.42 is included in the plot.

Some empirical results show standard errors on only one parameter, the other parameter being confined to specific values, such as integers. A plot of data of this kind is shown in Figure 5:10 (a). If both parameters have indicated standard errors, the plot must include these errors in two dimensions; i.e., and area bounded by the extremes of the two errors, as shown in Figure 5:10 (b). Neither the segments of the line for one-dimensional errors nor the areas for two-dimensional errors need be uniform over the graph, for the measurements may be made with different degrees of accuracy in different portions of the range. Furthermore, if the plot is made on non-linear paper, even errors of the same size will produce areas or segments of different sizes.

5.14 Interpolation[3]

A freehand sketch through the midpoints of the plotted ranges is first made, the object in this step being to plot the loci of these midpoints, rather than to produce a regular or smooth curve. This curve is slightly modified, as required, to remove discontinuities and fluctuations in radius of curvature. These modifications must be made in such a way as to hold the curve as near as possible to the midpoints of the plot, but not at the expense of logical form. The modified curve <u>must</u> pass through some portion of the plotted error range at each point; otherwise it does not conform to the data. The process of passing a curve through plotted points serves to define intermediate points, and is thus a form of interpolation.

---

3 The principles set forth here apply to the <u>graphical</u> interpretation of data, and should not be considered equivalent or preferable to a <u>statistical</u> interpretation. The latter is beyond the scope of this book.

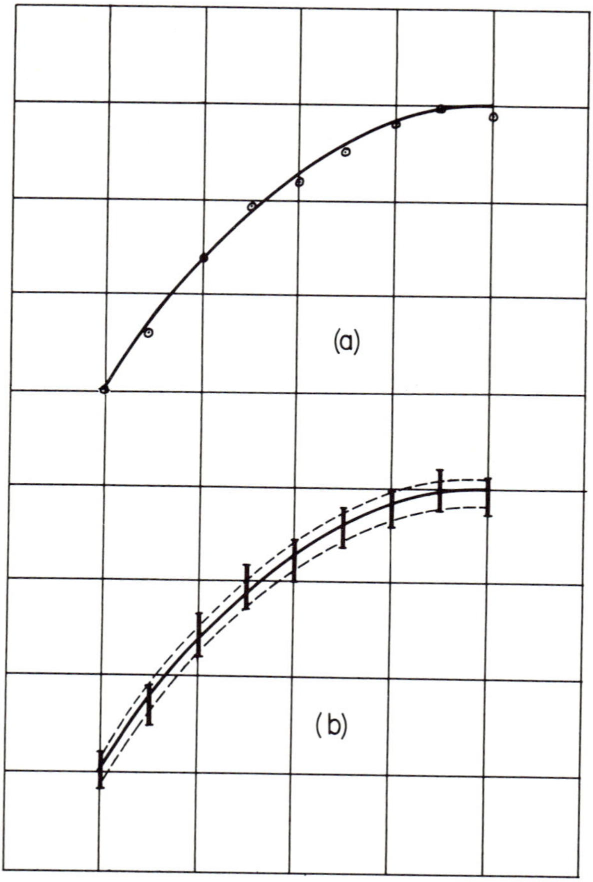

FIG. 5:11 USE OF POINTS IN PLOTTING DATA
a) Without Indicating Range of Error
b) Showing Ranges of Experimental
Error and Error Band

One may find in scientific and technical literature frequent accumulations of data accompanied by graphs in which the standard errors are not indicated in the data, or, if indicated, are not included in the plot. Most of these graphs show a curve passing through some of the plotted points, but lying slightly above some points and slightly below others so that the sum of the positive displacements approximately balances the sum of the negative displacements. This is in accordance with the precepts of many writers on graphical methods and is based upon a naive belief that experimental errors tend to balance out. Certainly, in many types of experimentation, the opposite is true; experimental errors tend to lie in the same direction, but to differing degrees. One may not, however, justify either assumption in a particular case; the error range is the only reliable guide, and one is completely justified in insisting that the curve satisfy each point within the error range of that point.

Obviously, not one curve, but many may be drawn for a given plot. The one drawn is the curve that, for its proximity to the midpoints and its conformity to the limits of error of the data, is considered the most probable form the interpolation would take. In the presentation of experimental data by graph it is frequently desirable to show not only the error range for each point and the interpolated curve, but also curves through the maximum and minimum errors. These latter curves are, in essence, representations of the maximum positive and negative displacements that the interpolated curve may experience and still stay within the limits of error of each point. The space between the extreme curves is called an error band. Figure 5:11 shows the contrast in information conveyed by graphs that include errors and error bands with that conveyed by graphs of the conventional sort.

5.15 Rectification

A more satisfactory interpolation can be made if a graph paper can be found on which the plotted error ranges may all be intercepted by the same straight line. In matching the straight line to the points, it is desirable to have the line pass through the midpoints, if possible, but it must certainly pass through each error range. The curve, thus represented by a straight line, is said to be rectified. Again, it is possible to construct an "error band," indicating the maximum positive and negative displacements of the straight line that still conform to the limits of error.

5.16 Analysis.[4]

From the equation of the straight line of analytical geometry and the values of two points (or a point and the slope) of the line, one may derive an equation corresponding to the functional relationship of the parameters represented. This is known as graphical analysis. One may, in fact, write an equation derived from the most probable position of the straight line, or separate equations for the maximum displacements of that line. The latter equations may be superimposed to show the limits of accuracy of each parameter. Thus, for the graph shown in Figure 5:12, one may write

$$\log y = 1.63\,x + \log 4.50,$$

or

$$y = 4.50\,e^{1.63x}$$

as the equation for the most probable position of the straight line. To include the error band, one may write

$$y = (4.50 \pm .12)\,e^{(1.63 \pm .05)\,x}$$

which is a somewhat truer picture of the status of our information. Even this form of the equation is somewhat misleading, however, for it suggests certain values of the coefficient and exponent that are incompatible. Thus, the exponent may not occupy every position in the range 1.63 ± 0.05 at the same time that the coefficient holds the specific value $4.50 + 0.12$. The two errors are not independent, although, if they are divorced from the graph, the equation might give that impression. For this reason, the graph is more informative than the equation.

---

[4] See also Sections 6.17 and 6.18.

There is a rather general opinion that graphical analysis is less reliable than a form of algebraic analysis known as the method of least squares.[5] Furthermore, it may be demonstrated mathematically that this method does give the

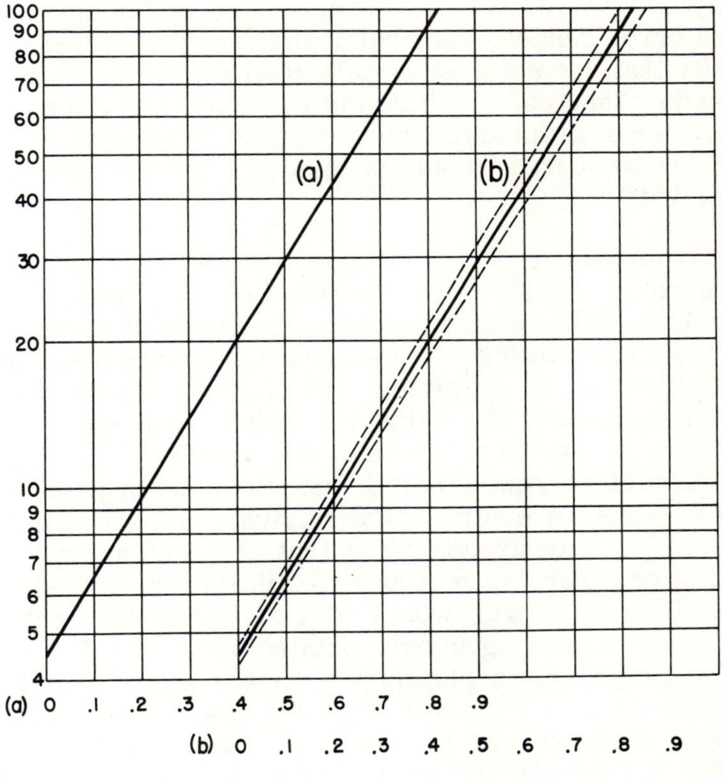

FIG. 5:12 RECTIFICATIONS OF CURVE ON SEMILOG GRAPH PAPER FOR PURPOSES OF ANALYSIS

equation of a straight line that more nearly satisfies a set of points than does a straight line effected by balancing positive differences between the points and the line against negative differences. It should be kept in mind, however, that this

[5] Lipka, Joseph, Graphical and Mechanical Computations, p. 124: John Wiley and Sons, Inc., New York, 1918.

method assumes the existence of a set of definite points, whereas experimental data are properly stated in error ranges rather than in points. The method of least squares is also based upon the assumption that experimental errors tend to balance out, whereas, in a given experiment, they might as easily tend to fall in the same direction. By taking cognizance of the error ranges, the graphical method of analysis leads to results that are not easily challenged by those from analytical methods. Some compromise in method, in which the foregoing principles of graphical representation are wedded to the method of least squares, should be applied for maximum accuracy in analysis.[6]

5.17 Extrapolation

There is seldom any justification for extrapolation of empirical data, although one encounters it frequently in engineering practice. It is usually in concealed form, the application of an empirical formula to computations that involve ranges of the variables which were not included in the data from which the formula was derived.

It should be remembered that there were probably very good reasons for confining the original data to certain ranges. Even if the only excuse were that of avoiding experimental difficulties, there is a strong likelihood that inclusion of the more difficult ranges would lead to vastly different results from those appearing in extrapolated data.

The graph shows plainly the ranges over which it is applicable; the formula, except in rare cases, does not. Nevertheless, care should be taken in every utilization of an empirical formula to ensure against the inclusion of ranges to which the formula does not apply.

COMPLEX GRAPHS

5.18 Three-Variable Expressions

Expressions such as f(x,y,z) = 0 in which each variable is a function of two other variables, actually represent three-

---

6 See a standard text on statistical analysis for a discussion of the Chi-square test for goodness of fit.

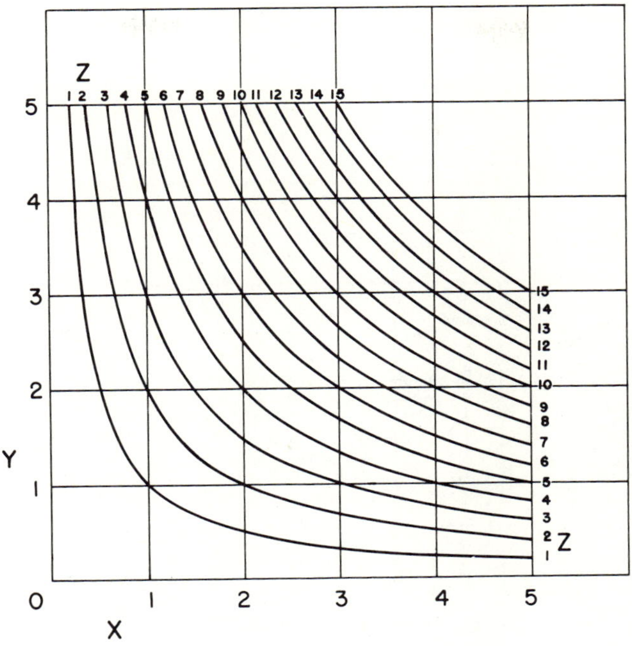

FIG. 5:13 TYPICAL PLOT OF A THREE-VARIABLE EXPRESSION, SHOWN AS A FAMILY OF CURVES

dimensional surfaces that cannot be graphed with complete satisfaction in a two-dimensional plane. The device generally adopted amounts to the projection of successive sections of the surface, corresponding to selected constant values of one variable, such as z, onto a plane that is parallel to those sections. The resulting graph, however, need not be regarded as a projection, for it consists of a family of two-dimensional curves, each member of which involves a constant value of one of the variables. Figure 5:13 shows a plot of $xy = z$, with a constant value of z shown on each curve. The equation of this family of curves is written $y = f(x)_z$, or "y is a function of x and z, with z being assigned constant values."

The enthalpy of a gas may be expressed either as a function of its entropy and temperature at constant pressure or as a function of its entropy and pressure at constant tempera-

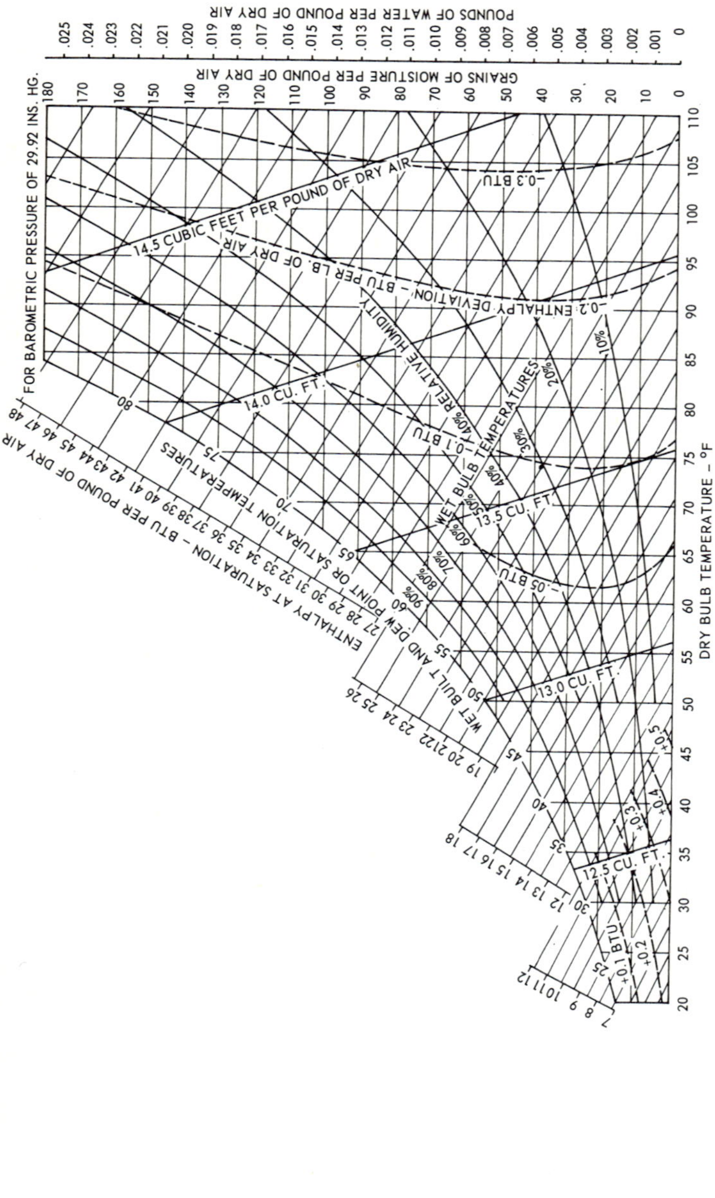

FIG. 5:14 PSYCHROMETRIC CHART
(From E. P. Palmatier and D. D. Wile, Refrigeration Engineering 52:31 1946) Note: Below 32°F the Vapor Pressure and Enthalpy are for Ice

ture. This is, $H = f_1(S,t)_p$ and $H = f_2(S,p)_t$. By assigning a series of constant values to t and obtaining the corresponding values of $f_1(S,t)_p$, one may plot a family of curves representing the H-S-t relationship. In similar manner one may obtain a family of curves for the H-S-p relationship. This amounts to plotting H against S first at a series of constant

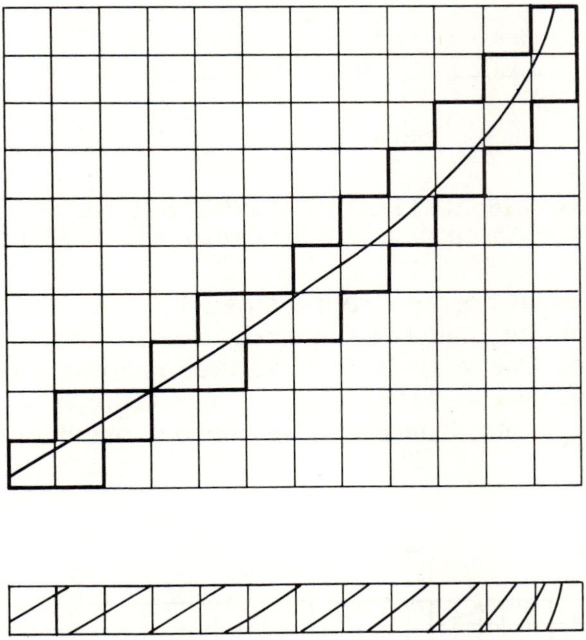

FIG. 5:15 CONDENSATION OF GRAPH BY ELIMINATION OF DEAD SPACE

values of p, and then at a series of constant values of t. The two families of curves intersect to form a grid superimposed upon the grid of the graph paper. This device is a highly informative one, for, given the value of any two of the variables, p, t, H, or S, one may immediately read the corresponding values of the other two variables. The extensively used Mollier diagram (Fig. 6:15) presents similar

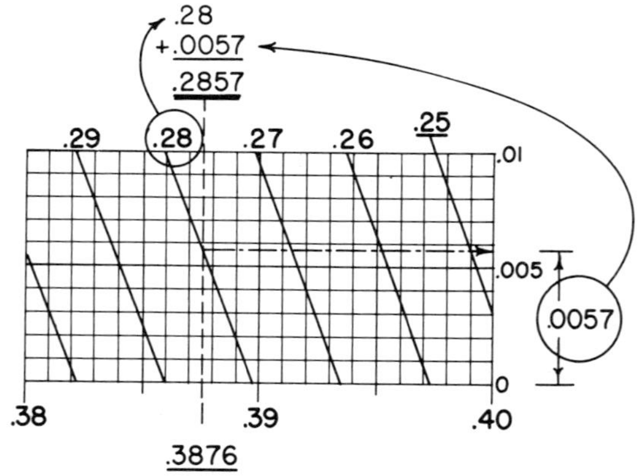

FIG. 5:16 METHOD OF CONDENSED GRAPH
(For Abscissa 0.3876 the Ordinate is 0.2857)

functional relationships represented by dual grid systems.

Indeed, more than two functional relationships may be so represented. One may in this manner represent the system of functions, $y_1 = f_1(x,u)_{v,w\ldots}$, $y_2 = f_2(x,v)_{u,w,\ldots}$, $y_3 = f_3(x,w)_{u,v,\ldots}$, etc. The resulting network provides for values

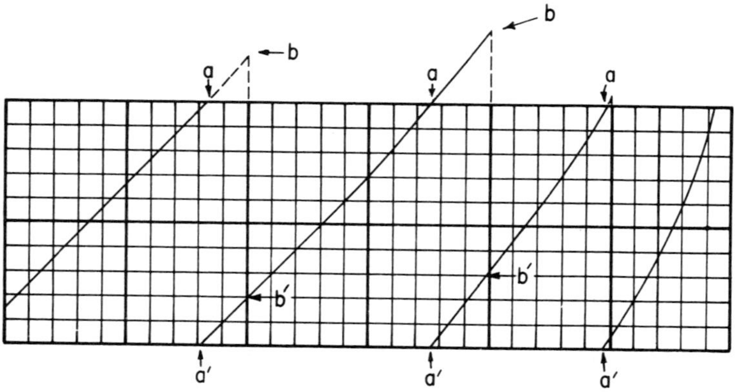

FIG. 5:17 INTERPOLATIVE METHODS FOR CONSTRUCTING CONDENSED GRAPHS
a and a', b and b' are Identical Points on Graph before and after Condensing

of any two variables the corresponding values of all of the remaining variables. A system of more than five variables, is, however, likely to be difficult to understand and interpolate. Figure 5:14 shows a psychrometric chart giving relationships of dry bulb temperature, wet bulb temperature, relative humidity, cubic feet per pound of dry air, total heat, vapor pressure, and ratio of sensible to total heat.

CONDENSED GRAPHS

5.19 Simple Condensed Graphs

One of the greatest drawbacks to the use of simple graphs in computational engineering is the large amount of space they consume for the amount of information they contain. There is a great expanse of dead space, empty space on either side of the line of the graph and between that line and the coordinates. This space is largely eliminated by condensing the graph, a process whose principle is illustrated in Figure 5:15:

a) The sheet is subdivided into small blocks of suitable size;
b) All blocks are eliminated except those through which the locus passes;
c) If the locus passes through more than one block in a vertical column, those blocks are superimposed:
d) The remaining blocks are arranged side-by-side in a horizontal row;
e) This row, if too long, is divided into segments, and the segments are arranged into a block with the first segment on top, successive segments in order under it.

Each portion of the locus is marked with the value of the ordinate on which lies the base of the block containing that portion. At the end of the row is a scale showing increments in these ordinates corresponding to the vertical displacement of the locus from the base of the block. A point on the locus corresponds to the abscissa given on the horizontal scale; its ordinate is the ordinate value (a) attached to the line segment, plus the vertical displacement (b) of the point from the base line (Figure 5:16).

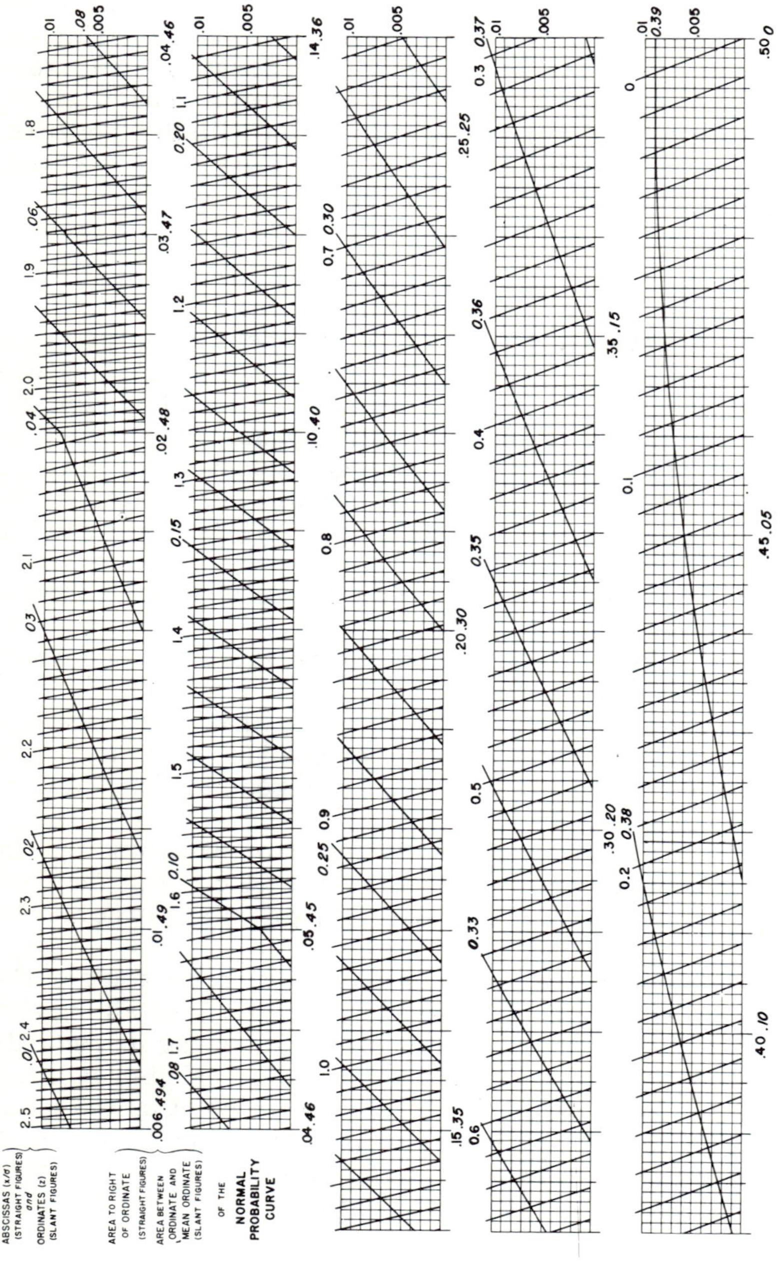

FIG. 5:18 CONDENSED GRAPH SHOWING PROPERTIES OF THE NORMAL DISTRIBUTION CURVE

The actual construction of the graph is quite simple. If possible, the block chosen is so small that each segment of the locus differs only negligibly from a straight line and may be located by two points. In some cases a slightly curved segment is unavoidable, and three or more necessary points are required. If the segments are virtually straight lines, it is necessary to calculate only one point for each segment, as a point calculated for an adjacent block may be used as one point, as shown in Figure 5:17. Indeed, there is occasionally sufficient regularity in the positions of the plotted points of adjacent blocks that only a few such points are actually computed, the remainder being plotted by graphical interpolation.

Figure 5:18 shows a portion of a condensed graph on which two properties are plotted simultaneously. Abscissas are the same for both loci, as are the values of their vertical displacements from the base line. Ordinates differ, and are marked on each line segment.

EXERCISES

1. For each of the following curves, calculate the value of y for a sufficient number of values of x, and plot the curve on linear graph paper: (a) $y = (\sin x)^n$ ($n = 1, 2, 3.$), (b) $y = 2/(e^x - e^{-x})$, (c) $y = x^3 - 2.3x^2 + 5.6x - 28$, (d) $y = 0.5 \sin x + 0.3 \sin 3x + 0.2 \sin 5x$.

2. In each of the curves of Problem 1 locate the position of each (a) intercept, (b) axis or point of symmetry, (c) maximum and minimum, and (d) flex point. Show how these may be employed as an aid to accurate plotting.

3. From the values of y and dy/dx at the points $x = 0$, $2\pi/3$, π, $4\pi/3$, 2π, make a plot of the function $y = \sin x$. Check interpolated portions of the curve with values from a trigonometric table.

4. From the moduli of the graph paper scales and the thickness of the locus plotted, estimate the maximum error

which should be observed in each of the computations of Problem 1.

5. Prepare a plot of the function $y = (\sin x)^n + 2$ on semi-logarithmic (linear-logarithmic) paper. The factor 2 is added to avoid negative and zero values of y, which cannot be plotted on a logarithmic axis. Compare this plot with the plot of $y = (\sin x)^n$ on linear paper made in Problem 1.

6. Plot each of the following curves:

$$\text{(a) } y^2 = \frac{x^2\,(0.1 + x)}{0.1 - x} \qquad \text{(b) } y^2 = \frac{x^3}{1 - x}$$

Note the occurrence of nodes (loops) and cusps in these curves.

7. Plot and construct the Sici spiral, for which $x = \text{Ci } x =$

$-\int_x^{\infty} \frac{\cos t}{t}\,dt$; and $y = \text{Si } x = \int_0^x \frac{\sin t}{t}\,dt$. Tables of values of

Ci x and Si x are available in Jahnke and Emde.* Portions of this curve may be interpolated with compass. Other portions are readily interpolated with French curves, as the Sici spiral is frequently employed as a basis for French curves.

8. Plot the evolute of each of the following curves, and construct the curve from the evolute: (a) $y = \log x$, (b) $y = \sin x$ ($x = 0$ to $\pi/2$), (c) $y = x^2$, (d) $y^2 = x^3/(1 - x)$ for positive values of y.

9. Plot the data of of the following table, indicating error ranges. Draw a curve which you consider best satisfies the data. Draw, also, an error band that satisfies the data.

---

* See footnote, Section 5:12

YIELDS OF PENICILLIN IN SHAKEN FLASKS*

Strain	Precursor Level, %	Mean Yield Units/ml
47-911	0	268 ± 13
	0.05	368 ± 8
	0.10	386 ± 24
	0.20	347 ± 8
47-1380	0	260 ± 21
	0.05	416 ± 9
	0.10	405 ± 15
	0.20	393 ± 24
47-1564	0	317 ± 5
	0.05	508 ± 14
	0.10	535 ± 30
	0.20	464 ± 52
48-701	0	372 ± 11
	0.05	589 ± 12
	0.10	558 ± 12
	0.20	487 ± 22
48-749	0	254 ± 9
	0.05	407 ± 22
	0.10	373 ± 18
	0.20	384 ± 16
49-133	0	648 ± 55
	0.05	1037 ± 14
	0.10	994 ± 58
	0.20	667 ± 28
49-2105	0	450 ± 0
	0.05	1030 ± 18
	0.10	963 ± 28
	0.20	755 ± 86

* Anderson, F. R. et al., Penicillin Production by Pigment-Free Molds, Ind. Eng. Chem., 45:768 (1953)

10. Plot each of the following sets of data on several types of graph paper. If a straight line is obtained, derive an equation to represent the data:

A

x	.0185	.0330	.054	.082	.168	.310
y	8.36	11.50	15.0	18.8	28.0	39.0

B

x	18.14	19.20	20.25	21.50	22.73
y	240	170	120	79.5	53

C

x	10.0	10.2	10.4	10.6	10.8	11.0
y	5.00	4.55	4.23	4.06	4.05	4.25

D

x	1.40	1.85	2.15	3.20	5.50
y	.470	.575	.620	.710	.787

11. Construct three-variable graphs to represent each of following formulas; (From <u>Orifice Meter Constants</u>, p. 20; American Meter Company):

(a) $h_u = \sqrt{100\ h_w/R_h}$ (b) $p_u = \sqrt{100\ (p + 14.4)/R_p}$

12. Prepare a condensed graph showing the squares of numbers between 35 and 45, inclusive, to ± 0.05 for the numbers and $\pm$ 1.0 for their squares.

chapter 6

CALCULATING WITH GRAPHS

Graphs are useful as a basis for many computations involving functions that are not readily represented by mathematical formulas. They are also applied to computations involving relationships that are not easily realized by any other form of graphical device. The variety of applications of graphs to computations is so great that it is impossible in a text of this sort to give more than a few outstanding applications; however, the principles underlying these applications are very general and suggestive of many other fields and methods of approach.

POSITION RELATIONSHIPS

6.1 Two-and Three-variable Expressions

In the preceding chapter the graph was discussed as a means of representing a functional relationship such as $f_1(x,y) = 0$ by means of a single curve on a graph paper grid, with x and y as abscissa and ordinate, or a relationship such as $f_2(x,y,z,) = 0$ by a family of such curves, one curve for each value of z. Functional relationships thus represented provide a simple and direct method of estimating the value of any one variable when the values of the remaining variables are known. The outstanding advantage of the graph over the numerical table is the greater accuracy with which interpolation is effected. Interpolation between successive values in a table is usually made on a linear basis, as any other basis poses quite a computational problem. The discrepancy between values thus interpolated and those read from a graph is often great (Fig. 6:1). In a table of the function f(x,y,z), even linear interpolation becomes laborious, and the error

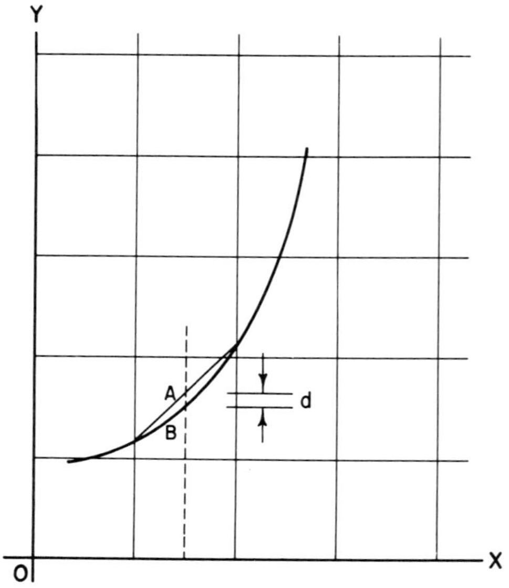

FIG. 6:1 DISCREPANCY BETWEEN LINEAR INTERPOLATION AND INTERPOLATION FROM A GRAPH

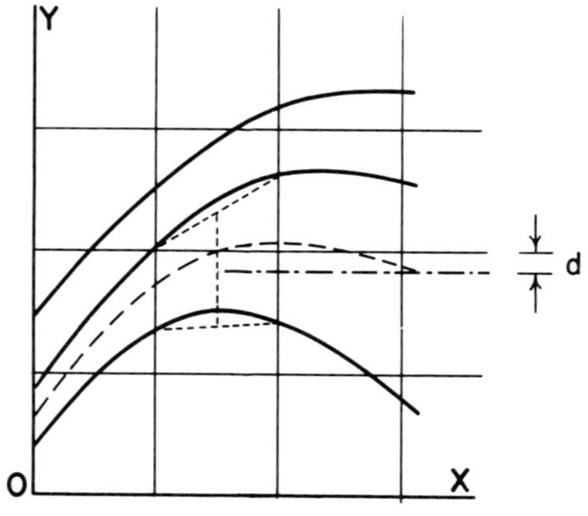

FIG. 6:2 INTERPOLATION IN A THREE-VARIABLE SYSTEM

resulting from the assumption of linearity may be even greater (Fig. 6:2).

The use of a graph to determine the value of one variable in a functional relationahip when the values of the other variables are known is often regarded as a form of computation. Actually, it is no more an act of computing than determining the value of y from a corresponding value of x in a table, or determining the telephone number of a person from the position of his name in the alphabetical listing of a telephone directory. The graph, in this sense, is an information storage unit. As such, it is an exceedingly valuable adjunct to computation, as is also the table or any other device for storing

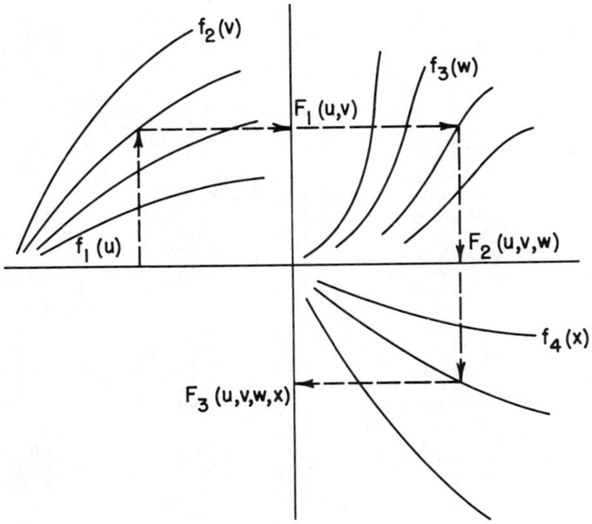

FIG. 6:3 SYSTEM OF SEVERAL VARIABLES ON A RUN-AROUND CHART

information. The graph does have certain properties which permit computations to be performed. These are discussed in part in subsequent portions of this chapter.

6.2 Expressions of More Than Three Variables

An expression of the sort, $F(u,v,w,x, \ldots) = 0$, which cannot be simply represented on a single grid in the manner of

Figure 6:2 can usually be represented by a series of such grids. The functional relationship must be such that it can be reduced to a series of simpler relationships,

$$f_1\ (u,v,t_1) = 0,$$

$$f_2\ (t_1,w,t_2) = 0,$$

$$f_3\ (t_2,x,t_3) = 0, \text{ etc.},$$

where t_1, t_2, t_3, etc., are functions of an extraneous parameter t. The first of these expressions is represented on a grid

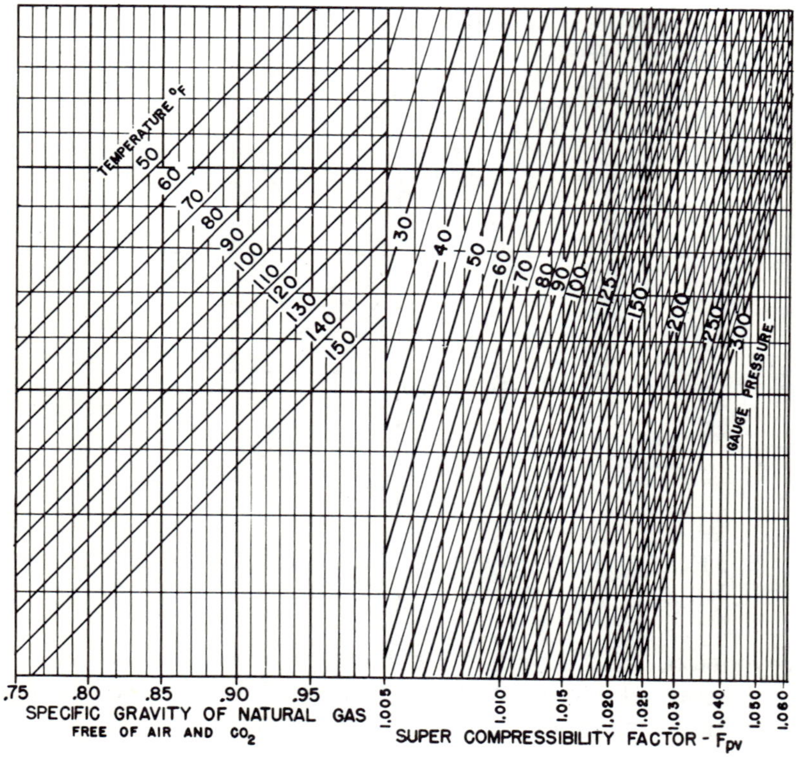

FIG. 6:4 RUN-AROUND CHART FOR SUPERCOMPRESSIBILITY FACTOR
(After NSGMA Data Book, 1951, p. 16)

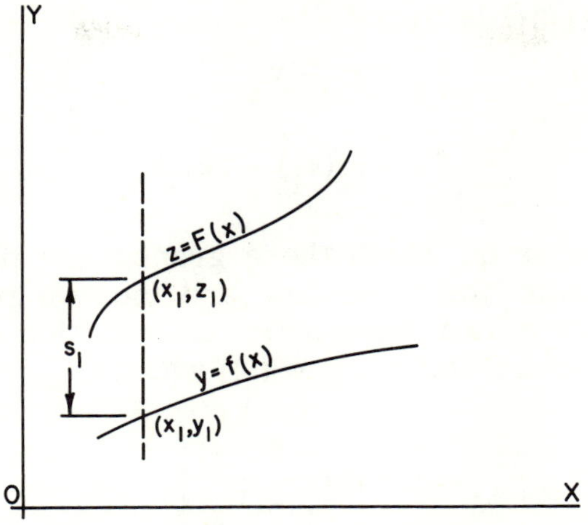

FIG. 6:5 COMPUTATION BASED UPON VERTICAL DISPLACEMENT OF TWO CURVES

having abscissas u and ordinates t_1, with a separate curve for each value of v. The second expression is represented on a grid placed adjacent to the first and sharing a common ordinate scale; its abscissas are t_2, and there are curves for successive values of w. The third grid has a common abscissa scale with the second grid, and so on. This system is represented diagrammatically in Figure 6:3. Figure 6:4 shows its application to the formula for supercompressibility factor (see Sec. 2.12). This chart, sometimes known as a "run-around" chart, is similar to the device described in Section 4.18 (Fig. 4:21).

VERTICAL AND HORIZONTAL DISPLACEMENTS

6.3 Vertical Displacements.

Several types of computation are based upon the superposition of one graph upon another and the measurement of the distance between points on the two graphs that have equal abscissas or equal ordinates. The two graphs may be plotted on separate paper and superimposed, or both curves may be plotted on the same paper.

Figure 6:5 illustrates this point. The two curves are $y = f(x)$ and $z = F(x)$. At abscissa x_1 the distance between the points

(x_1,y_1) and (x_1,z_1) is s_1, or

$$s_1 = z_1 - y_1$$

$$= F(x_1) - f(x_1).$$

Should the paper on which these graphs are plotted have a logarithmic scale for ordinates, and the two graphs represent a property of two items under examination, u and v, then $F(x) = \log\ \phi(u)$ and $f(x) = \log\ \phi(v)$; whence,

$$s_1 = \log\left[\frac{\phi(u_1)}{\phi(v_1)}\right]$$

or

$$\text{antilog}\ s_1 = \frac{\phi(u_1)}{\phi(v_1)}$$

This formula may be applied to the measurement of luminosity of lamps at various angles from their axes of symmetry. The logarithm of luminosity is plotted as ordinate against the angle as abscissa for the lamp under examination. The differences between the ordinates of this curve and those of the corresponding curve for a standard lamp are read on a logarithmic scale, giving immediately the value of the ratio of luminosities of the two lamps.

6.4 Analysis of Absorption Spectra

The mathematical relationships of absorption spectroscopy may be given as

$$T_s = 10^{-a_s bc} \qquad \text{(a)}$$

$$k_s = a_s c = \log_{10} \frac{1/T_s}{b} \qquad \text{(b)}$$

$$A_s = \log_{10}\left(1/T_s\right), \qquad \text{(c)}$$

in which the symbols are

$$T_s = \text{transmittancy} = \frac{\text{transmittance of solution}}{\text{transmittance of solvent}}$$

a_s = absorbency index (1000 sq cm/gm)

c = concentration (g/1000 cc)

b = cell thickness (cm)

k_s = absorbency coefficient

A_s = absorbency

From equations a and c, it is apparent that

$$\log A_s = \log a_s + \log b + \log c; \qquad \text{(d)}$$

consequently, if $\log A_s$ is plotted as ordinate against the wave length as abscissa, the shape of the curve is determined entirely by a_s, that being the only term on the right of Equation d which varies with the wave length of light incident upon the solution. If a different thickness of solution (b) or a different concentration (c) is used, the curve is displaced upward or downward (this displacement may be used to calculate the change in concentration), but its shape remains the same.

The value of a_s—and consequently, the shape of the curve—are determined by the nature of the absorbing component in the solution. Matching the shape of the photometric plot of an unknown solution with that of known solutions is sufficient to identify the chemical composition of the unknown solute. Were the shape of the curve subject to variations due to thickness of the cell or concentration of solution, this method of identification would be impossible.

More significantly, the presence of an unknown dye or other absorbing compound in a mixture whose other components are known may be detected and the identity of the unknown substance established by an extension of this method. If several components, u, v, and w are present in the same solution, the absorbency of the mixture is related to the absorbencies of the components by[1]

$$\log A_s = \log (A_s)_u + \log (A_s)_v + \log (A_s)_w$$

Figure 6:6 shows the method by which the logarithmic graphs of absorbencies of two substances may be combined to obtain the graph of the absorbency of the mixture. A logarithmic reference scale having a modulus equal to that of

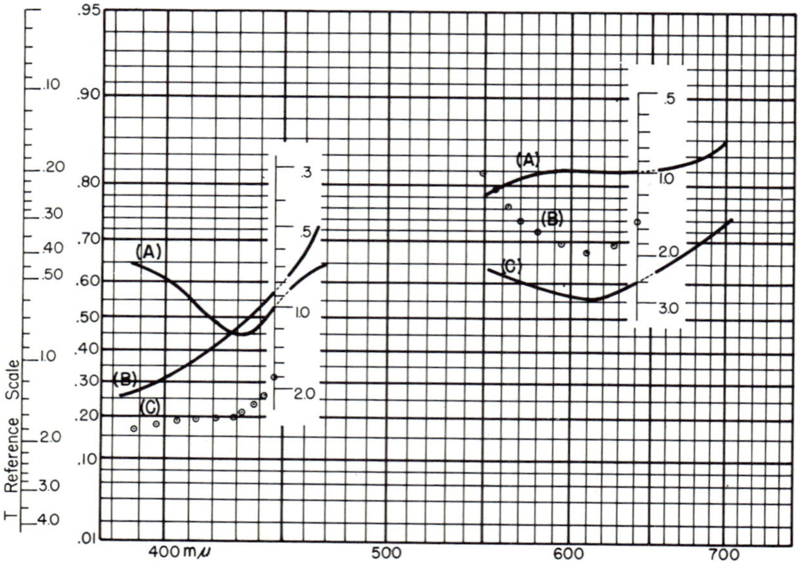

FIG. 6:6 CURVE OF SPECTRUM USED IN ANALYZING DYE FOR UNKNOWN COMPONENT

the ordinate scale is placed in a vertical position so that the 1 of the reference scale is on one curve. The value of the intersection with the other curve is read, the two values are added, and their sum is plotted to the reference scale still in the same position. The process is repeated for enough wave lengths to define the curve for the mixture accurately.

[1] Mellon, M. G. (ed.), Analytical Absorption Spectroscopy, p. 310; John Wiley & Sons, Inc., New York, 1950.

Also in Figure 6:6 is the method for obtaining the plot of an unknown component from that of the mixture, when remaining components are known. Curve C is the plot for the mixture, Curve A the combined plot of the other components. The reference scale is placed vertically so that 1 is on Curve A. The value of Curve C is read, 1 is subtracted from it, and the difference is plotted. Again, enough points are plotted for different wave lengths to obtain an accurate plot of the com-

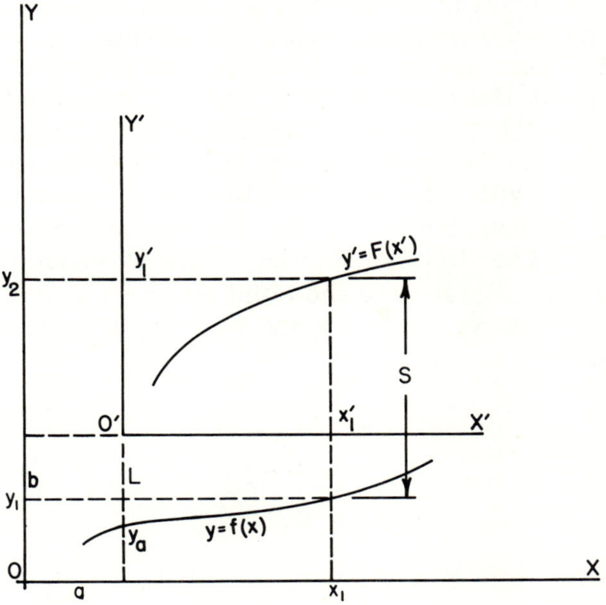

FIG. 6:7 RELATIONSHIP OF POINTS ON SUPERIMPOSED GRAPHS

ponent. The identity of the unknown component can then be established from the shape of its curve.

In preparing plots of this kind, it is convenient to employ a graph paper with log-log ordinates, rather than log ordinates, as the data to be plotted are generally given in transmittancy, rather than in absorbency. Log A_S = log log $1/T_S$. Figure 6:6

shows both log and log-log ordinates for the sake of comparison.

The excellent treatment of spectrophotometric data methods given by Stearns [2] includes, also, examples of computation of concentrations of solutions from their log absorbency charts.

6.5 Vertical and Horizontal Displacements

In Figure 6:7 the curve $y = f(x)$ is plotted on the coordinate system XOY, and the curve $y' = F(x')$ on the coordinate system X'O'Y'. The two coordinate systems are superimposed so that the Y and Y' axes are parallel, as are the X and X' axes. The origin O' is at the point (a,b) of the XOY coordinate system. The point on the curve $y' = F(x')$ whose coordinates with respect to the X'O'Y' axes are x'_1 and y'_1 has the coordinates x_1 and y_2 with respect to the XOY axes. The point on the $y = f(x)$ curve having the same abscissa has the ordinate y_1. The distance of the origin O' above the $y = f(x)$ curve is L; the distance of the point (x'_1, y'_1) above that same curve is S. Then $x_1 = x'_1 + a$ and $y_2 = y'_1 + b$. The value of S is, therefore,

$$S = y_2 - y_1$$

$$= y'_1 + b - y_1,$$

or, as

$$b = y_a + L,$$

$$S = y'_1 + y_a - y_1 + L. \tag{6.1}$$

6.6 Freight Rate Differentials

A most interesting application of this device is the computation of freight rate differentials. When a commodity, such as cotton, is shipped, the shipper is billed separately for the goods from point of origin to warehouse (inbound) and from warehouse to destination (outbound). Subject to certain limitations, however, such as date, route, etc., the shipper may file a claim with the railroad and obtain application of a

---

[2] In Mellon, op. cit., Chapter 7.

through rate to both bills. This results in a saving, for the rate per mile decreases as the mileage increases (Table 6:1). Thus, a shipper who paid 23 cents a hundred pounds for an inbound shipment of 110 miles and 21 cents a hundred pounds for an outbound shipment of 87 miles—a total of 44 cents—is permitted to apply, instead, the rate for 197 miles, which is only 39 cents. The saving on a hundred pounds is 5 cents, which in the volume shipment of cotton amounts to a good sum.

There are other savings, too. The rate for carload lots may be applied to both shipments if one of them was 25,000

TABLE 6:1

MILEAGE SCALE OF RATES ON COTTON

(Rates in Cents/cwt)

Miles	Rates	Miles	Rates	Miles	Rate
35	11	85	19	170	33
40	11	90	21	180	34
45	11	95	22	190	36
50	12	100	22	200	39
55	12	110	23	210	40
60	14	120	24	220	42
65	15	130	26	230	45
70	15	140	29	240	46
75	15	150	30	250	48
80	18	160	32		

pounds or more, although the other shipment might have been less than a carload. There is another source of saving which is somewhat more subtle and requires the application of a bit of computation. The decrease in freight rates is not uniform with increase in mileage; certain combinations of inbound with outbound bills yield greater savings than other combinations. It is to the shipper's advantage, in matching groups of inbound and outbound bills for filing claims, to know those combinations for which the greatest difference is realized between through rate and combined inbound and outbound rate. The enormity of the problem may be realized by considering the matching of ten inbound and ten outbound bills. There are ten different choices of outbound bills that may be matched with the first inbound bill. That match completed, there remain

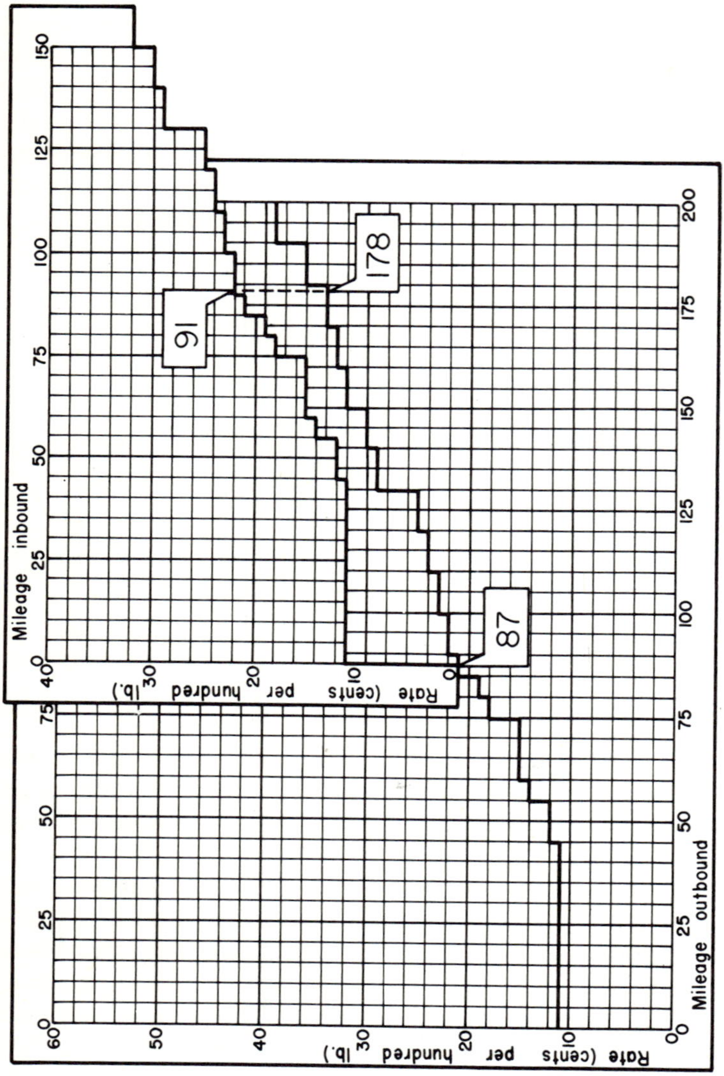

FIG. 6:8 FREIGHT RATE DIFFERENTIALS BY METHOD OF SUPERIMPOSED GRAPHS

nine different choices for matching the second inbound bill, eight for matching the third, etc. The total number of choices possible is 10 x 9 x 8 x . . . x 1 = 10! = 3,628,800. In actual practice, it is more common that the shipper is faced with the problem of matching some fifty or more inbound bills with outbound bills.

There are many practices that may be introduced to simplify the problem a bit, but one indispensable step in solving it is that of determining at a glance the combinations that yield the greatest savings. If I is the inbound mileage and O the outbound mileage, the through mileage is I + O; R_I is the inbound mileage rate, R_O the outbound mileage rate, R_{I+O} the through rate, and D the rate differential, where

$$D = R_I + R_O - R_{I+O}$$

This is of the same form as Equation 6.1 and may be identified with that equation by setting $S = D$, $y' = R_O$, $y_a = R_I$, $y_1 = R_{I+O}$ and $L = O$. In short, the solution is this: Two identical plots of the values of the freight rates as ordinates are made against miles as abscissas. The origin of one graph is placed on the other graph at the point corresponding to the inbound mileage. In this position, the vertical displacements between the two graphs indicate immediately the freight rate differentials obtainable by combining that inbound mileage with all possible outbound mileages, as shown in Figure 6:8. Here it is apparent that an outbound bill for 87 miles may be matched with an inbound bill of 91 miles for a rate differential of 9 cents. This match is obviously better than a match with a bill for 110 miles, where the differential is only 5 cents, or 60 miles, where the differential is only 4 cents. With the aid of this device it is possible to prepare rapidly a table showing rate differentials for all combinations of the bills under examination in a given set. Those that yield the greatest sum of differentials for the entire set may then be selected by inspection from the table.

GENERAL DISTANCE RELATIONSHIPS

6.7 Distance Formula

The distance between two points (x_1, y_1) and (x_2, y_2) is

$$s = \sqrt{(x_2 - x_1)^2 + (y_2 - y_1)^2} \qquad (6.2)$$

There is no restriction as to definition of the two points; they may lie on a common curve or on separate curves. If the two points are on separate curves defined by functions of the variables u and v, then

$$s = \sqrt{\left[f_3(u_1) - f_1(v_1)\right]^2 + \left[f_4(u_1) - f_2(v_1)\right]^2} \qquad (6.3)$$

is the distance between the points u_1 and v_1 on the two curves.

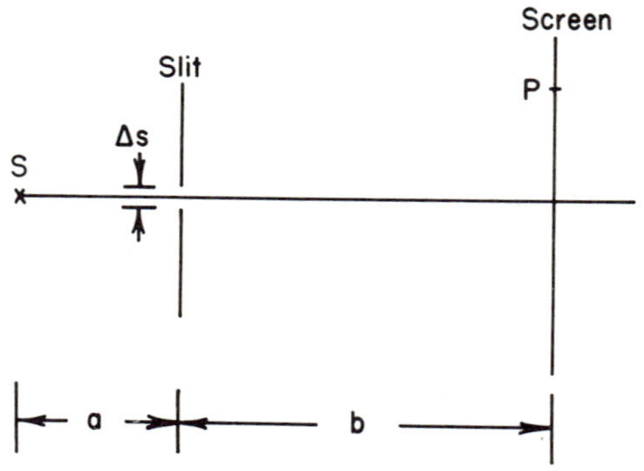

FIG. 6:9 DIFFRACTION OF ELECTROMAGNETIC RADIATION PASSING THROUGH A SLIT

If the two points lie on the same curve, defined by the functions, $x = f_1(u)$, $y = f_2(u)$, the value of s is

$$s = \sqrt{\left[f_1(u_2) - f_1(u_1)\right]^2 + \left[f_2(u_2) - f_2(u_1)\right]^2} \qquad (6.4)$$

Equations 6.3 and 6.4 are not readily reduced to the form of defining equation commonly employed for nomographs (see

Chap. 9); consequently, application of the distance formula is a most convenient method for handling formulas of this type.

6.8 Fresnel Diffraction

The latter relationship is applied to the problem of determining the intensity at various points in a field illuminated by electromagnetic radiation which has been passed through a slit or otherwise subjected to the phenomenon of diffraction. A full discussion of this measurement would require more space than can be given here,[3] but an example will serve to illustrate the method.

Radiation from a distant source S (Fig. 6:9) passing through a slit of width Δs produces at the point p a resultant oscillation whose amplitude is R, where

$$R^2 = \text{const.} \times (x_2 - x_1)^2 + (y_2 - y_1)^2.$$

Here

$$x_2 = \int_0^{v+\Delta v} \frac{\cos \pi v^2 dv}{2}, \quad x_1 = \int_0^{v-\Delta v} \frac{\cos \pi v^2 dv}{2}$$

$$y_2 = \int_0^{v+\Delta v} \frac{\sin \pi v^2 dv}{2}, \text{ and } y_1 = \int_0^{v-\Delta v} \frac{\sin \pi v^2 dv}{2}.$$

Δv is related to the elements of Figure 6:10 by

$$\Delta v = \frac{\Delta s}{\sqrt{\frac{2ab\lambda}{(a+b)}}}$$

λ is the wave length of the radiation.

The integrals in these definitions are the Fresnel integrals employed in constructing the Cornu spiral (Sec. 5.10-5.12). The length of a chord of this spiral from $v - \Delta v$ to $v + \Delta v$ is, therefore, identical with $(x_2 - x_1)^2 + (y_2 - y_1)^2$ in the above

[3] Jenkins, A. J., and White, H. E., Fundamentals of Optics, 2nd Ed., Ch. 18; McGraw-Hill Book Company, Inc., New York, 1950.

expression for R^2, as may be observed by substitution in Equation 6.4.

The problem is generally one of plotting the intensity (I, proportional to R^2) over the surface of the screen, MM'. The value of Δv is fixed by the distances a and b, the wave length λ, and the slit width Δs. An arc whose length is $2\Delta v$ is measured at any position on the spiral. The midpoint of the arc

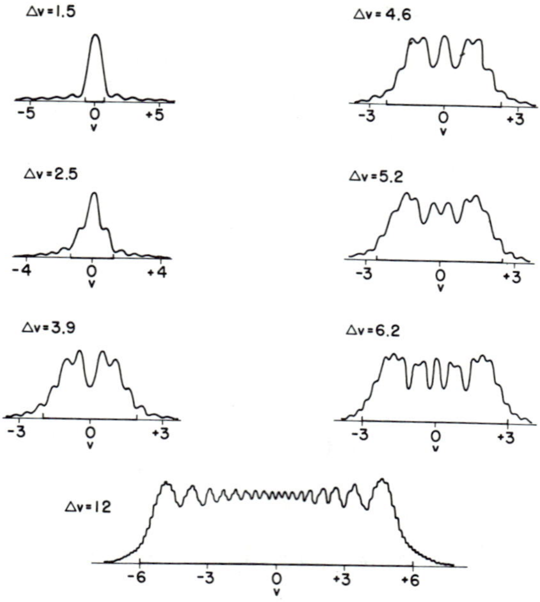

FIG. 6:10 PLOT OF TYPICAL DIFFRACTION PATTERNS OBTAINED FROM THE CORNU SPIRAL
After Jenkins, F. A., and White, H. E., Fundamentals of Optics, 2nd. ed., p. 372. New York, McGraw-Hill Book Co.

is the value of v; the length of the chord subtending this arc is R (or const. × R depending upon the units employed), the square of the chord length is I (or const. × I). This value of I is plotted as ordinate against the value of v as abscissa for this and a series of equal arcs along the length of the spiral. The resulting plot of I against v shows the intensity of light falling on each part of the screen, as shown in Figure 6:10.

As an aid to plotting I, a tape for reading the length of the chord may be graduated in R^2, rather than in R. If graduated in R, unit length is taken as the length of the arc of the spiral from the point $v = 0$ to the point $v = 1$; this is also the distance from the origin to 1 along either axis. If the scale is graduated in R^2, the same modulus is used. An alternate method employs a linear scale (graduated in R) for measuring the length of the arc, and this length is plotted on a linear-square graph paper, abscissas being linear, and ordinates, squares.

If the tape on which R or I^2 is read is to have the same modulus as the axes on which the spiral is constructed, (the units which must be employed) are such as to normalize the measurement. The total illumination passing through the slit has unit value; consequently, the total illumination falling on the screen, regardless of its distance from the slit, also has the value 1. This is, $\int_{-\infty}^{+\infty} I dv = 1$. This may be used as a check on the accuracy of the measuring and plotting; the area under the curve must equal 1. From a practical standpoint, it is rarely necessary to plot more than half of the curve, as the other half is an inverted mirror image of the plotted portion; here the area under the curve is 0.5.

The same principles of diffraction that apply to visible light and X-rays also apply to microwave radiation. Many types of microwave antennas emit radiation which is diffracted in a pattern analogous to that of light passing through a slit, and its field intensities may be plotted by reference to the Cornu spiral (see 5.12; Fig. 5:7). Here is, then, an extremely modern application of the brilliant researches of Fresnel, dating from 1815. The work of Cornu is a classic among graphical methods of computation.

ISOMETRIC TRANSLATION

6.9 Isometrics

Figure 6:11 is a typical pressure-temperature-volume diagram for a gas, consisting of a network of three families of isometric lines. The vertical lines are isometric temperature lines, or isotherms; the horizontal lines are isometric volume lines, or volume isometrics; the diagonal lines are

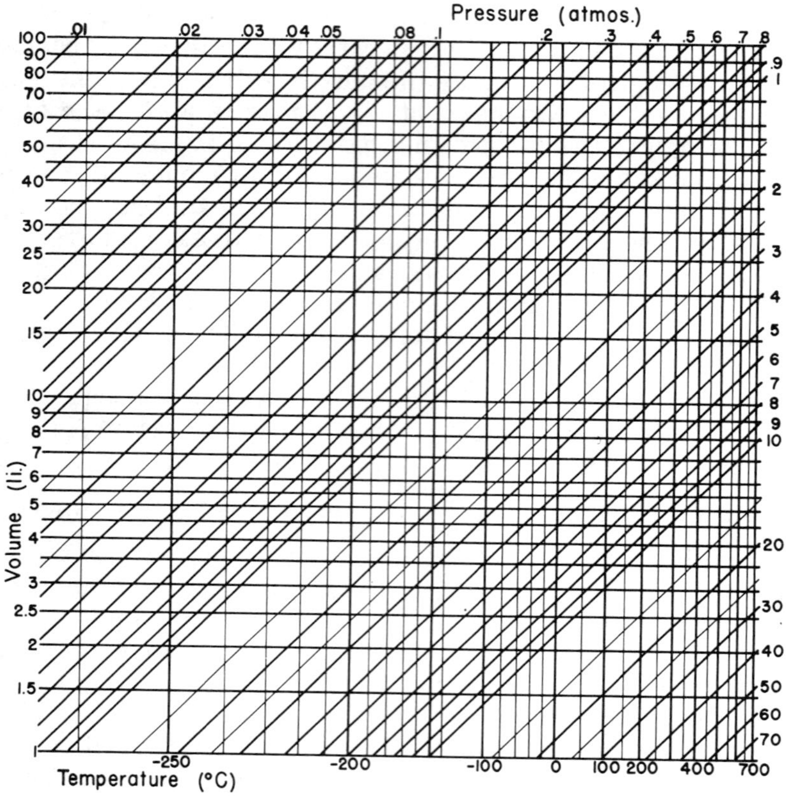

FIG. 6:11 SIMPLE PRESSURE-VOLUME-TEMPERATURE CHART FOR IDEAL GASES

isometric pressure lines, or isobars. The present diagram represents one mole of an ideal gas, which is completely defined by the expression, PV = RT, where P is the pressure, V the volume, and T the absolute temperature, degrees Absolute. R is the molar gas constant, which has the value 0.08206 when P is measured in atmospheres and V in liters.

6.10 Translation

The isobars represent the relationship $V_2/V_1 = (T_2/T_1)_p$, where the subscript p has the meaning "at constant pressure." This is an expression of Charles' Law, relating gas volumes and temperatures. A typical problem is this: The volume of a mole of an ideal gas at 85° is 20 liters; what volume does it occupy at -150° if no change in pressure takes place? The solution is: Locate the intersection of the isometrics for 85° and 20 liters. Follow the isobar to its intersection with the -150° isotherm, and read the value of the volume by interpolating between the volume isometrics (Fig. 6:12).

The isotherms represent the relationship $V_2/V_1 = (P_1/P_2)_T$, which is Boyle's Law, relating gas volumes and pressures at constant temperatures. In solving a problem that involves this law, one would locate the intersection of

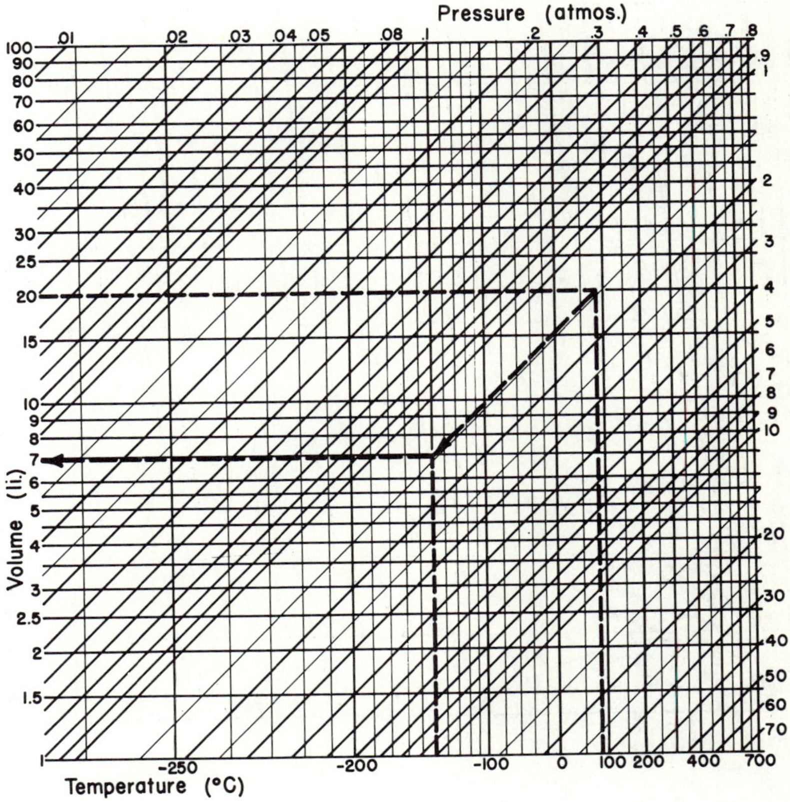

FIG. 6:12 COMPUTATION BY ISOMETRIC TRANSLATION

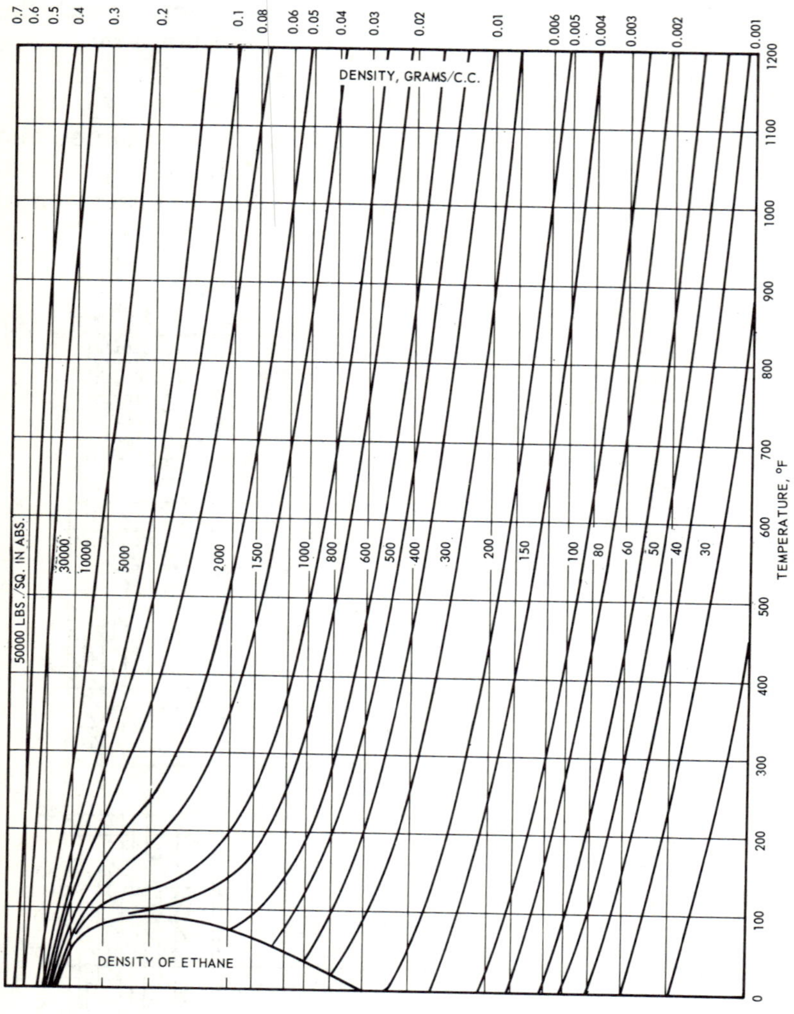

FIG. 6:13 DENSITY-TEMPERATURE-PRESSURE DIAGRAM FOR ETHANE
(After NSGMA Data Book, 1951, p. 129)

the isometrics for the original volume and pressure, then follow a vertical isotherm to the intersection representing the new pressure and volume.

Problems involving changes in both temperature and pressure are solved by a sequence of the two foregoing steps, one step for the change in volume at constant pressure, the other for the change in volume at constant temperature. The result is the same, regardless of which change is effected first.

The volume isometrics represent Gay-Lussac's Law, $P_1/P_2 = (T_1/T_2)_V$. Problems involving changes in pressure and temperature of gases at constant volume are solved by following one of these lines. There are, therefore, in this diagram all the necessary relationships for solving every type of problem involving the ideal gas laws.

The problems solved by Figure 6:11 can be solved more easily by ordinary slide rule, graph paper, or nomograph computation. There is, therefore, little need for a device of this kind in problems as simple as these. On the other hand, a diagram representing the relationships of pressure, temperature, and volume of a real gas has great value in calculating the results of changes in conditions, such as the one given. In this case the relationships are mathematically much more complex than those of the ideal gas laws, and cannot be solved by any slide rule routine or other simple multiplication procedure. Yet, the graphical solution is identical with that illustrated in the foregoing problems.

Figure 6:13 is a density-temperature-pressure diagram for ethane. Density is employed, rather than volume of one mole, as it more readily lends itself to further calculations involving weight and volume of any quantity of gas. In this diagram, one may solve a problem involving the change in density with pressure at constant temperature by following a vertical isotherm from the intersection of the original density and pressure isometrics to the intersection corresponding to the final conditions. A problem involving change of density with temperature at constant pressure is solved by following an isobar (i.e., one of the pressure curves) in the diagram, from the original conditions to the final conditions.

6.11 Defining Equation

The general relationship represented by a diagram such as Figure 6:11 or 6:13 is f(u,v,w,) = 0. One family of lines represents $f(u_i,v,w) = 0$, where u_i is a series of constant values of u. Similarly, the other families of lines are $f(u,v_i,w) = 0$, and $f(u,v,w_i) = 0$. The problems solved by isometric translation on a diagram of this kind are of the type $f(u_1,v_1)_w = f(u_2,v_2)_w$.

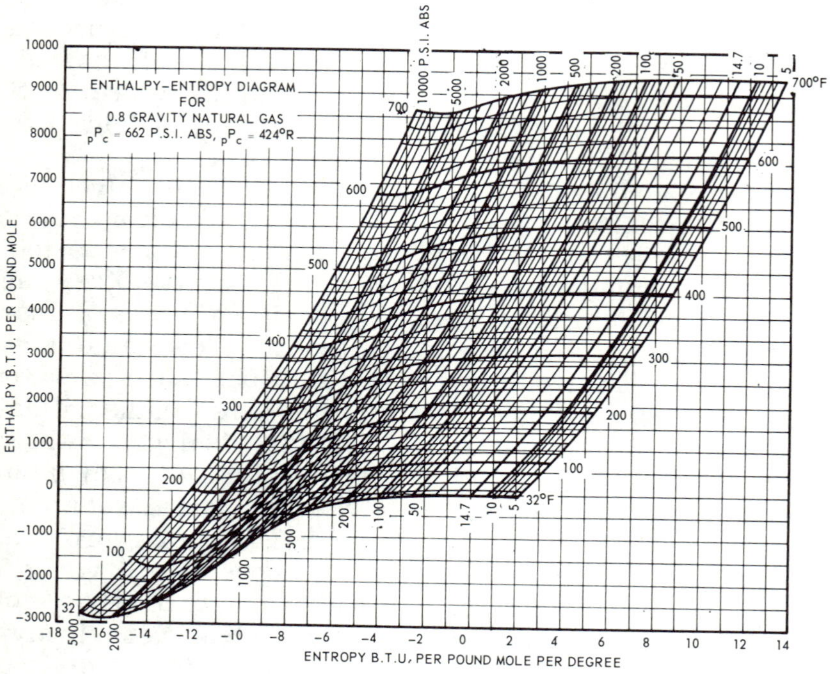

FIG. 6:14 TYPICAL ENTHALPY-ENTROPY DIAGRAM FOR GAS
(After NSGMA Data Book, 1951, p. 90)

6.12 Four-variable Systems

The method is easily extended to more complex graphs, such as the entropy-enthalpy diagram (Fig. 6:14) and the pressure-enthalpy (Mollier) diagram (Fig. 6:15). Translational problems are solved on these diagrams just as on the simpler diagram for pressure-temperature-volume. For ex-

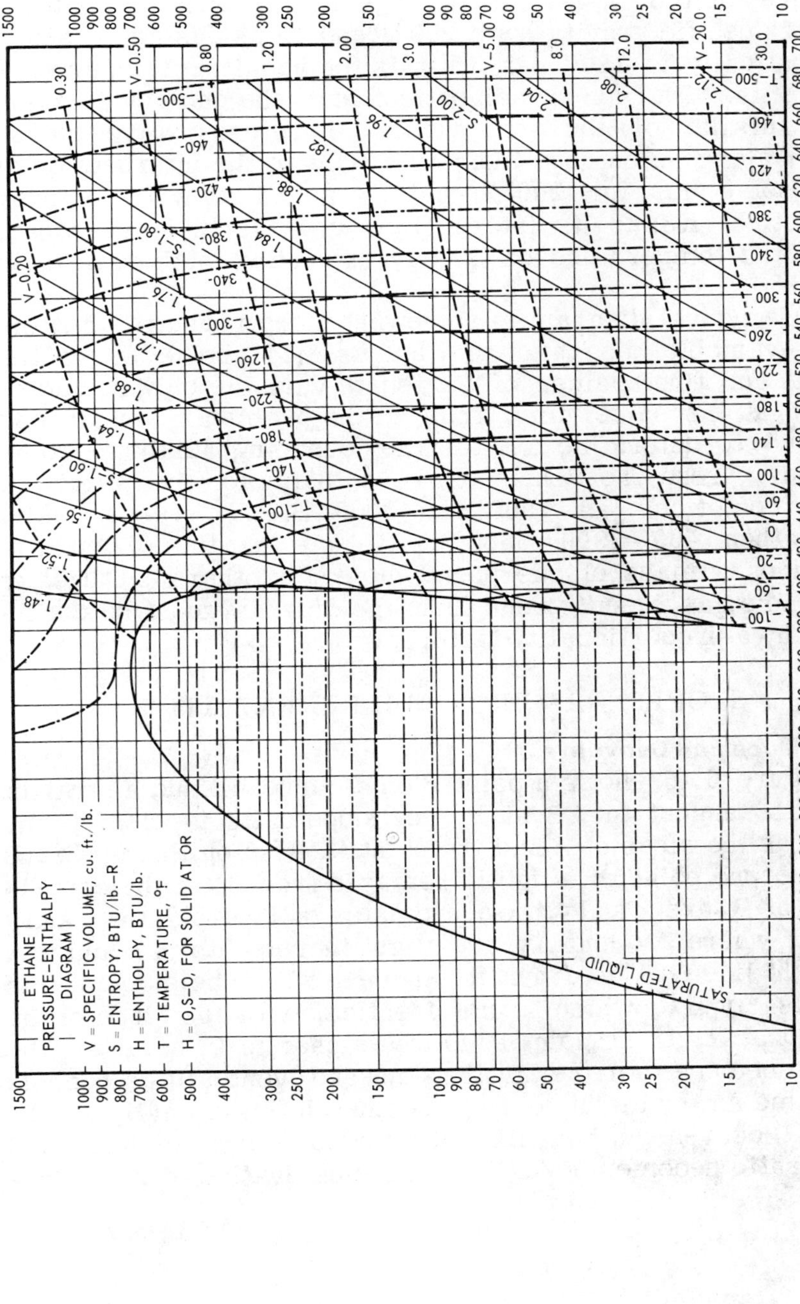

FIG. 6:15 MOLLIER (PRESSURE-ENTHALPY) DIAGRAM FOR ETHANE
(After NSGMA Data Book, 1951, p. 100)

ample, one may employ Figure 6:14 for the solution of a problem involving change of temperature during adiabatic compression of a gas, as follows. If, for example, natural gas of specific gravity 0.8 at 200°F and 40 psi, is compressed adiabatically to 1200 psi, what is the resulting temperature of the gas? During an adiabatic change the entropy remains constant; the solution of the problem, therefore involves translation along an entropy isometric. Locate the intersection of the isometrics for 200°F and 33 psi. Follow the entropy isometric (1.8) to its intersection with the 1200 psi isobar. Read the temperature on the corresponding isotherm, which is 600°F.

This type of diagram is somewhat more versatile than the three-variable diagrams of Figures 6:11 and 6:13, for the initial and final values of the fourth variable may be determined at the same time. Thus, in the problem given, if it is desired to determine ΔH for the foregoing adiabatic compression, read the values of the enthalpy for the initial and final conditions as 1900 and 8000 B.T.U., respectively. ΔH is then 7400 - 2000 = 5500 B.T.U. Note that this diagram is based on a mass of <u>one pound mole</u>; for other quantities of the same gas, results read from the diagram must be multiplied by a proportionality factor.

SLOPES; GRAPHICAL DIFFERENTIATION

6.13 Slope and Derivative

Figure 6:16 shows a point P on a curve AB and a straight line CD tangent to AB at P. The slope of CD is, clearly, the same as the slope of AB at the point P. Also on the curve AB is a second point P', slightly removed from P. If the coordinates of P are x and y, the coordinates of P' are $x + \Delta x$ and $y + \Delta y$, where Δx and Δy are slight increments in x and y. A straight line C'D' is drawn through P and P'. The slope of this line is $\Delta y/\Delta x$, which is not identical with the slope of the tangent at P. If the point P' moves closer to P, however, the slopes become more nearly the same and eventually do become the same as the distance PP' becomes infinitesimally small. As PP' decreases, both Δx and Δy become infinitesimal, and their ratio becomes dy/dx. Thus, the derivative of y with respect to x, is still, however, the slope C'D which now is identical with CD; consequently, the slope and the derivative are identical.

Differentiation is the process of determining the value of the derivative of a function; either its numerical value at a point on the curve, or its value as an algebraic expression applying to a length of the curve. Where the curve represents a well-defined function, the algebraic expression for

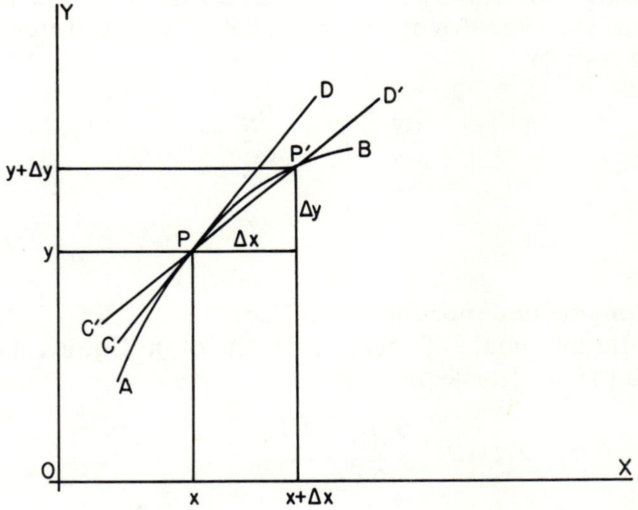

FIG. 6:16 RELATIONSHIP OF SLOPE OF A LINE TO ITS DERIVATIVE

its derivative may be derived by the principles of differential calculus. Where the curve has no analytical counterpart, one has recourse to graphical methods of differentiation; in short, to measurement of its slope at a point, or its slopes over a length of the curve.

6.14 The Derivative at a Point

The derivative of a function at a point is the numerical value of the slope of a graph of that function at that point. In a plot of y against x on a linear graph paper, the slope is simply dy/dx. A plot on semilogarithmic graph paper, however, is a plot of log y against x, and the slope is d(log y)/dx. Given y = f(x), the value of df/dx, obtained by measuring the slope, depends not only upon the nature of f, but also upon the type of the graph paper on which the plot is made.

In determining the value of the derivative from the slope, one must, also, take into account the moduli of the two axes. A function plotted on graph paper on which the modulus of the Y axis is twice that of the X axis has at any point a slope that is twice that of the same function plotted on paper with equal moduli of the two axes. The value of the derivative at any point is, therefore, the product of slope times the ratio of m_x to m_y, or

$$\frac{dy}{dx} = M \times \frac{m_x}{m_y}$$

6.15 Clapeyron Equation for Latent Heat

The latent heat of vaporization of a liquid, L_V may be estimated from the expression,

$$\frac{dP}{dT} = \frac{PL_V}{RT^2}$$

where P, T, and R are, respectively, pressure, temperature, and molar gas constant, and dP/dT is derivative of vapor pressure with respect to temperature. The value of this derivative may be estimated by measuring the slope of a vapor pressure graph (P as ordinates against T as abscissas, on linear graph paper) at the value of T given in the problem. The rest of the arithmetic is a simple slide rule operation.

In the derivation of the given expression for latent heat, it was necessary to assume (a) that L_V is constant, and (b) that the volume of the liquid is neglible in comparison with the volume of the gas. Both of these assumptions introduce slight errors; consequently, it is quite easy to obtain, by graphical methods, a degree of accuracy entirely in keeping with the limits of error of the formula.

6.16 The Derivative of a Curve

The derivative of a curve may be plotted by measuring the slope of the curve at frequent intervals and plotting these slopes, as ordinates, against the abscissas of the points at

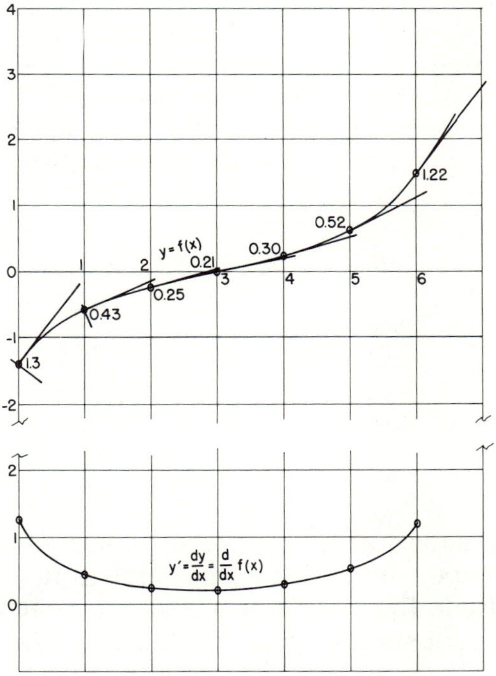

FIG. 6:17 DERIVATIVE OF CURVE OBTAINED FROM PLOT OF ITS SLOPES

which the slopes are measured (Fig. 6:17). As the derivative of a curve is often another curve; the same process may be applied to it to obtain a second derivative, to the second derivative to obtain a third derivative, and so on.

6.17 Analysis of a Curve

A discussion of graphical analysis of empirical data in Section 5.16 deals with methods whereby an algebraic expression representing the data can be derived from a rectified graph of those data. Rectification with common types of graph papers is not always possible, however. Polynomial expressions, such as $y = ax^2 + bx + c$, and $y = ax^3 + bx^2 + cx + d$, cannot be analyzed by rectification, except with graph papers especially prepared for the problem being analyzed. Lipka,[4] for instance, in analyzing the parabola $y = ax^2 + bx + c$, achieves rectification by plotting $(y - y_k)/(x - x_k)$ against x, where (x_k, y_k) defines some convenient point on the curve. The ordinates of this graph paper must be calculated from the empirical data or estimated from a plot of those data; they are then substituted for the conventional ordinates of linear graph paper before the final plot is made. If the final plot is a straight line, its slope and intercepts provide the values of the constants, thus completing the analysis. If, however, the curve differs only slightly from a parabola, as might a third degree curve, a hyperbola, or an exponential curve, the plot obtained by this method fails to provide the desired analysis. It is then necessary to attack the problem by a method especially designed for the hyperbola,[5] the exponential curve, etc.

TABLE 6:2

x	y	x	y
-1.0	8.25	0.6	4.06
-0.8	7.55	0.8	4.05
-0.6	6.84	1.0	4.25
-0.4	6.17	1.2	4.66
-0.2	5.55	1.4	5.32
0	5.00	1.6	6.25
0.2	4.55	1.8	7.46
0.4	4.23		

[4] Lipka, J., Graphical and Mechanical Computation, p. 145; John Wiley and Sons, Inc., New York, 1918.

[5] Lipka, J., op cit., pp. 149-152.

A more general approach is by way of the derivatives. Consider for instance, the data of Table 6:2. A plot of these data (Fig. 6:18, curve A), although suggestive of a parabola or hyperbola, fails to respond to standard measures for rectification of either of those forms. The slopes of the curve at each value of x in Table 6:2 are measured and plotted (Fig. 6:18 curve B) against the same abscissas. This curve

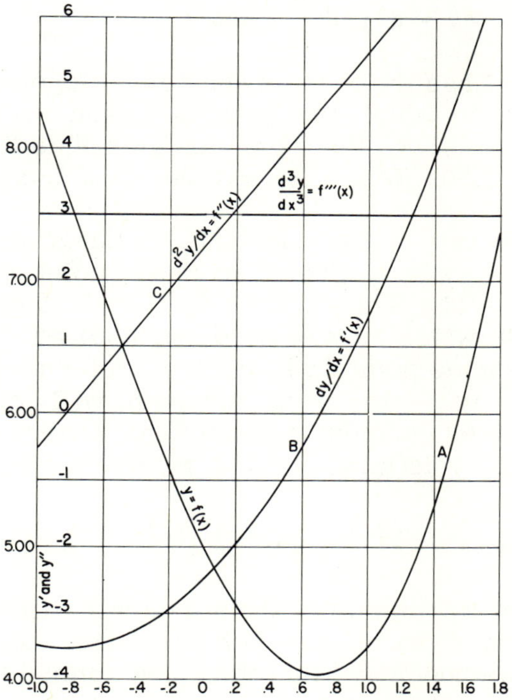

FIG. 6:18 SUCCESSIVE DERIVATIVES OF THIRD DEGREE CURVE

is a plot of the first derivative, dy/dx. The slopes of curve B are measured and plotted against the same abscissas (Fig. 6:18 curve C). This locus, which is a straight line, is a plot of the second derivative, d^2y/dx^2.

As the second derivative of y is linear with respect to x, one may assume for y the value

$$y = ax^3 + bx^2 + cx + d \tag{6.5A}$$

The first derivative is, then,

$$y' = dy/dx = 3ax^2 + 2bx + c \tag{6.5B}$$

and the second derivative,

$$y'' = d^2y/dx^2 = 6ax + 2b. \tag{6.5C}$$

Equation 6.5C has the form of the point-slope equation of the straight line, in which the slope is 6a and the y intercept is 2b. In Figure 6:18, the slope of line C is 3.00, and the y intercept is 2.50. Hence, $6a = 3.00$, $a = 0.50$; and $2b = 2.50$, $b = 1.25$.

These are two of the constants of Equation 6.5A; a third constant, c, may be evaluated from any point on locus B, for $c = y' - 3ax^2 - 2bx$. The point $x = 0$, $y' = -2.50$ gives the value of c as -2.50, which may be used, together with the values of a and b and some point on locus A, to determine the value of d: $d = y - ax^3 - bx^2 - cx$. Any point will do, but the point $x = 0$, $y = 5.00$ is most satisfactory, whence $d = 5.00$. The expression for y defining curve A is, therefore,

$$y = 0.50x^3 + 1.25x^2 - 2.50\,x + 5.00$$

Every true polynomial of the nth degree will by successive differentiation yield an $(n - 1)$th derivative that is a straight line. This method may, therefore, be applied to all polynomials. It is not essential, however, that a straight line derivative be obtained; the following method clarifies this point somewhat.

6.18 Series Expansions

In the preceding example, the derivation of a value for y is actually an application of the Maclaurin series,

$$f(x) = f(0) + \frac{f'(0)x}{1!} + \frac{f'\,(0)x^2}{2!} + \frac{f'''(0)x^3}{3!} + \ldots + R_1,$$

where f(0), f'(0), f''(0), etc., are the values of y, dy/dx, d^2y/dx^2, etc., at the point x = 0. These are, in other words, the values of the y intercepts of curve A (Fig. 6.18) and its derivatives. From Figure 6:18, these intercepts are f(0) = 5.00, f'(0) = -2.50, f''(0) = 2.50, and f'''(0) = 3.00. All remaining derivatives are zero; hence, the value of the series is

$$f(x) = 5.00 - 2.50x + \frac{2.50x^2}{2!} + \frac{3.00x^3}{3!}$$

$$= 5.00 - 2.50x + 1.25x^2 + .50x^3.$$

This is, clearly, the same expression previously found for this curve. The process is virtually the same, for the y intercepts were used in both cases in evaluating b, c, and d. Note that the y intercept of f'''(0) is equal to the slope of f''(x) in this example.

As pointed out in Section 6.17, however, any point on each curve may be employed to evaluate a constant, the other constants being known. So, also, in series expansion, the values of f(x), f'(x), f''(x) at some abscissa x = a, rather than 0, may be employed. The series here is Taylor's series:

$$f(x) = f(a) + \left(\frac{f'(a)}{1!}\right)(x-a)$$

$$+ \left(\frac{f''(a)}{2!}\right)(x-a)^2 + \left(\frac{f'''(a)}{3!}\right)(x-a)^3 + \ldots + R_2$$

Returning to the example cited in Section 6.17, one may consider the expansion at the axis, x = 0.6. At this point, y = 4.06, y' = -.46, y'' = 4.30, and y''' = 3.00. Hence,

$$f(x) = 4.06 - 0.46\ (x - 0.6) + \frac{4.30}{2}(x - 0.6)^2 + \frac{3.00}{6}(x - 0.6)^3$$

$$= 4.06 - 0.46x + 0.28 + 2.15x^2 - 2.58x + 0.77 + 0.5x^3$$

$$- 0.90x^2 + .54x - .11$$

$$= 5.00 - 2.50x + 1.25x^2 + 0.50x^3.$$

This, again, is identical with the expression previously derived.

The successful application of this method depends very largely upon the accuracy of the original plot and the care with which it is interpolated. True, the expansion is made only at one point of the curve, but the correct value of its derivative at that point can be obtained only if an appreciable portion of the curve is known on both sides of the point. Thus, in curve A, if one used only the values of x in the vicinity $-1.0 = x = -0.6$, one would be likely to take y' as zero and miss entirely the true character of the curve.

For values of x in the region $(a - 0.7) \leq x \leq (a + 0.7)$, terms higher than the third power become fairly small; consequently, this method may be satisfactorily applied to limited ranges, whether or not higher derivatives of the original function vanish. If the curve is an extensive one, however, it is advisable to divide it into portions such that the range of x in each portion is about 1.5 units, and derive separate expressions for y in each portion. If the curve is highly regular, as is the cubic equation just derived, the various expressions will be virtually identical; if not, the relationship may be left in the form of separate expressions or united into a composite expression in which one or more of the coefficients is also a function of x.

The series form of the function is very satisfactory for many purposes, such as the computation of tables and the derivation of defining equations for nomographs. It is also a desirable means of defining individual curves in a family of curves, for it is frequently possible to derive a relationship between the values of the coefficients and those of the variable that, in the family, has a series of constant values. If, however, one prefers a generic form (i.e., a form characteristic of certain trigonometric, geometric, or other elements) in preference to the expanded form of the function, the series may be compared with the series obtained by expansion of various functions, as listed in certain texts or handbooks.[6] Data can be analyzed by this method as well as by rectifica-

---

[6] See, for instance, Perry, J., Chemical Engineers' Handbook, 2nd Ed. p. 220, McGraw-Hill Book Company, Inc., New York, 1941.

tion. The series method has the following advantages over rectification methods:

1. The method is the same, regardless of the character of the curve. Trial-and-error methods are avoided.
2. It is more readily adapted to the analysis of families of curves.
3. It produces the same results for standard forms, often in a fraction of the time required to try various rectifying plots.
4. No computation of rectifying ordinates is required. The method is completely graphical.
5. It may be applied to irregular forms, for which no rectification methods are available.

6.19 Measurement of Slope

From the preceding applications of derivatives, one may readily realize that accurate, rapid methods of measuring slopes are essential. As has been pointed out, the slope of a locus at a given point is identical with the slope of its tangent at that point. The slope of the normal (perpendicular) to the curve that point is the negative reciprocal of the slope. Either the tangent or the normal may, therefore, be used to measure the slope; the problem is to construct the tangent or normal accurately, and this can be done with the proper equipment.

a) Method of Tangents: A simple tangent meter may be constructed as follows (Fig. 6:19). A claw spring (a clothes pin or paper clamp may be adapted) is equipped on one jaw with a pair of parallel steel bars, placed close together, and, on the opposing jaw, with a point that presses midway between the bars. When this spring is clamped onto a flexible metal strip, with the bars perpendicular to the length of the strip, the centers of the bars align themselves with the tangent to the strip at that point. The smaller the space between the bars, the more accurate this alignment is. A scale of slopes is attached by which the slope is read directly. The flexible metal strip (a piece of spring steel) is moved to the point at which a reading is to be made, and is bent to the exact shape of the curve at that point. With the strip in that position, the slope is read on the scale.

The instrument must be carefully calibrated, and checked from time to time. Even so, it is a difficult device to use, unless one has three or four hands. It is possible, however, to increase ease and accuracy in the use of it by including in the equipment a number of preshaped steel strips, with which almost any curve can be matched without much force on the strip. The shape of strip shown in Figure 6:19 is de-

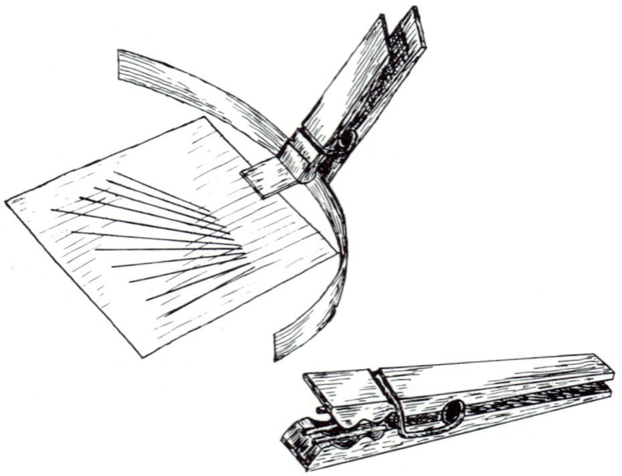

FIG. 6:19 METHOD OF MEASURING SLOPES OF TANGENTS

sirable, as the grippers for thumb and fore finger expedite adjustment of its curvature.

b) Method of Normals: The normal to a curve is found by placing a small mirror on the curve perpendicular to the plane of the curve and rotating it until the curve and its reflection appear to be perfectly continuous. A glass mirror mounted on a square block may be used, but, as there is some refraction of light by the glass, a polished metal surface is preferable. Excellent metal mirrors of this sort may be salvaged from a toy kaleidoscope a day or two after Christmas. The normals may be marked with a pencil stroke along the edge of the mirror, and their slopes measured later, or a slope meter may be attached to the mirror to read each slope separately. By either method, the slope meter should be calibrated to read the slope of the tangent, rather than that of the normal; the difference between the two slopes is 90° and may be realized by rotating the slope meter through that angle. Alternatively, the slope of the normal may be read on the meter with reference to the horizontal lines of the grid,

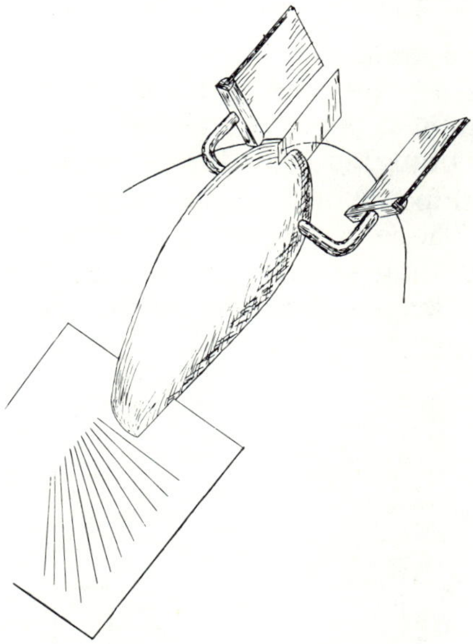

FIG. 6:20 METHOD OF MEASURING SLOPES OF NORMALS

and the slope of the tangent with reference to the vertical lines.

At each point the slope should be determined from both sides of the curve. If there is no discontinuity in the curve, the two readings should agree. If they do not, it is possible that sightings are not being made quite parallel to the surface of the paper, or that the mirror is not quite perpendicular to the surface.

Figure 6:20 shows a convenient device for locating normals. A thin metal plate, highly polished on both sides, is held in position by a cast steel block. Mirrors on both sides of the plate are tilted so that reflections of the curve on both sides of the plate may be observed simultaneously. The mirrors also have the effect of placing the line of sight parallel to the board. A slope meter attached to the casting provides values of the slope of the tangent. With this device, the derivatives of a curve at a number of points may be accurately and rapidly determined.

AREAS; GRAPHICAL INTEGRATION

6.20 Integrals and Areas

Figure 6:21 shows one method for estimating the area between a segment of a curve and the X axis. The area is subdivided into rectangles, the lower ends of which lie on the X axis. The midpoint of the upper end of each rectangle lies on the curve. There are errors in this estimate, for some parts of the rectangles project beyond the curve, and some portions of the area under the curve are not included in the

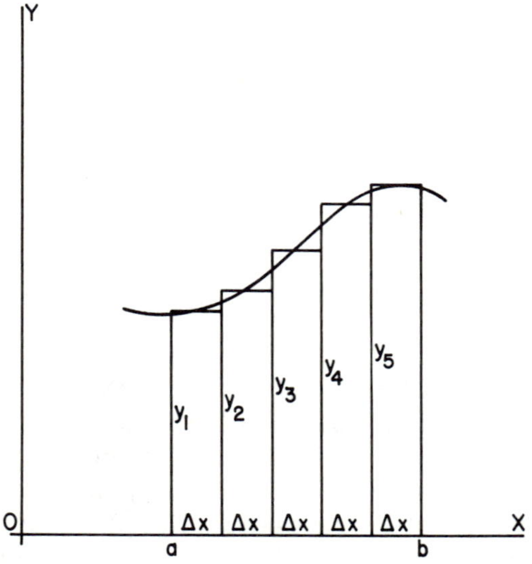

FIG. 6:21 MEASUREMENT OF AREA UNDER A CURVE; METHOD OF RECTANGLES

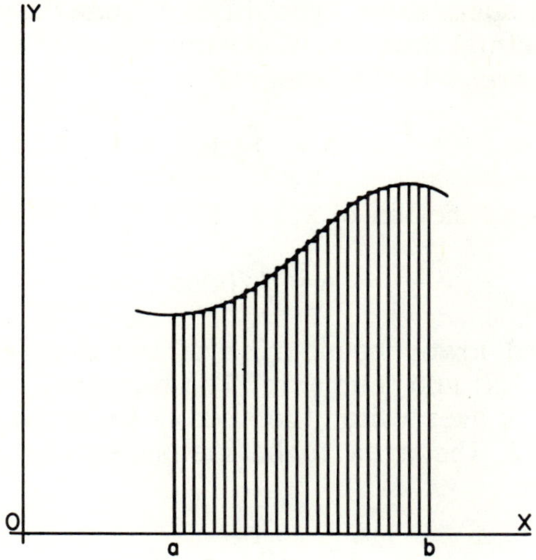

FIG. 6:22 USE OF SMALLER INCREMENTS OF x TO INCREASE ACCURACY OF METHOD OF RECTANGLES

rectangles. These errors partially balance out, but not entirely. If the rectangles all have equal bases, Δx, and their altitudes are y_1, y_2, y_3, and so on, the total area included in the rectangles is

$$\text{Area} = (y_1 + y_2 + y_3 + \ldots)\Delta x$$

$$= \sum_{i=1}^{n} y_i \Delta x$$

To reduce the errors mentioned previously, one might use thinner rectangles, as in Figure 6:22. The number of such rectangles increases as Δx decreases. If the errors are to be reduced beyond measurable proportions, however, it is necessary that Δx become infinitesimal. The infinitesimal increment is dx, and the process of summation, instead of being applied to discrete areas, becomes a continuous process in which each successive area differs from the preceding one by an infinitesimal amount. The summation of a continuum of infinitesimal areas is the integral

$$A = \int y dx,$$

or, since y is a function of x,

$$A = \int f(x) dx \tag{6.6}$$

The upper and lower boundaries of the area are defined by the curve y = f(x) and the x axis. The terminal boundaries are the limits of x over which the area is to be measured. If, as in Figure 6:22, the area to be measured lies between x = a and x = b, then

$$A = \int_a^b f(x) dx$$

If f(x) is a common function, such as cos x, log x, x^2, and so on, the integral may be evaluated analytically. Extensive integral tables[7] are available, showing the algebraic expressions obtained by integrating a large variety of functions.

If the curve is derived from empirical data, its integral may be derived from a series expression obtained by the method of Section 6.19. Thus, if the curve is found to be equivalent to

$$y = f(x) = ax^3 + bx^2 + cx + d,$$

its integral (by the rules of integral calculus) is

$$\int f(x) dx = \frac{ax^4}{4} + \frac{bx^3}{3} + \frac{cx^2}{2} + dx + e$$

The area under the curve between x_1 and x_2 is

---

[7] For example, Peirce, B., Table of Integrals, Ginn and Company, Boston, 1910.

$$A = \int_{x_1}^{x2} f(x)dx = \frac{a}{4}\left(x_2^{\ 4} - x_1^{\ 4}\right) +$$

$$\frac{b}{3}\left(x_2^{\ 3} - x_1^{\ 3}\right) + \frac{c}{2}\left(x_2^{\ 2} - x_1^{\ 2}\right) + d\left(x_2 - x_1\right).$$

Unless there is reason to analyze the curve for other purposes, this is a rather tedious method for obtaining the area. As a general rule, any one of the following direct measurements is simpler, easier, and equally accurate.

6.21 Counting Method

If the curve is plotted on linear graph paper, a fairly close estimate of its area may be obtained by counting squares under a curve. A systematic approach is in order, however.

A plot made on paper divided into inches and tenths or half-inches and twentieths is recommended. If the division is in inches and tenths, the area should be viewed as composed of columns one inch wide. At each end of the area and at the top of each column are areas that must be estimated by smaller subdivisions; but the estimate should proceed from the larger to the smaller subdivisions. A typical column is shown in Figure 6:23. The number of one inch squares (3) lying completely under the curve is counted first, and their total area is written down. The number of squares (22) with 0.1-inch sides, other than those in the one inch squares, lying wholly under the curve is counted next. Their area is 1/100 times their number; this is added to the first area. The small squares intersected by the curve are then examined, and the fraction of their area lying under the curve is estimated. The total of these fractions, multiplied by 1/100 is the area; added to the other areas, this completes the estimate of area for the column. This process is repeated for other columns and fractions of columns.

The only source of error is in estimating the fractions of small squares lying under the curve. This error is less if paper divided into twentieths is used. The procedure is the same, the counting being done by half-inch and twentieth-inch squares. The total count is then divided by four to obtain square inches, as the area of each of these squares is only one-fourth that of the inch squares. Errors, as well as the tedium, of estimating the small portions of squares lying under the curve may be reduced considerably by using a reading glass. With these small squares, there is very little error involved if the segment of each small square under the curve is regarded as a triangle or trapezoid, and its area is estimated accordingly.

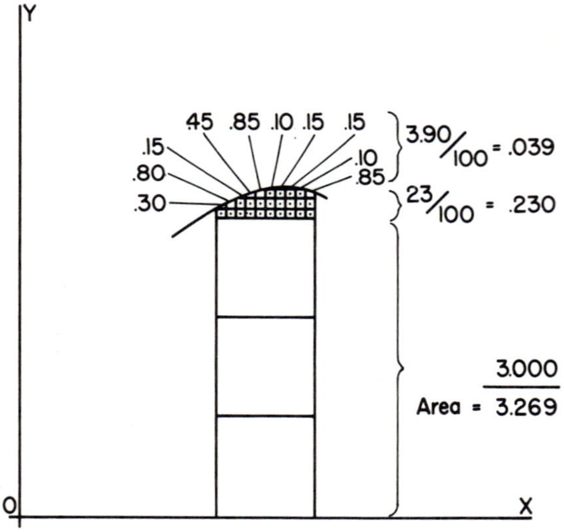

FIG. 6:23 METHOD OF ESTIMATING AREA BY COUNTING SQUARES

Other, more accurate methods are described in the following sections, but the method of counting has its advantages. No special equipment is required, and for non-repetitive computations the time factor is not of great importance.

6.22 Weight Methods

For small-scale repetitive computations, areas are conveniently measured by weighing them. The curve is traced out on metal foil, rather than on paper; the foil is cut along the boundaries of the area, and the cut-out portion is weighed.

The foil used for this purpose must be uniform in thickness, within very close tolerances. Percentage deviations in thickness are less in the thicker foils, which, also, are much more easily manipulated and cut with scissors. The foil must also be resistant to corrosion, as oxidation increases its weight. A heavy aluminum foil is recommended. The weight of a standard area, such as 100 square inches, should be checked frequently. In some laboratories the foil is purchased in a standard size; all the pieces should weigh the same. As a check, both the cut-out portion and the scrap are weighed separately; if the sum of their weights differs from standard, a correction is introduced to compensate for the factor of thickness.

If the curves to be integrated are on graph paper, they may be traced onto the foil, using a hard carbon and a very fine stylus, or without carbon, using a toothed-wheel stylus. If the curves are plots produced by a mechanical recording device, the foil may be substituted for the paper of the recording device, and the curves traced out directly on the foil by the machine with a fine stylus. If the tracing thus made is difficult to see, the foil may be coated with soap film or water soluble color for the stylus to trace through. After cutting, those materials may be washed off and the foil dried.

An accurate analytical balance, sensitive to 0.1 mg., should be used, and all the usual precautions for handling material to be weighed should be observed. The foil should not come into contact with the fingers, because of the weight and corrosive effects of fingerprints. The foil may be rolled into a cylinder by wrapping it around a clean glass rod, thus providing a more compact form for weighing. Unrolled, it may be filed as a more or less permanent record.

6.23 Photometric Methods

Several methods are available for measuring areas photometrically. The area may be painted black, leaving the remainder of the sheet white, and the reflection of light from the sheet thus prepared is then compared with that from blank sheets and from sheets blackened over standard areas. Alternatively, the area to be measured may be kept white, and the remainder of the sheet blackened.

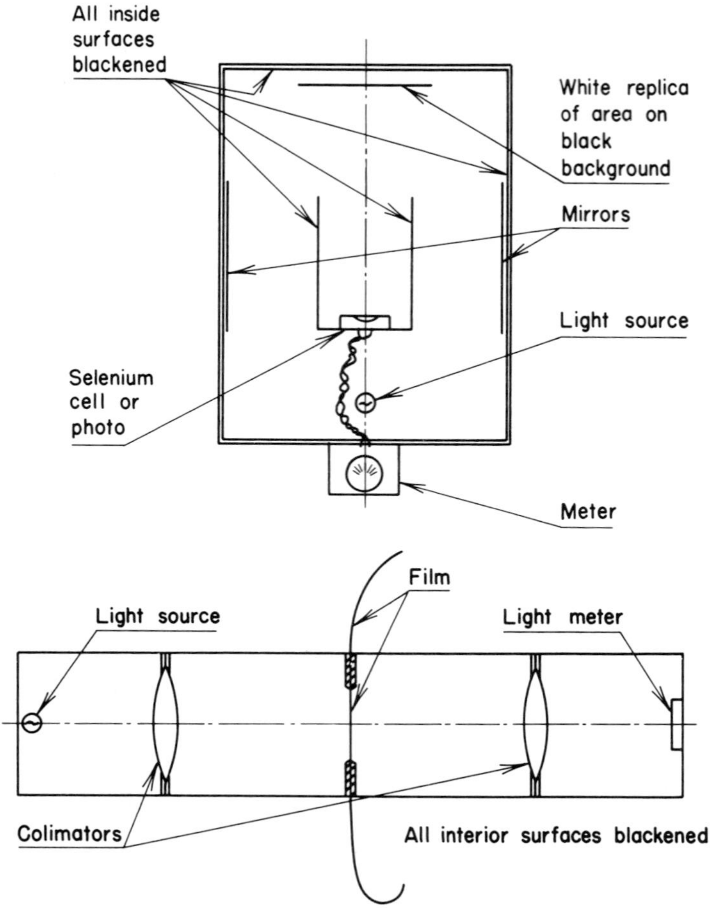

FIG. 6:24 PHOTOMETRIC METHOD OF MEASURING AREA
a) By Reflectance; b) By Transmission through a film

If preferred, the area to be measured may be cut out of opaque white paper and mounted on a black background, or vice versa. In all of these methods, the reflected light is measured by a standard photometer, which must be so illuminated by the reflected light as to be equally sensitive to all portions of the field. It is also essential that all portions of the field be equally illuminated by the source of light. One such arrangement is shown in Figure 6:24.

As uniform illumination of large areas is both difficult and expensive, it may be avoided by the use of film. The area to be measured is either painted black or cut out of black paper and mounted on white. It is then photographed under studio-type fluorescent flood lights onto 16mm. or 32 mm. film. A large number of areas can thus be photographed very rapidly. The developed film is opaque except for apertures that are exact miniatures of the area to be measured.

The film is passed, frame by frame, through a device such as that in Figure 6:24 (b). Uniform illumination is much simpler with the small field, and this may usually be achieved by the use of a distant light source without a colimator. The photosensitive meter may be calibrated to read sq inch or sq cm, directly, or it may even be rigged to record readings mechanically.

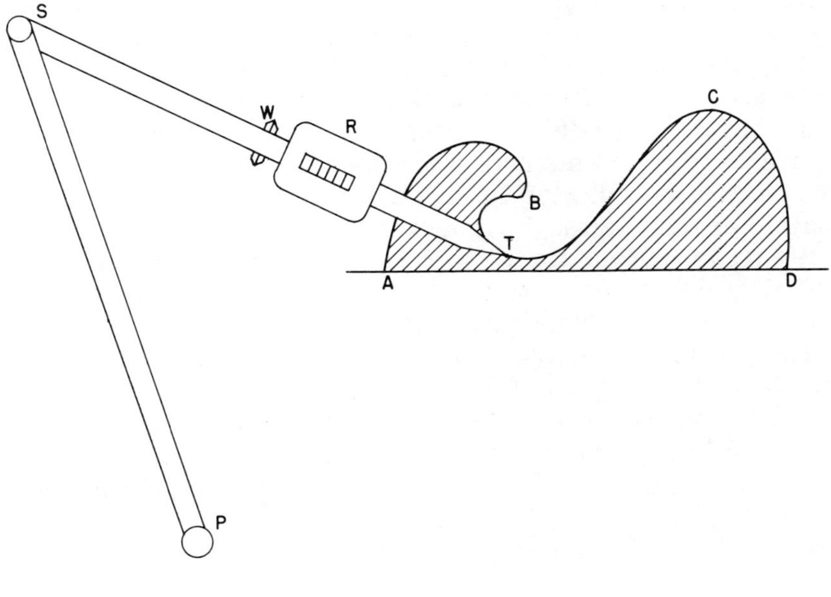

FIG. 6:25 MEASURING AREA BY PLANIMETER
P, Pivot Point; T, Tracing Point; W, Wheel;
R, Recorder; ABCD, Area to be Measured

6.24 Planimeter Methods

The planimeter[8] is a device for measuring areas mechanically (Fig. 6:25). The point P' is stationary; the point P traces the boundary of the area to be measured. The wheel W partly slides and partly rolls as a result of the motion of the arm on which it is mounted and its traction against the surface of the paper. The wheel is so calibrated that, starting at zero, its reading at the end of one circuit of the area S by the point P is exactly equal to that area.

As the tracing point starts at A, outlines the perimeter of ABCD, and returns to A, the wheel combines the motions of sliding, positive rotation, and negative rotation, resulting in a reading, n, on the recorder, which is the algebraic sum of its rotations. Area ABCD is then equal to $2\pi rln$, where r is the radius of the wheel, and l in the length of the bar, ST.

The planimeter may be used, by hand, equally well for isolated measurements or for repeated readings. When operated by hand, it should be used with the same care one would use in inking a curve with a ruling pen. Wherever possible, a straight edge or standard curve should be used as a guide, the curve being carefully fitted to the boundary of the area before the tracing point is moved.

In the large-scale measurement of strip chart areas it is desirable to avoid manual operations altogether. The curves are plotted automatically by a recording device, and the planimeter may often be adapted in such a way as to determine areas automatically. If it is not feasible to have the recording device operate the planimeter, the planimeter may be attached to a curve-following device.

The preferable method is a linkage between planimeter and recording device. The recording device has two moving elements, usually the paper and the scriber. The paper may be a roll that is passed under the scriber by rotation of the drums on which it is wound (or a feeding mechanism), or it may be a disk rotated on center. As a general rule, the rate at which the paper advances is constant and it is synchronized

---

[8] For a discussion of the theory of the planimeter, see Granville, W. A., P. F., Smith, and W. R. Longley, Elements of the Differential and Integral Calculus, pp. 429-3; Ginn & Company, New York, 1929.

with an independent variable of the system that is being measured, such as the wave length of light passed by a spectrophotometer or the angle of rotation of a microwave antenna. The motion of the scriber is at right angles to that of the paper; both its speed and its direction are subject to wide fluctuations. The extent of the displacement of the scriber from zero position is determined by a dependent variable, such as the intensity of light reaching a photosensitive tube or of radiation attenuated by the antenna.

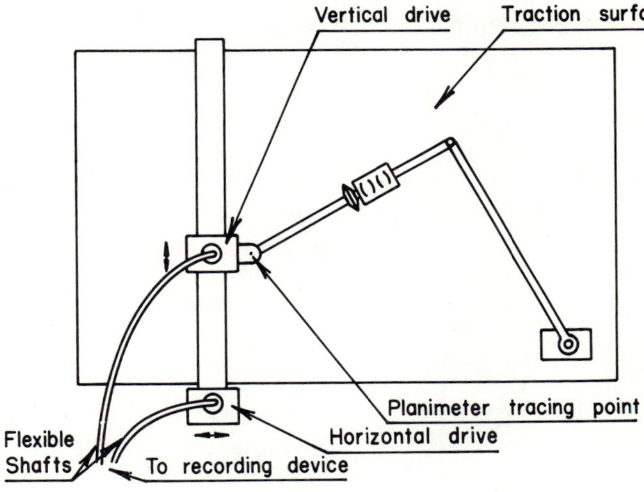

FIG. 6:26 PLANIMETER ATTACHED TO RECORDING DEVICE

Figure 6:26 shows a planimeter linked to these two motions. The motion of the paper is geared to an element that moves horizontally across the board on which the planimeter is mounted. The motion of the scriber is translated to an element that travels vertically on the first one. The tracing point of the planimeter is attached to the second element, thereby being constrained to move with both of them.

At the start of a recording, both the recording device and the planimeter are set at zero. During the recording, the scriber of the recording device describes a curve from x_1 to x_2; and the tracing point of the planimeter describes an identical curve on the board. To enclose the area completely, the tracing point must now move, in zero position, from x_2 to x_1. This may be done by reversing the feed of the recording device or by disengaging the horizontal element on the board from the recording device and moving it back to x_1. The reading on the planimeter, which is the desired area, is noted and, as a permanent record, attached to the chart of the recording device.

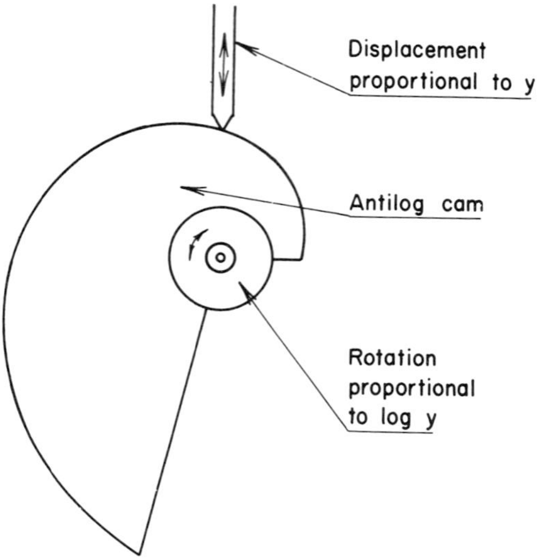

FIG. 6:27 ANTILOG CAM FOR CONVERTING LOGARITHMIC DISPLACEMENTS TO LINEAR SCALE

One difficulty occasionally encountered in putting this arrangement into operation is the non-linearity of the dependent variable to which the scriber is sensitive. This must be corrected by a linkage or cam. An example is the measurement of attenuation at various angles by a microwave antenna. The intensity of the signal received by the antenna is measured and plotted by the recording device in decibels, which is a most convenient unit for comparative purposes. The intensity thus recorded (db) is related to the power (watts) of the signal (P) as

$$db = \log_{10} P/P_s,$$

where P_S is a standard of power, such as 6 microwatts. The integral of the db curve would have little meaning; but the integral of P over the same range is a valuable part of the data. The units are transferred by inserting an antilog cam (Fig. 6:27) in the drive of the vertical element that guides the tracing point of the planimeter.

In recent years, most of the manufacturers of strip chart recorders have developed commercial models of automatic integrators based on principles similar to those described. Many of these print the area measured on an auxiliary tape.

EXERCISES

1. The accompanying graph shows plots of spectrographic data for (Curve C) a mixture of three dyes, and (Curves A and B) two of the component dyes. (a) Plot the curve for the mixture of A and B. (b) Use the plotted curve and Curve C to

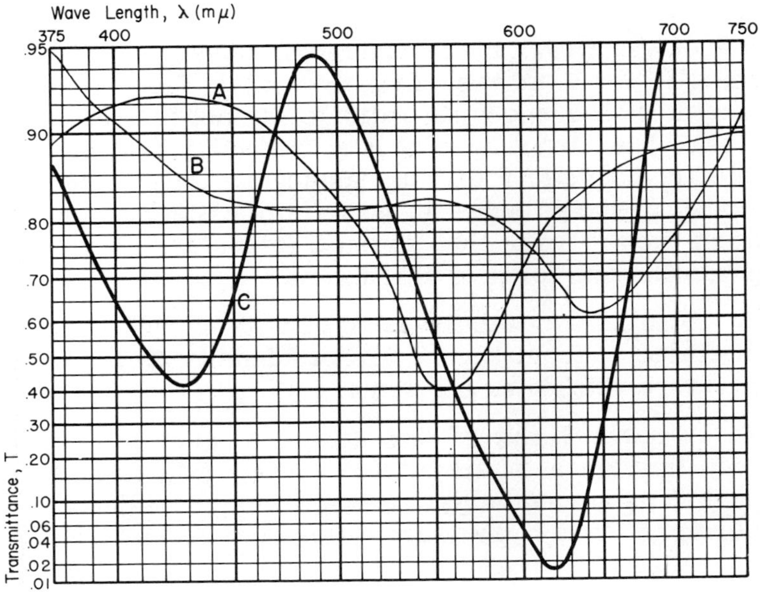

plot (by difference) the curve for the third component. Vertical displacements may be made to correct for changes in concentration of the components and the mixture.

2. From Figure 6:8 (Sec. 6.6), read the values of the freight rate differentials for the following outbound mileages, when the inbound mileage is 88 mi: 75, 83, 87, 102, 113, 128. Check each reading by first adding inbound mileage to outbound mileage to obtain through mileage; and then reading the rates for inbound, outbound, and through mileage from Table 6:1.

3. Using arcs of lengths 0.5, 1.0, 1.25, and 1.5 units, measure the lengths of chords subtending these arcs as the arcs are moved along the Cornu spiral (5:12). Plot the square of the length of the chord as ordinate against v as abscissa, v being the midpoint of the arc. The resulting plots are Fresnel diffraction patterns at different distances from a slit. Calculate these distances, taking the width of the slit as ten wave lengths for each of the following wave lengths; (a) 5×10^{-5}cm (visible light), (b) 10 cm, (microwave radiation), (c) 10^4 cm (A-M radio).

4. Taking the point P' at (1,1) as one end of the arc and a point P on the spiral as the other end, measure the lengths of chords PP' as P moves along the spiral. Plot the square of the length of the chord as ordinate against the value of P as abscissa. The resulting plot is a Fresnel diffraction pattern for electromagnetic radiation passing a straight edge (i.e., a slit, one of whose edges is an infinite distance). Derive the equation for I from the distance formula and the Fresnel integrals for this form of diffraction.

5. A sample of ethane has a density of .005 g/cc at 200 psi. Determine from Figure 6:13 its density at 2000 psi, the temperature remaining the same.

6. A natural gas of 0.8 sp. gr. at 5000 psi and 500°F expands through a small orifice without gain or loss of heat (i.e., at constant enthalpy). From Figure 6:14 determine its temperature if its pressure after expansion is 100 psi.

7. Plot the curve $y = x^4 - 5.5x^2 + 15x - 27$ for the values $-5 \leq x \leq +5$. Measure the slopes of the curve at $x = -3, 0$, and $+4$. Derive dy/dx from the expression given and calculate its value for the same values of x. Compare the values of the slopes obtained graphically with those obtained from the expression for dy/dx.

8. Plot the curve defined by the following points:

x	-2	-1	0	+1	+2	+3
y	-15.6	-13.2	-10.0	-3.8	+3.6	+16.4

Using a power series expansion, derive an expression for this curve.

9. Measure the areas under the curves prepared in Exercises 3 and 4.

10. Plot the curve $dy/dx = 3x^2 - 2x + 5$. Measure the area under the curve between the limits of x = 2.5 to 5.0. This area should equal the value of the integral $\int_{2.5}^{5.0} (3x^2 - 2x + 5)dx$. Evaluate this integral analytically and compare its value with that obtained graphically.

11. Compute the values of $\pi v^2/2$ for v = 0, 0.1, 0.2, 0.3, etc. Using these values with a table of trigonometric functions, determine and plot $\cos \pi v^2/2$ and $\sin \pi v^2/2$. Measure the areas under these curves at intervals of $\Delta v = 0.1$. Compare the values obtained with a table of Fresnel integrals (such as in Jahnke and Emde) prepared by direct computation from a series expansion of the integral.

12. Compare the Mollier diagram, Figure 6:15, with the Nomograph for Enthalpy of Pure Hydrocarbons, by Schiebel and Jenny,* as to (a) ease of reading, and (b) ease of making computations involving changes in pressure and temperature.

---

* Schiebel, E. G., and F. J. Jenny, Ind. Eng. Chem. 37:990 (1945); also in Nomographs for Chemical Engineers by O. P. Kharbanda, p. 29; Heywood & Company, London, 1958.

chapter 7

NOMOGRAPHY

The inception and development of the ingenious method of graphical representation called nomography is indisputably attributable to Maurice d'Ocagne. It is interesting to follow the growth of an idea from d'Ocagne's account of the development of his method[1] which traces its beginnings to the success of solving complex relationships by means of a straight index line that intersects three circles. Inspired to pursue the matter in broader applications, d'Ocagne was led to substitute for the circle a cross-sectional matrix representing values of two variables. In this he derived benefit from the work of Lelanne (1842) and Massau (1884), who developed general methods for the selection of axes that produce rectified families of curves.

The outstanding contribution in this development was the method for realizing solutions of relationships among as many as six variables, each of which could be represented by a pencil of scales, the pencils paired into three cross-section matrices. (Fig. 7:1). The defining equation of such a system is so limited, however, that few engineering formulas of five or six variables may be adapted to it. Consequently, one seldom encounters in practice formulas of more than four variables; some of this type are shown in Chapter 14 the charts consisting of two linear scales and a network scale representing two variables.

The word nomograph (Gr., nomos, law) was originated by d'Ocagne. It was a most fortunate selection, for the nomograph is a graphical representation of a mathematical law and it is more properly regarded as such than as merely a device for solving of problems. The term nomogram is sometimes used, also. As applied by many, the nomograph includes

[1] d'Ocagne, M., Traité de Nomographie, Gauthier-Villars, Paris, 1921.

cross-section charts and combinations of cross-section charts of the type discussed in Chapter 6 as well as alignment charts, in which solutions are realized by the intersections of a straight index line with systems of scales. As has been demonstrated, the cross-section chart is more properly considered a graph; the alignment chart employs an entirely different principle of representation of which d'Ocagne was the originator. It is to charts of this type that the term nomograph is applied in this book (Fig. 7:2).

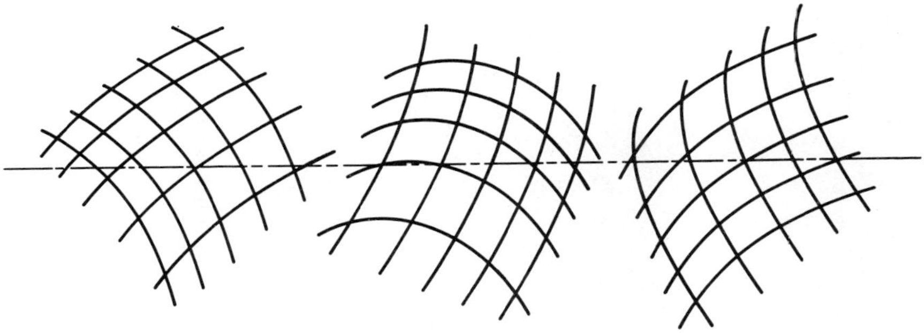

FIG. 7:1 d'OCAGNE'S ARRANGEMENT OF NOMOGRAPH FOR SIX VARIABLES

NOMOGRAPHIC METHODS

7.1 Fundamental Problems of Nomography

The design of a nomograph to provide solutions of a given formula for specified ranges of each of its variables requires:

1. An arrangement of scales and index line(s) that can realize graphically the numerical relationships inherent in the formula.

It is usually possible to find one such arrangement; sometimes, several are available. The arrangement, however, may be such that an intricate method of operation is required, or the index lines may intersect the scales at such angles as to make reading difficult or inaccurate, or too large a chart is required in order to achieve the desired degree of accuracy. In such a case, construction of a nomograph is inadvisable. An alternate arrangement may be found, a different graphical

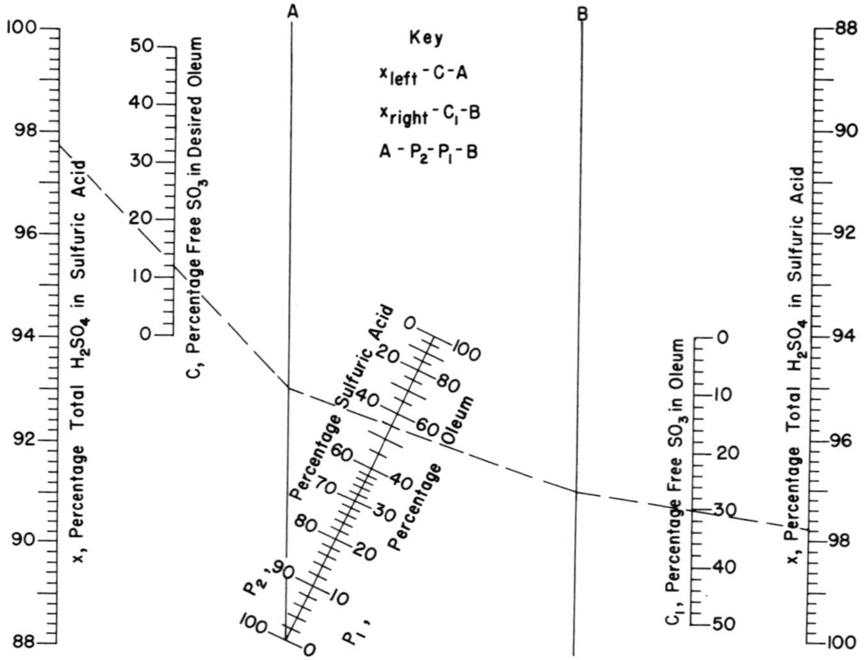

FIG. 7:2 CALCULATION OF MIXTURES OF OLEUM AND SULFURIC ACID
A Modern Nomograph Showing a Relationship of Six Variables
(From Davis, D. S., Chemical Engineering Nomographs, McGraw-Hill Book Company, Inc., New York, 1944)

method of solution may be substituted, or a numerical table may be constructed. Indeed, it is the existence of such possibilities that makes necessary an intimate knowledge of all graphical and numerical methods. The nomograph is, however, a very versatile device, which can be fitted to a surprisingly large variety of problems.

2. Adjustments in moduli and distribution of the scales to accommodate the particular ranges and degrees of accuracy required of the variables to be presented by the nomograph.

There is not so much freedom of choice in this matter in the construction of nomographs as in other graphical devices. In the graph, for instance, the moduli of the X and Y axes may be chosen independently; there is no dependent modulus. In the nomograph, which always involves at least three scales, the modulus and position of the third scale depend upon the moduli and positions of the other two scales. If these characteristics are poorly chosen for two of the scales, the third scale might have so great a modulus as to exceed the allowable dimensions of the nomograph. On the other hand, its modulus might be so small as to reduce the accuracy of the scale below that required.

Over the years various systems of nomography have been developed to provide satisfactory solutions of these two problems. The methods employed may be classed as synthetic or analytic, depending on the nature of the geometrical approach (see Chapter 3).

7.2 The Synthetic Method

In the synthetic method the analysis of scale relationships is based upon the occurrence of the nomographic scales as corresponding sides of similar triangles, parallelograms, or other familiar geometrical figures. d'Ocagne investigated a large number of possible arrangements, some of which are discussed in Chapter 8, and derived the characteristic defining equation of each.

The thesis of the synthetic method is simple, but very broad in its scope. It is virtually impossible, given a particular problem, to derive from the relationships of the variables in the problem a suitable arrangement of scales to realize solutions of the problem. Therefore, one must, as a preliminary step, investigate all possible arrangements of scales and their corresponding defining equations. These must be catalogued by some system that will enable one to locate those defining equations to which a particular formula can be adapted.

When a suitable defining equation has been found, its variables, which are geometrical elements, are equated to corresponding variables in the formula to be represented. These, in turn, are inserted in the defining equations of the corresponding scales to provide values for the moduli, zero points, and defining functions of the scales.

The scales are constructed in accordance with their scale equations and their positions as given in the geometrical figure from which the defining equation of the nomograph was derived. If the resulting nomograph shows defects not previously anticipated, such as illegibility of a portion of one scale, it should be revised. This can be accomplished only by revising certain scale moduli and computing from the defining equation the effect of these revisions on remaining moduli in the nomograph. Considerable experience in the art of the synthetic method is required to avoid excessive "backtracking" of this sort.

7.3 The Analytic Method

Examination of all of the scale arrangements investigated and accepted by the synthetic method indicates a surprising similarity. All consist of units composed of three scales intersected by a straight index line or have defining equations equivalent to those of such units. There is, therefore, justification for accepting this unit as universal in nomography and approaching the problems of nomography through the equation of the straight line of analytic geometry.

In essence, the analytic approach contends that the defining equations of all nomographs may be reduced to units that are special cases of the equation of the straight line.

If the three scales of the nomograph represent the variables U, V, and W, the points of intersection of the index line and scales may be designated as (x_u, y_u), (x_v, y_v), and (x_w, y_w), respectively. The condition that these points lie on a straight line is expressed by the following relationship:

$$y_u(x_v - x_w) - y_v(x_u - x_w) + y_w(x_u - x_v) = 0. \qquad (7.1)$$

The formula that is to be expressed as a nomograph has a form similar to this relationship, which is loosely called the equation of the straight line. Corresponding elements of the two expressions are equated to arrive at values of the coordinates of these points in terms of the variables of the formula. The resulting values of the x's and y's determine the form and position that each scale takes on the grid of a cartesian coordinate system.

Because of the single type of defining equation applicable to all nomographs, this approach is much simpler than that of the synthetic method. The simplification does not end there, however. As each x occurs twice in the relationship shown it is occasionally a bit difficult to determine the best method of expressing the x's and y's as functions of the variables in the formula. A quicker approach is made through use of the determinant form of the equation of the straight line,

$$\begin{vmatrix} y_u & x_u & 1 \\ y_v & x_v & 1 \\ y_w & x_w & 1 \end{vmatrix} = 0, \qquad (7.2)$$

in which each x and y occurs only once. The procedure for finding the functions that may be substituted for these x's and y's is then quite direct. Suppose that the formula to be represented is the reduced cubic equation, $U^3 + VU + W = 0$. This is expressed in determinant form as

$$\begin{vmatrix} U^3 & U & 1 \\ -V & 1 & 0 \\ -W & 0 & 1 \end{vmatrix} = 0 \qquad (7.3)$$

This determinant may be manipulated by standard procedures to obtain a determinant whose third-column elements are all 1's:

$$\begin{vmatrix} \dfrac{U^3}{U+1} & \dfrac{U}{U+1} & 1 \\ -V & 1 & 1 \\ -W & 0 & 1 \end{vmatrix} = 0. \qquad (7.4)$$

This last determinant is called the defining equation of the nomograph. Each element in this equation is set equal to the corresponding element in Equation 7.2; the resulting equations indicate the functional content of the defining equations of the scales of the nomograph. From these equations it may be observed that the V scale has the constant abscissa 1 and ordinate equal to the values of -V; it is a straight scale parallel to the Y axis at a distance of unity from that axis, and the values of V increase towards the bottom of the scale. The W scale is similar to the V scale, but it lies on the Y-axis, because its abscissa is 0. The U scale is curved, because both its abscissas and its ordinates are nonlinear functions of U. For all positive values of U, the U scale lies between the V and W scales, because the abscissa U/(U + 1) vanishes when U vanishes and approaches 1 as U approaches infinity.

The analytic approach thus provides two simplifications towards solving the first problem of nomography, that of finding an arrangement of scales capable of realizing solutions of a given formula. The determinant form of defining equation affords further simplifications in the second problem of nomography, the modification of scale moduli and positions to accommodate the particular ranges encountered in the variables to be represented. One approach is algebraic, involving modification of the defining equation; the other is geometrical, involving modification of the coordinate system on which the nomograph is constructed.

The algebraic method consists in multiplying the determinant of the defining equation by a non-vanishing matrix containing nine (or sometimes fewer) constants. The product is then a defining equation whose elements contain these constants, and the values of the constants are determined by solving an equal number of simultaneous equations derived from the desired dimensions of the nomograph and the limiting values of the U, V, and W.

The necessity for this algebraic step arises from the fact that the defining equation is given in terms of cartesian coordinates which lack flexibility, whereas, in order to represent all possible ranges of its variables, the form of the nomograph must be very flexible. The geometrical approach involves the use of a system of hyperbolic coordinates with which it is possible to make all necessary adjustments of moduli and positions of scales by selecting an optimum value for a single constant r of the coordinate system, or, occasionally, two such constants. The selection of this value is usually quite simple; the value of r may be chosen as an integer or simple fraction, and, as it does not enter into the defining equation, its value may be changed at will without affecting the computation of scale factors.

SOME CHARACTERISTICS OF NOMOGRAPHS

7.4 Versatility of Nomographs

It would appear that limiting the nomograph to the form of three scales and a straight index line would unduly limit its applicability, but this is not so. The scales may be parallel, concurrent, N-shaped, or H-shaped combinations of straight scales; or they may lie on hyperbolic, parabolic, elliptical, exponential, or other curves. By the methods of synthetic nomography, the various shapes of nomographs fall into different categories, whereas by analytic methods they are simply special cases of a general defining equation.

The versatility is increased by the method of subdividing scales to represent linear, logarithmic, hyperbolic, exponential, or other regular functions as well as a host of special functions. Also, a scale may represent more than one variable if it is a network scale (see Sec. 11.15), or if it is a common element of two units of a compound nomograph.

7.5 Advantages Over Other Graphical Methods

Unusual versatility is only one advantage that the nomograph holds over other graphical methods of computation. A comparison of defining equations indicates clearly that the nomograph can represent a far wider variety of formulas than can the slide rule. The scope of cross-section charts (graphs) is as great as that of nomographs, but these charts necessitate interpolation between irregular curves as well as the construction of difficult curves. The graph requires two index lines (one for each coordinate) as opposed to one index line for the nomograph.

Both graphs and nomographs are much simpler and less costly to construct and duplicate than are slide rules. Their printed copies occupy very little space, making them ideal for inclusion in handbooks and data books.

Unfortunately many printed copies of what were at one time very good nomographs are virtually useless because they have been reduced to a fraction of their original dimensions. In some cases, scale points have been removed to avoid crowding in the reduced copy. Such nomographs are useful primarily in indicating the method by which a more accurate nomograph may be prepared. If a nomograph is to be used for routine calculations in a plant or laboratory, it should be specially prepared for that particular use, drawn to cover only the ranges involved and in sufficiently fine detail to give all the accuracy required, or all that is justified by the data from which the basic formula of the nomograph is evolved. Prepared in this manner, the nomograph can be considerably more accurate than the slide rule of average length, and much faster in yielding results.

7.6 Limitations and Precautions

The nomograph should not be "oversold," however. Versatile as it is, there are still many computations that are more satisfactorily handled by other graphical means; some computations of Chapter 6 are good examples in point.

One should be cautious in applying a nomograph whose origin and ancestry are vague unless the nomograph is checked against known values of the variables involved. All too frequently, nomographs are printed without their basic formulas; or, if the formula is included, it is stated without reference to its origin in the literature, the ranges of the variables from which the relationship was derived, or the limits of error in the original data. Consequently, a formula or its nomographic representation may be misapplied to data that lie outside the ranges of the original data, or used to obtain results whose apparent accuracy is entirely unjustified.

chapter 8

NOMOGRAPHS BY SYNTHETIC METHODS

A large number of nomographs used in engineering fields today have been designed and constructed by the methods of synthetic geometry—the "plane geometry" of the preparatory school. The arithmetic is more cumbersome than that of analytic methods, the most satisfactory nomographic form is not always so clearly indicated, and some types possible with analytic methods are all but impossible by these methods. The synthetic methods, however, are more easily understood by the beginner or by any person not thoroughly versed in analytic geometry and are valuable to anyone engaged in the construction of nomographs, regardless of other nomographic methods that are also employed.

SYNTHETIC METHODS

8.1 Geometrical Arrangements

A large variety of arrangements of axes and index lines may be investigated by synthetic methods. The axes may be parallel, concurrent, N-shaped, or otherwise, with a straight index line. There may also be pairs of parallel or perpendicular axes intersected by two parallel index lines, two perpendicular index lines, and so on.

Some satisfactory point must be chosen as the origin of each axis. This choice is entirely arbitrary, and it is generally made in the interest of simplifying the derivation of geometrical relationships among the axes and index lines. If the axes are parallel, the origins are generally chosen as colinear points on those axes. If the axes are concurrent, their point of intersection is a convenient origin.

With the position and origin of each axis determined, the index line(s) and auxiliary lines are then drawn, giving rise to familiar geometrical figures, such as triangles, and parallelograms. The geometrical relationships among the various displacements of the index line(s) from the origins of the axis are then derived from their positions as corresponding elements of similar triangles or other convenient figures.

8.2 Defining Equation

The displacements of index lines from the origins of the axes of the nomographic figure are related by means of an equation. These displacements are also related to the values of the scale functions by the defining equations of the respective scales plotted on those axes. If the arrangement of scales is to serve as a means of arriving at solutions of a given formula, it is necessary that the defining equations of the scales, substituted in the equation that represents the geometrical arrangement of the axes, give rise to the same relationship among the variables as exists in the formula that is to be solved.

This is, clearly, not a situation in which one may easily proceed from the formula to the geometrical figure required for solving it. It becomes necessary, therefore, to investigate a number of possible geometrical arrangements and derive their characteristic defining equations in general terms. Once this has been done, the only necessary step in fitting the figure to a formula is to select suitable defining equations for the scales of the nomograph. The following examples may serve to clarify this procedure.

FIGURES WITH ONE STRAIGHT INDEX LINE

8.3 Three Parallel Scales

Let U, V, and W be three axes on which scales representing the variables u, v, and w are to be plotted. Let these three axes be parallel and the scales so arranged that a straight index line realizes relationships among the variables u, v and w. Let the origins of the three axes (O_u, O_V and O_W) lie on a line (Fig. 8:1), the origin of the U axis at distance d_1 and that of the W axis at distance d_2 from the origin of the V axis. Draw any index line I, intersecting the three axes at dis-

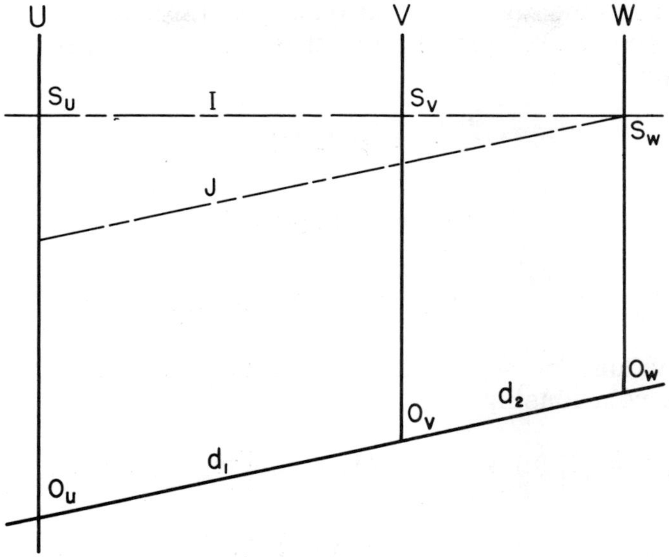

FIG. 8:1 GEOMETRICAL RELATIONSHIPS OF THREE PARALLEL AXES

tances s_u, s_v and s_w from their respective origins. Also draw an auxiliary line through the origins and another one, J, parallel to it through the intersection of the index line and one of the other scales, say W. The length of axis subtended by these parallel lines on the three axes are all equal to s_w. I and J intersect V and W to form a pair of similar triangles whose bases are $s_v - s_w$ and $s_u - s_w$, respectively. From the relationships of similar triangles,

$$\frac{s_v - s_w}{s_u - s_w} = \frac{d_2}{d_1 + d_2},$$

or

$$d_2 s_u - (d_1 + d_2) s_v + d_1 s_w = 0. \tag{8.1}$$

This is the fundamental relationship of the axes.

Let scales be plotted on these axes as defined by the equations

$$s_u = m_u f_1(u) + l_{01},$$

$$s_v = m_v f_2(v) + l_{02},$$

$$s_w = m_w f_3(w) + l_{03}.$$

These values of s_u, s_v and s_w, substituted in Equation 8.1 give the relationship

$$d_2\left[m_u f_1(u) + l_{01}\right] - (d_1 + d_2)\left[m_v f_2(v) + l_{02}\right]$$
$$+ d_1\left[m_w f_3(w) + l_{03}\right] = 0. \qquad (8.2)$$

This can be simplified. The constants l_{01}, l_{02} and l_{03} may be set equal to zero, or, if desired, one constant may be retained. Let $l_{01} = l_{03} = 0$. To be completely general, however, the defining equation of the form must be independent of the geometrical elements (scale moduli and distances). Therefore let

$$\frac{m_v}{m_u} = \frac{d_2}{d_1 + d_2}$$

$$\frac{m_v}{m_w} = \frac{d_1}{d_1 + d_2}$$

$$l_{02} = m_v.$$

Equation 8.2 then becomes

$$f_1(u) - f_2(v) - 1 + f_3(w) = 0, \qquad (8.3)$$

which is the basic defining equation of this form of nomograph, subject only to the foregoing limitations as to scale moduli and distances. From the above ratios of moduli, one obtains

$$d_2/d_1 = \left(\frac{m_v}{m_u}\right) \Big/ \left(\frac{m_v}{m_w}\right) = \frac{m_w}{m_u} \qquad (8.4)$$

Equation 8.3 may be matched to a number of engineering formulas; among them is the simplified formula for flow of fluids in pipes[1]

$$D_i = 2.2w^{0.45}/\rho^{0.31},$$

where D_i is the internal diameter of the pipe, w is the weight flow and ρ is the fluid density. In logarithmic form, this becomes

$$\log D_i - \log 2.2 - 0.45 \log w + 0.31 \log \rho = 0$$

In matching this formula with the basic defining equation one may define the scales as follows:

$$s_u = 0.45 \log w,$$

$$s_v = \log D_i - \log 2.2$$

$$s_w = -0.31 \log \rho.$$

The ranges of the variables and their functions must be taken into account in arriving at the moduli to be employed for each scale. These are

---

[1] Perry, John H., Chemical Engineers Handbook, 3rd Ed., p. 385; McGraw-Hill Book Company, New York, 1941.

w	1000 / 0.001	.45 log w	1.35 / -1.35	$m_w = \frac{l_w}{2.7}$	(U axis)
D_i	100 / 0.1	log D_i	2 / -1	$m_{D_i} = \frac{l_{D_i}}{3}$	(V axis)
ρ	200 / .0001	.31 log ρ	.71 / -1.24	$m_\rho = \frac{l_\rho}{1.95}$	(W axis)

The w and ρ scales may be made 10 inches long; the resulting moduli are $m_w = 10/2.7 = 3.70$ inches; and $m_\rho = 10/1.95 = 5.14$ inches. It is clear, however, that the actual length of the scale is not so critical a factor as to require rigid adherence to moduli that may be a bit difficult to use. The scales constructed with these moduli would be

$$s_u = 3.70\,(.45 \log w) = 1.67 \log w$$

and

$$s_w = 5.14\,(.31 \log \rho) = 1.59 \log \rho.$$

The moduli for the construction (including the constants) are 1.67 and 1.59.

The construction can be greatly simplified by using a common modulus, say 1.60 for both scales and the appearance and accuracy of the nomograph are not appreciably affected. If the value $m' = 1.60$ is used for the construction, the corresponding values of the moduli without the constants are $m_w = 1.6/.45 = 3.56$ and $m_\rho = 1.6/.31 = 5.165$. It is these values that determine, by Equation 8.4, the ratio d_2/d_1; thus

$$\frac{d_2}{d_1} = \frac{5.165}{3.56} = 1.45$$

If the total width of the nomograph is 8 inches, then

$$d_1 = 8\,\frac{(1.00)}{2.45} = 3.27 \text{ inches}$$

and

$$d_2 = 8\,\frac{(1.45)}{2.45} = 4.73 \text{ inches}$$

These values, in turn, establish the value of the modulus of the third scale, for

$$\frac{m_v}{m_u} = \frac{d_2}{d_1 + d_2} = \frac{4.73}{8.00};$$

$$m_v = 3.56 \times \frac{4.73}{8.00} = 2.10.$$

The nomograph may now be constructed. Two logarithmic scales with moduli 1.60 inches are constructed parallel to one another at a distance of 8 inches. The left hand scale increases in value towards the top ($s_u = .45 \log w$), and is designated with values of w ranging from 10^{-3} to 10^3; the right hand scale increases in value towards the bottom ($s_w = -.31 \log \rho$), with values of ρ ranging from 10^{-4} to 2×10^2. A light line is drawn through the origins ($w = 1$, $\rho = 1$) to determine the origin of the D_i scale, and a light vertical line is drawn parallel to the other scales at a distance of 3.27 inches to the right of the left hand scale. A logarithmic scale of modulus 2.10 inches is constructed on this line with its value 2.2 on the line of origins ($s_v = \log D_i - \log 2.2$). A range of D_i from 10^{-1} to 10^2 is drawn; the completed nomograph is shown in Figure 8:2.

The distinction between the modulus of the scale equation and the modulus for construction of the scale is an important matter in analytic as well as synthetic methods of design. By omitting the constant coefficient from the scale modulus, the basic defining equation for the form, as well as the relationship of moduli to distance between scales, may be generalized on a much simpler basis. By including the coefficient in the modulus for construction, the computation required for laying off the scale is simplified.

The scale functions need not be so simple as those in the foregoing example. For example Gess and Irwin[2] include a correction factor for pipe taps

$$Y_1 = 1 - \left[0.333 + 1.145\,(\beta^2 + 0.7\beta^5 + 12\beta^{13})\right]\frac{x}{k}.$$

The parameters may be taken as Y_1, β, and x/k, each of which is defined in the foregoing reference. Corresponding scale functions are $f_1(Y_1) = Y_1 - 1$, $f_2(\beta) = 0.333 + 1.145\,(\beta^2 + 0.7\beta^5 + 12\beta^{13})$, and $f_3(x/k) = x/k$.

The method for other forms follows the procedure shown here in detail. The geometrical equation is derived first, and the general defining equations of the scales are inserted. For most purposes, the scale equation may take the simplified form $s_i = m_i f(i)$; the resulting equation is then more readily converted to the required form of the general defining equation. Relationships are next established among the moduli of the scales and any other geometrical elements in the equation, so that these factors vanish from the equation.

The resulting equation is the general defining equation of the form from which it was derived. In fitting it to a mathematical formula it is necessary that the functions of the general defining equation be so defined as to give rise to an expression identical with that mathematical formula. In selecting moduli for the scales, the following factors must be considered:

a) Relationships imposed by the derivation of the general defining equations,
b) Dimensions of the nomograph,
c) Ranges of the variables and their functions,
d) Practical aspects of scale construction.

8.4 Other Forms With Straight Scales

Various other geometrical arrangements of straight index line and straight scales may be investigated in similar fashion. A popular type consists of two parallel axes and a

---

[2] Gess, L., and Irwin, R. D., Flow Meter Engineering Handbook, 2nd Ed., p. 36; The Brown Instrument Company, Philadelphia, 1946.

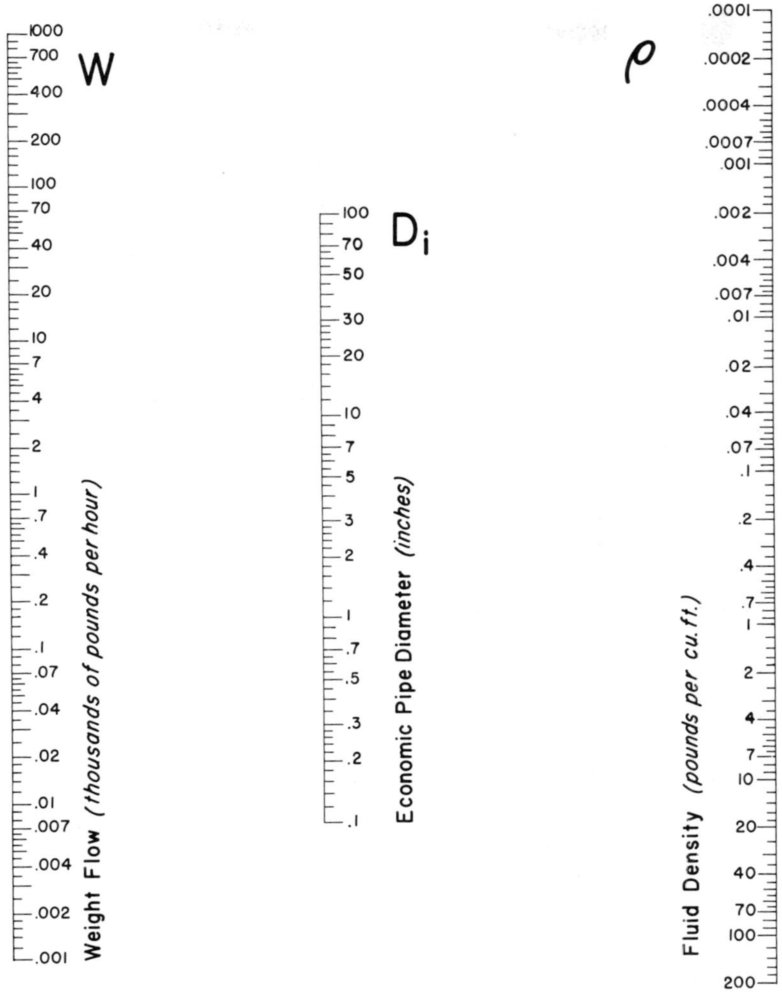

FIG. 8:2 NOMOGRAPH FOR FLOW OF FLUIDS IN PIPES

$$D_i = 2.2w^{0.45}\rho^{0.31}$$

diagonal axis, the familiar N- or Z-type (Fig. 8:3). The intersection of the diagonal axis with one of the parallel axes serves as the origin of both axes; the other parallel axis has its origin at its intersection with the diagonal axis.

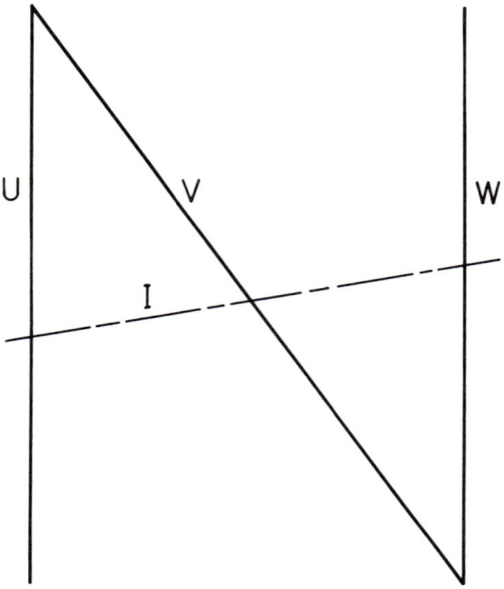

FIG. 8:3 GEOMETRICAL RELATIONSHIPS OF N OR Z CHART

Without loss of generality, the length of the diagonal axis may be taken as 1. Any index line cuts off a pair of similar triangles from which the geometrical relationship may be derived:

$$s_u - s_w\left[s_v/(1 - s_v)\right] = 0 \tag{8.5}$$

If, instead of the linear factor s_v one introduces the hyperbolic factor p_v, so that $s_v = p_v/(p_v + 1)$, this equation becomes

$$s_u - p_v s_w = 0 \tag{8.6}$$

This form is rather extensively used for multiplication, especially in combination-type nomographs, wherein it is desirable to have a linear scale for one term in the product. Generally, however, logarithmic multiplication using parallel axes is preferable; the index line makes less acute angles with the scales, and the logarithmic scales are characterized by uniform relative rather than uniform absolute errors.

More extensively used are double Z charts of the types shown in Figure 8:4; the diagonal axis is not scaled, but two index lines intersect at the diagonal axis to realize on the parallel scales solutions of the proportionality, $u_1/w_1 = u_2/w_2 = u_3/w_3 = \ldots$

Nomographs with concurrent scales may be examined in similar manner; however, this type is much more effectively treated with general hyperbolic coordinates (Chapter 12).

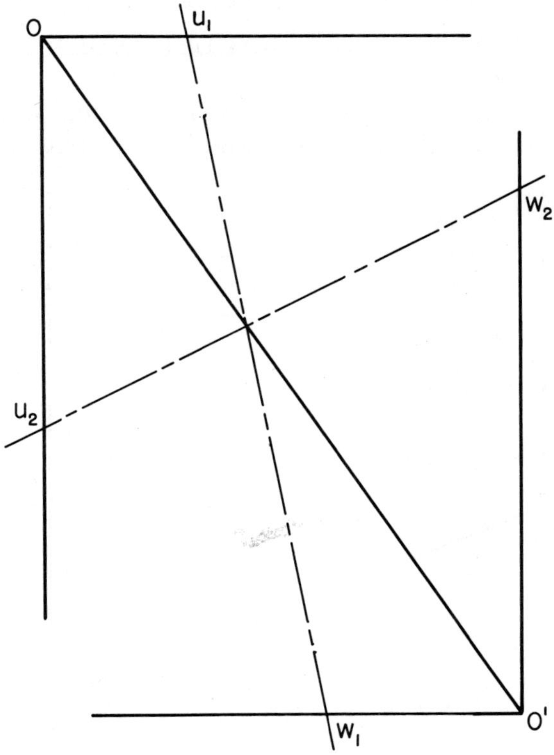

FIG. 8:4 DOUBLE-N or DOUBLE-Z CHART

8.5 Nomographs With Curved Scales

Formulas are at times encountered which may be fitted to Equation 8.1 if d_1 is taken as a function of the variable represented on the v axis rather than as a constant. The distance between extreme parallel scales then remains constant; but the distance between the middle scale and an outer scale varies with the value of the parameter it represents. The middle scale in such nomographs is quite likely to be curved.

Notable among nomographs of this kind are the polynomials; these are discussed in detail in Chapter 11 wherein certain generalities are employed which are more readily associated with analytic than with synthetic methods.

Nomographs involving more than two curved scales are very intricate by synthetic methods; hence, the discussion of such charts is deferred to the section on analytic methods.

FIGURES WITH TWO PARALLEL INDEX LINES

8.6 Parallel Axes

Figure 8:5 shows two parallel axes, one bearing the scales T and U, the other, the scales V and W. Let one index line I_1

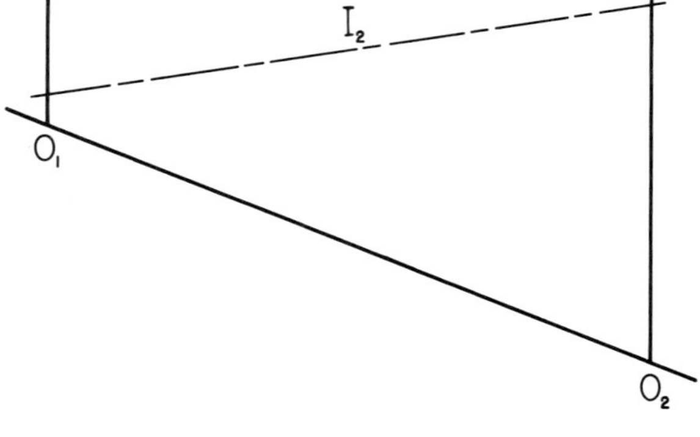

FIG. 8:5 RELATIONSHIPS OF CHART WITH PARALLEL SCALES AND INDEX LINES

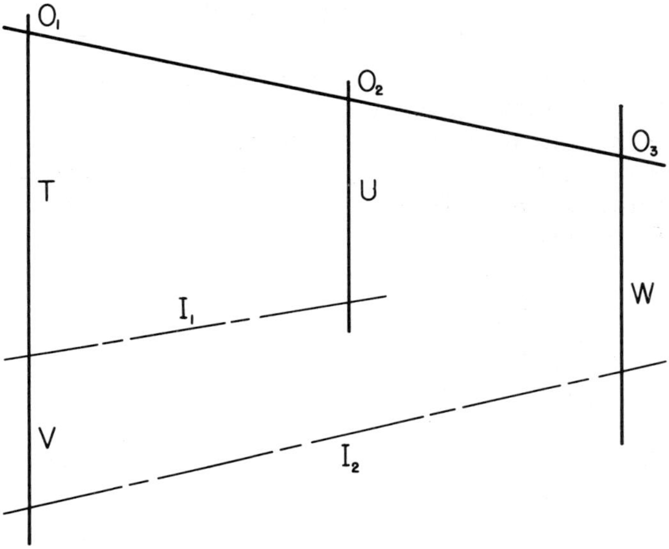

FIG. 8:6 MORE VERSATILE FORM OF FIGURE 8:5

intersect T at a distance s_t from its origin O_1 and V at a distance s_v from its origin O_2. Let I_2 intersect U at s_u and W at s_w, where s_u and s_w are distances from O_1 and O_2 respectively. As the parallel index lines subtend equal lengths on the two axes,

$$s_u - s_t = s_w - s_v \tag{8.7}$$

is the basic geometrical relationship of this form.

From Equation 8.7, by substitution of scale equations, there is

$$m_u f_2(u) + m_v f_3(v) = m_t f_1(t) + m_w f_4(w) + k, \tag{8.8}$$

where $k = l_{01} - l_{02} - l_{03} + l_{04}$. There being no geometrical elements other than the scale moduli and k in this equation, if the general defining equation is to be free of geometrical elements (except k), it is necessary that the moduli all be equal:

$$m_t = m_u = m_v = m_w.$$

This restriction limits application of this form, as very few practical problems involve functions whose ranges are so nearly equal. It is desirable, therefore, to modify this form by introducing of additional geometrical elements. One such modification is a difference in the spacing of the scales (Fig. 8:6), so that

$$(s_v - s_u)/d_2 = (s_w - s_t)/d_1,$$

or

$$d_1 (s_v - s_u) = d_2 (s_w - s_t) \tag{8.9}$$

Introduction of the scale equations leads to the general defining equation,

$$f_3 (v) - f_2 (u) = f_4 (w) - f_1 (t) + k, \tag{8.10}$$

subject to the restrictions that $m_u = m_v$, $m_w = m_t$, and $m_u/m_t = d_2/d_1$.

8.7 V-Axes

In another modification the axes are not all parallel; instead, there are two pairs of parallel axes and each index line intersects one scale from each pair (Fig. 8:7). If these axes be extended and the points of intersection of correspond-

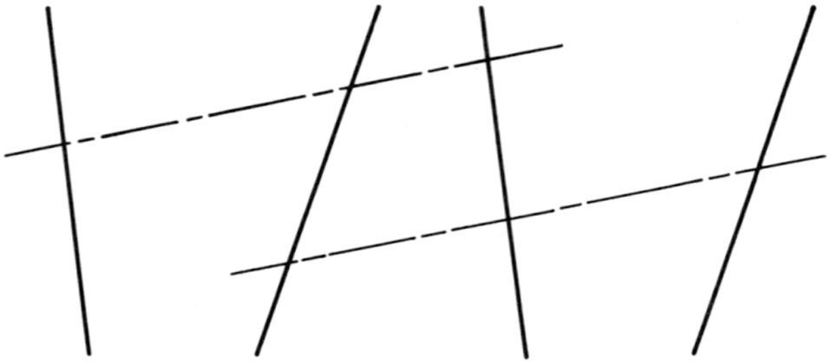

FIG. 8:7 FORM OF NOMOGRAPH WITH V AXES AND PARALLEL INDEX LINES

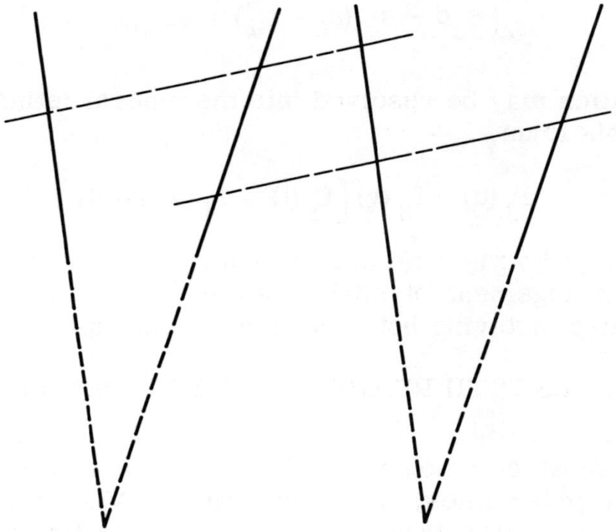

FIG. 8:8 GEOMETRICAL RELATIONSHIPS OF THE DOUBLE-V AXIS FORM

ing axes be taken as origins (Fig. 8:8), it is apparent that

$$s_t : s_u = s_v : s_w \tag{8.11}$$

which becomes the general defining equation

$$f_2(u) \times f_3(v) = f_1(t) \times f_4(w), \tag{8.12}$$

provided $l_{01} = l_{02} = l_{03} = l_{04} = 0$ and $m_t : m_u = m_v : m_w$.

Either Equation 8.10 or Equation 8.12 may be used for products that are conveniently expressed as proportionality. With 8.12, linear scales are employed, whereas with 8.10, logarithmic scales are required. Each has its advantages, as is pointed out in Section 8.3.

Still another modification is that shown in Figure 8:9, in which there is one pair of parallel axes; the second pair of axes forms a V with one leg of the V superimposed on one of the parallel axes. From similar triangles in this figure one may establish the basic geometrical relationship,

$$s_u d - s_v (s_t - s_w) = 0. \tag{8.13}$$

This equation may be resolved into the general defining equation for this form,

$$f_2(u) - f_3(v)\left[f_1(t) + f_4(w)\right] = 0, \tag{8.14}$$

provided $m_t : d = m_u : m_v$, and $m_u = m_w$.

This arrangement of scales is a desirable one to employ for formulas involving both addition and multiplication.

FIGURES WITH PERPENDICULAR INDEX LINES

8.8 V-Axes

In the most common type of nomograph employing perpendicular index lines, the axes form two V's with mutually perpendicular sides (Fig. 8:10 a). Each index line intersects

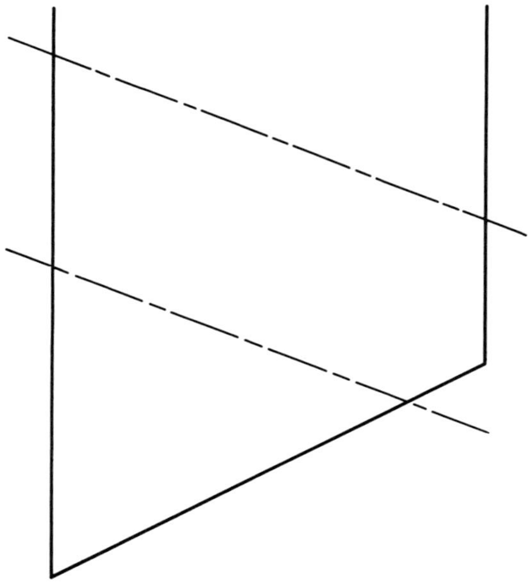

FIG. 8:9 COMBINED FORM; V AND PARALLEL AXES

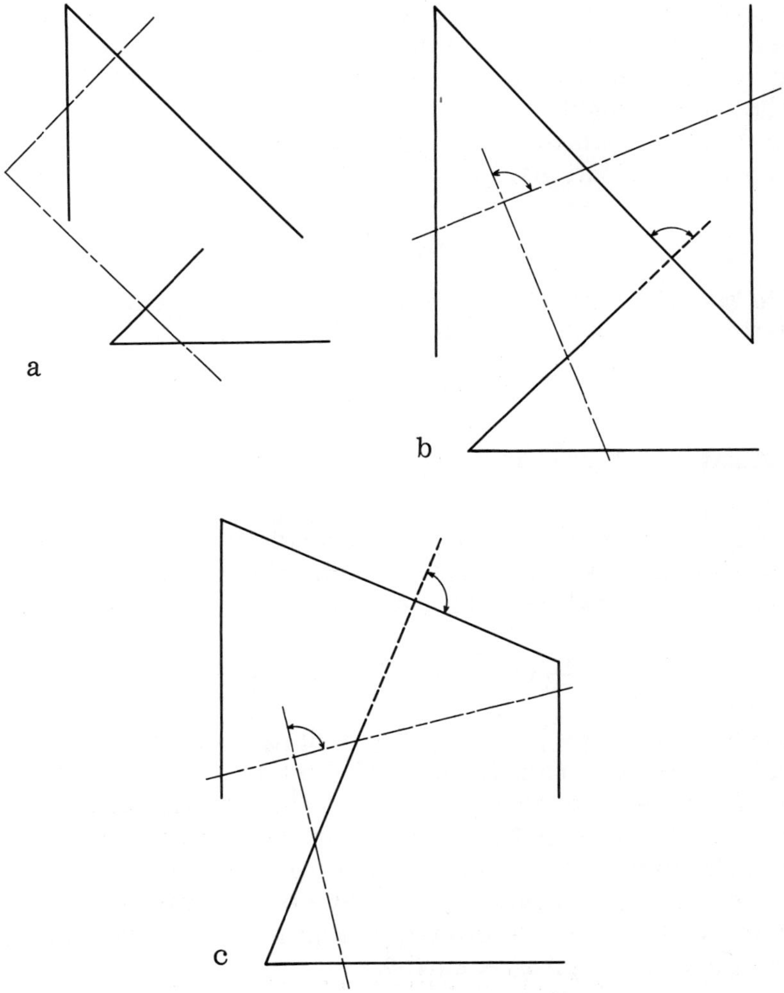

FIG. 8:10 FORMS OF NOMOGRAPH WITH PERPENDICULAR INDEX LINES

the axes of one V, the only restriction being that the two index lines be perpendicular to one another. A simple method of reading the nomograph is this: Place the edge of a transparent straightedge on the values of two variables on one V. Place the edge of one leg of a right angled triangle against the first straight edge and slide the triangle along until its other leg lies on the value of the third variable. In that position it will also lie on the value of the fourth variable, which is the value sought.

From the similar triangles of Figure 8:10, the basic geometrical relationship is

$$s_t : s_u = s_v : s_w. \tag{8.15}$$

Scale equations may be inserted to obtain the general defining equation for this form, which is

$$f(t) : f(u) = f(v) : f(w), \tag{8.16}$$

provided $$m_t : m_u = m_v : m_w.$$

8.9 Other Forms

Other forms that employ perpendicular index lines are shown in Figures 8:10 b and c. For the device shown in Figure 8:10 b, the general defining equation is

$$f_2(u) + f_3(v) = f_1(t)/f_4(w)$$

provided $m_u = m_v$, and $m_v m_t = d m_w$. Neither the defining equation nor the relationship of moduli differs from that of the device shown in Figure 8:9. The use of parallel index lines being somewhat simpler than the use of perpendicular lines, the former is recommended for this relationship.

For the same reasons, the device shown in Figure 8:6 is recommended over the device of Figure 8:10 c, which has the same general defining equation.

REVIEW OF SYNTHETIC METHODS

The foregoing methods have been extensively used in the construction of engineering nomographs for many years. They are still in use, even by some engineers who are familiar with analytic methods, for they present a simple, direct method of approach to numerous practical examples. The more complicated formulas, such as polynomials, are more readily handled by analytic methods, as are cases involving difficult projective transformations. Where complicating factors such as these do not arise, the simple method of approach is usually the best.

Not all possible arrangements of axes and index lines have been examined in this chapter. Instead, the general method for examining such arrangements has been discussed. Those interested in pursuing these methods further will find it a fascinating pastime to devise additional geometrical arrangements and derive their general defining equations.

The desirable approach for any method is from the general to the particular. Following this approach we would derive a general equation for say, any three axes, straight or curved, intersected by a straight index line, or for any four axes intersected in pairs by two parallel index lines, and so on. Unfortunately, without a reference frame, the geometry becomes too complicated for practical use. It is, therefore, necessary to examine particular problems and avoid generalities altogether.

It is significant that all of the general defining equations for nomographs that employ parallel or perpendicular index lines may also be realized in combinations of three-scale forms employing straight index lines (see Chapter 9). Certainly it is easier to draw two straight index lines with only a pivotal point in common than to draw either parallel or perpendicular index lines. Usually the relationship of moduli is more flexible with the combined straight line forms than with forms employing parallel or perpendicular index lines. It is for these reasons, and for the simplicity of the equation of the straight line, that the analytical methods are almost wholly confined to straight line forms.

EXERCISES

1. Construct a nomograph for the Gess and Irwin formula. Ranges of the independent parameters are $\beta \Big|_{.10}^{.75}$, $\frac{x}{k}\Big|_{0.1}^{1.0}$.

2. Derive the defining equation for each of the nomographic forms shown in the accompanying figure.

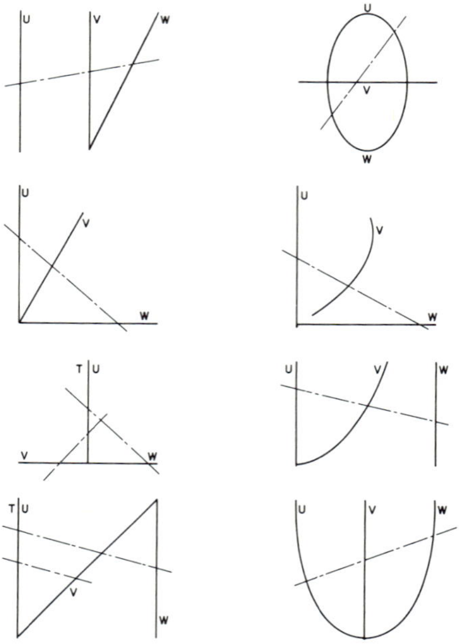

3. The volume of a bushing, washer, or other cylinder is $\frac{\pi}{4}(D^2 - d^2)L$, where D is outside diameter, d inside diameter and L length. Using Equation 8.14, construct a nomograph for this formula, using the ranges D$\left|\begin{matrix}2.00\\ \\ 0.50\end{matrix}\right.$, d$\left|\begin{matrix}1.75\\ \\ 0.25\end{matrix}\right.$, and L$\left|\begin{matrix}2.00\\ \\ 0.10\end{matrix}\right.$, all in inches. On each scale indicate the subdivisions in tenths of inches and in sixteenths.

4. Microwave radiation from a point source P strikes a distant reflector of diameter D. The difference in phase of radiation reaching the point b and that reaching the point a is $\Delta\phi$, where $\Delta\phi = 360\,\Delta R/\lambda$, λ being the wave length of the radiation. From the accompanying figure, it is obvious that

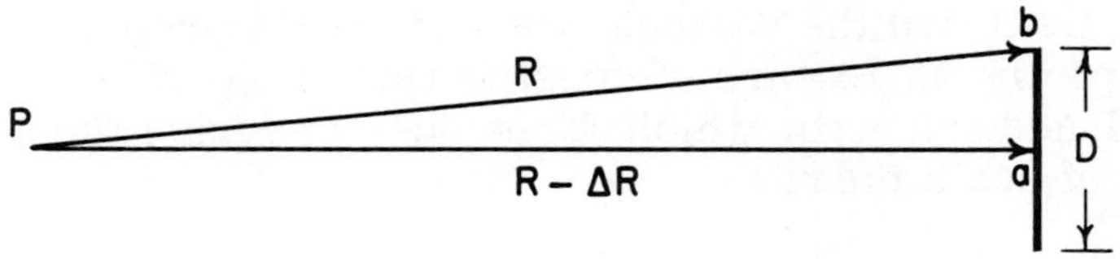

$(R - \Delta R)^2 = R^2 - (D/2)^2$; however, for practical purposes, ΔR is so small compared with the other terms that $(\Delta R)^2$ may be neglected, whence $2R\Delta R = (D/2)^2$. Eliminate ΔR between this expression and the expression for $\Delta\phi$ to obtain a formula for $\Delta\phi$ in terms of D, R and λ. Using Equation 8.16, construct a nomograph for this formula. Ranges are $D\Big|_{0}^{12}$ ft, $R\Big|_{0}^{1200}$ ft, $\lambda\Big|_{0.6}^{10}$ cm, and $\Delta\phi\Big|_{0°}^{360°}$. A conversion factor (cm/ft) must be introduced if the dimensions are to be compatible.

5. Derive a defining equation for a nomograph having two parallel scales with a curved scale between them.

6. Demonstrate that formulas that can be solved by nomographs employing parallel or perpendicular index lines can also be solved by nomographic forms employing two straight index lines, if the resulting nomograph is a composite of two parallel scale charts, two N-type charts, or an N-type chart and a parallel scale chart.

7. Prepare a chart for the formula of Problem 4, using a pair of parallel index lines.

8. Prepare a chart for the formula of Problem 4, using a double N-type chart. If desired, the D and $\Delta\phi$ scales may be placed on one leg of the N, the $\lambda\phi$ and R scales on the other leg. The two index lines intersect at the diagonal.

9. Express the formula for $\Delta\phi$ in logarithmic form and construct a nomograph for four parallel logarithmic scales and two index lines for solving it.

10. Compare the various types of charts prepared for the $\Delta \phi$ formula as to ease of construction, ease of reading, and overall accuracy. In which form does one have the greatest choice of scale moduli?

chapter 9

NOMOGRAPHS WITH CARTESIAN COORDINATES

In the preceding chapter it was apparent that without the aid of a reference frame it would be exceedingly difficult to approach the construction of nomographs from a general standpoint. This would be true even if the form were limited to a single index line intersecting three axes. Yet this form, which is so versatile that it may be used instead of forms with parallel and perpendicular index lines, may be quite readily generalized by the use of analytic geometry.

RELATIONSHIP OF DEFINING EQUATION TO TYPE

9.1 Defining Equation

The condition that three points (x_u, y_u), (x_v, y_v), and (x_w, y_w) lie on a straight line is expressed in the identity,

$$y_u(x_v - x_w) - y_v(x_u - x_w) + y_w(x_u - x_v) = 0 \qquad (9.1)$$

Consider now three axes, U, V, and W, in a cartesian plane (Figure 9:1). Let (x_u, y_u) be a point on the U axis, (x_v, y_v) a point on the V axis, and (x_w, y_w) a point on the W axis. If a scale representing some parameter, u, be plotted on the U-axis, the values of x_u and y_u will be determined entirely by the value of u for each point on the axis. That is,

$$\left.\begin{aligned} x_u &= f_1(u), \\ y_u &= f_2(u) \end{aligned}\right\} \qquad (9.2a)$$

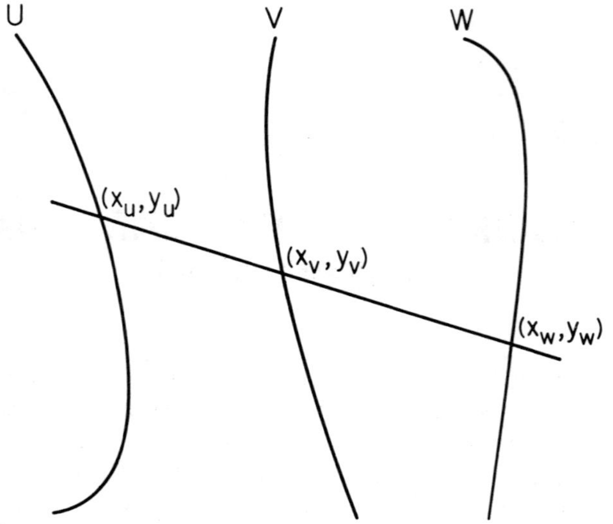

FIG. 9:1 STRAIGHT INDEX LINE INTERSECTING THREE SCALES, U, V, AND W

are the functional relationships defining x_u and y_u. In similar manner, for scales plotted on the V and W axes,

$$\left.\begin{aligned} x_v &= f_3(v), \\ y_v &= f_4(v), \end{aligned}\right\} \qquad (9.2b)$$

$$\left.\begin{aligned} x_w &= f_5(w), \\ y_w &= f_6(w), \end{aligned}\right\} \qquad (9.2c)$$

relate points on the axes to values of v and w.

There is no restriction on the nature of $f_1, f_2, \ldots f_6$; they may be variable or constant, even zero. In fitting a formula to a nomograph of this form, it is necessary only to define these f's in such a way that, substituted for the corresponding x's and y's of Equation 9.1, they give rise to an expression identical with that which it is desired to represent in nomographic form.

Suppose, for example, one wished to construct a nomograph for the hyperbola, $T \times R = S$. If logarithmic scales were desired, one would write

$$\log T + \log R - \log S = 0,$$

and define the functions as follows:

$$x_u = 0, \qquad y_u = \log T,$$

$$x_v = 1/2 \qquad y_v = 1/2 \log S,$$

$$x_w = 1, \qquad y_w = \log R$$

From the constant values of the abscissas, it is apparent that the T, R, and S scales are straight and parallel; the T scale lies on the Y axis of the cartesian coordinate system, with the S scale at a distance half as great as the distance to the R scale. The modulus of the S scale is half that of the T and R scales (Fig. 9:2).

If, on the other hand, linear scales were desired, one might retain the expression as $T \times R = S$ and write the function as

$$x_u = 0 \qquad y_u = S,$$

$$x_v = T/(T + 1) \qquad y_v = 0,$$

$$x_w = 1, \qquad y_w = -R$$

From their constant abscissas, it is clear that the S and R scales are linear, straight and parallel, the S scale on the Y axis and the R scale at unit distance. The R scale, being negative, increases towards the bottom. The T scale is hyperbolic and lies on the X axis; the point $T = 0$ lies at the origin of the S scale, the point $T = \infty$ at the origin of the R

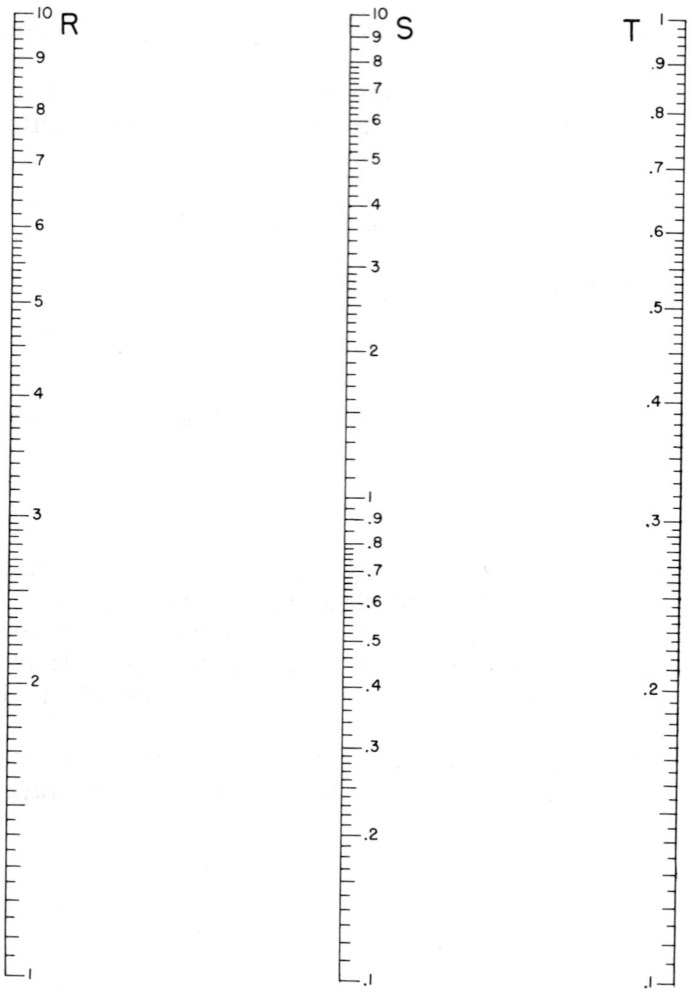

FIG. 9:2 LOGARITHMIC NOMOGRAPH FOR T × R = S

scale. In Figure 9:3 the axes are placed at an acute angle to bring the R and S scales to the same vertical position.

9.2 Defining Equation in Determinant Form

The selection of suitable functions for the x's and y's of the defining equation is not always as simple as in the foregoing examples. Indeed, the selection of the value of x_V in the second example is not an obvious one. A complicating factor is the dual appearance of each y in Equation 9.1. A very helpful step in matching a mathematical formula to a nomographic defining equation is to express both in determinant form; it

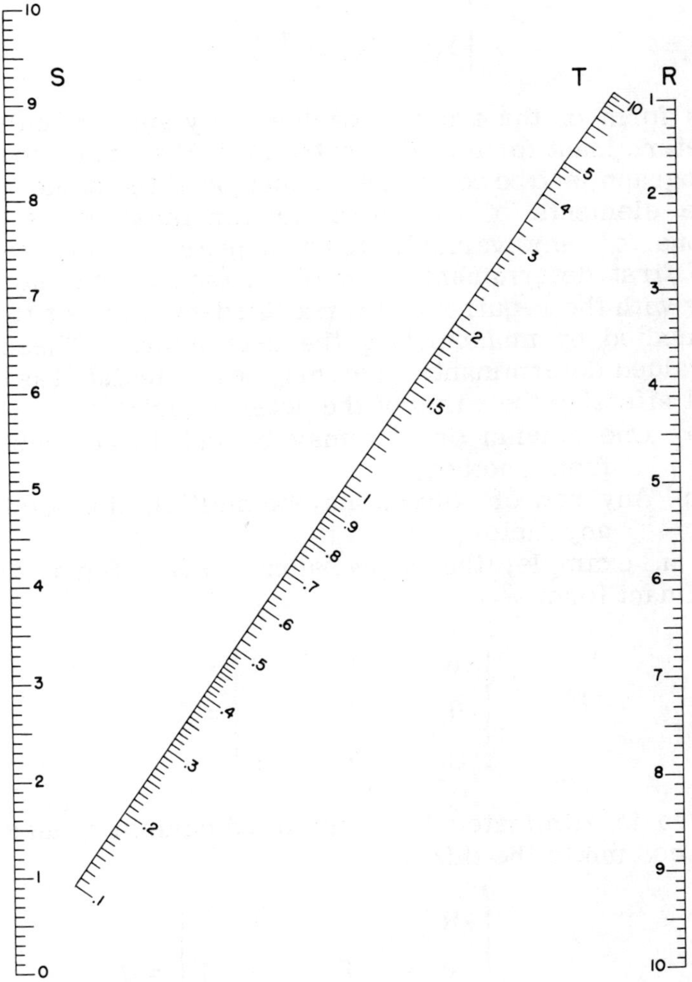

FIG. 9:3 N-TYPE CHART FOR T × R = S

is then necessary only to equate corresponding elements of the two determinants.

Thus, Equation 9.1 may be written,

$$\begin{vmatrix} y_u & x_u & 1 \\ y_v & x_v & 1 \\ y_w & x_w & 1 \end{vmatrix} = 0 \qquad (9.3)$$

In this form of the equation each x and y appears only once. The determinant form of the mathematical formula for which a nomograph is to be constructed, must be of the same pattern; all the elements of the third column must be 1's, and all functions of any variable must appear in only one row.

The first determinant form of an expression may fail to comply with the requirement for a third-column but that may be remedied by manipulating the determinant. These being zero-valued determinants, they may be manipulated as follows without affecting the value of the determinant:

a) One column or row may be added to or subtracted from another.

b) Any row or column may be multiplied or divided by any factor.

As an example, the expression $T \times R = S$ may take the determinant form

$$\begin{vmatrix} -R & 1 & 0 \\ 0 & T & 1 \\ S & 0 & 1 \end{vmatrix} = 0.$$

The zero is eliminated from the third column by adding the second column to the third,

$$\begin{vmatrix} -R & 1 & 1 \\ 0 & T & T+1 \\ S & 0 & 1 \end{vmatrix} = 0$$

and dividing the second row by T + 1,

$$\begin{vmatrix} -R & 1 & 1 \\ 0 & T/(T+1) & 1 \\ S & 0 & 1 \end{vmatrix} = 0$$

A little experience enables one to convert formulas rapidly to their determinant forms and manipulate the determinants into suitable defining equations. This, however, is but one advantage in the use of determinant forms. Methods for the introduction and balancing of scale moduli are based upon the determinant, as manipulation is greatly facilitated by that array of terms.

9.3 Introduction of Scale Moduli

The defining equation is a general expression applicable to all nomographs of a certain type that represent a given formula, regardless of the ranges of the various parameters of the formula. A particular nomograph, however, is limited to specific ranges of its variables, and its scales must be constructed with that factor in mind.

The quadratic

$$W^2 + VW + U = 0$$

is an example. In determinant form this equation is

$$\begin{vmatrix} -U & 0 & 1 \\ -V & 1 & 0 \\ W^2 & W & 1 \end{vmatrix} = 0$$

By the manipulation previously described, a defining equation is obtained in the form

$$\begin{vmatrix} -U & 0 & 1 \\ -V & 1 & 1 \\ \dfrac{W^2}{W+1} & \dfrac{W}{W+1} & 1 \end{vmatrix} = 0 \qquad (9.4)$$

The U and V scales are straight and parallel, the U scale on the Y axis, and the V scale at unit distance. Both scales increase in a negative direction. The W scale is curved, for both its abscissa and its ordinate are non-linear functions of W and of each other.

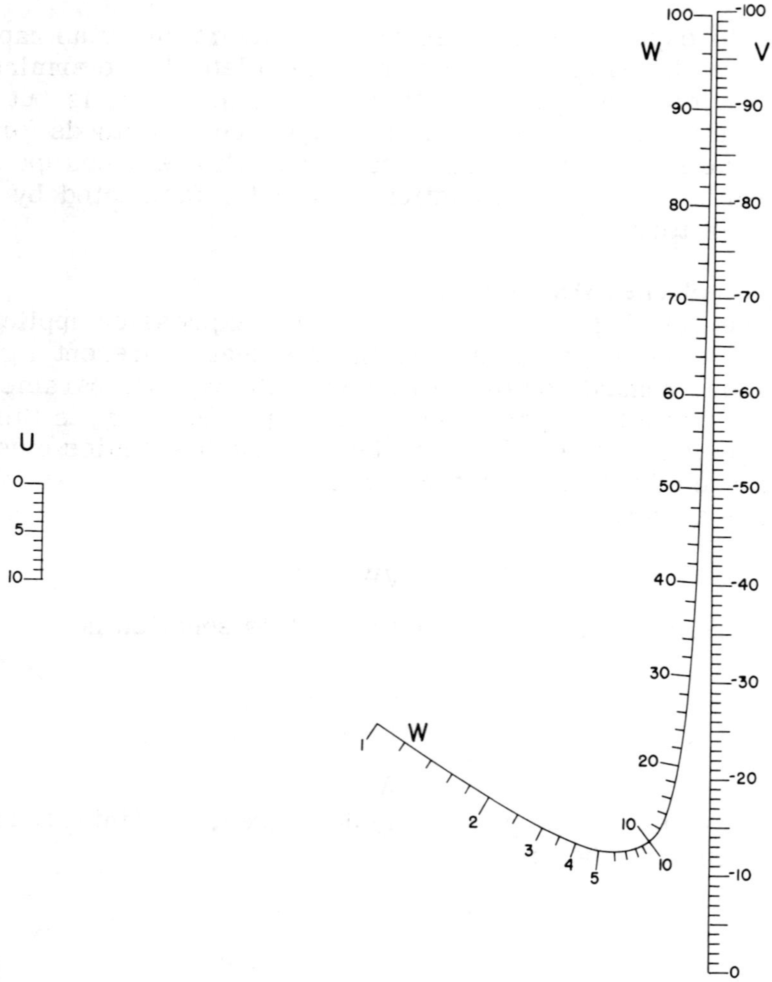

FIG. 9:4 CHART FOR $W^2 + VW + U = 0$, SHOWING DISTORTION DUE TO UNEQUAL RANGES

If U and V have equal ranges and are to be represented by scales 10 inches long and 8 inches apart, one may multiply the elements of column 1 by 10 and those of column 2 by 8 to obtain the actual values of the ordinates and abscissas of all scale points:

$$\begin{vmatrix} -10U & 0 & 1 \\ -10V & 8 & 1 \\ \dfrac{10W^2}{W+1} & \dfrac{8W}{W+1} & 1 \end{vmatrix} = 0 \qquad (9.5)$$

These factors do not alter the value of the determinant, for its expansion is identical with that of Equation 9.4, which is based upon unit dimensions rather than actual dimensions and is the normalized defining equation.

If U and V have dissimilar ranges as is generally the case, some additional manipulation of the determinant is required. Suppose, for example, that the range of U is 0 to 1, while that of V is 0 to -100. If the U and V scales are constructed with equal moduli, the V scale is one hundred times the length of the U scale, which is clearly undesirable (Fig. 9:4). It is preferable to choose the moduli so that these scales are of approximately the same length; this, in turn, shifts the position of the W scale and alters the moduli of its abscissa and ordinate scales. The exact change in the W scale is indicated by the determinant.

Returning to Equation 9.4, to include the diverse ranges of U and V, one may make the following changes:

(1) Divide row 2 by 100:

$$\begin{vmatrix} -U & 0 & 1 \\ -.01V & .01 & .01 \\ \dfrac{W^2}{W+1} & \dfrac{W}{W+1} & 1 \end{vmatrix} = 0$$

(2) Multiply column 2 by 99 and add to column 3:

$$\begin{vmatrix} -U & 0 & 1 \\ -.01V & .99 & 1 \\ \dfrac{W^2}{W+1} & \dfrac{99W}{W+1} & \dfrac{100W+1}{W+1} \end{vmatrix} = 0$$

(3) Divide column 2 by 0.99:

$$\begin{vmatrix} -U & 0 & 1 \\ -.01V & 1 & 1 \\ \dfrac{W^2}{W+1} & \dfrac{100W}{W+1} & \dfrac{100W+1}{W+1} \end{vmatrix} = 0$$

(4) Divide row 3 by $(100\,W + 1)/(W + 1)$:

$$\begin{vmatrix} -U & 0 & 1 \\ -.01V & 1 & 1 \\ \dfrac{W^2}{100W+1} & \dfrac{100W}{100W+1} & 1 \end{vmatrix} = 0 \qquad (9.6)$$

This determinant, in normalized form, is suitable for the construction of nomographs for all ranges of U, V, and W in which the range of V is approximately one hundred times the range of U. A similar procedure may be followed to derive from Eq. 9.4 a determinant for a nomograph in which the ratio of these ranges has some other value.

Introduction of the actual dimensions of the nomograph, as in Eq. 9.5, is a geometrical, not an algebraic step, and is left to the discretion of the individual. Many prefer to work from the normalized form of the determinant, assuming whatever actual dimensions are chosen to be unit dimensions. A nomograph constructed from Eq. 9.6 is shown in Figure 9:5.

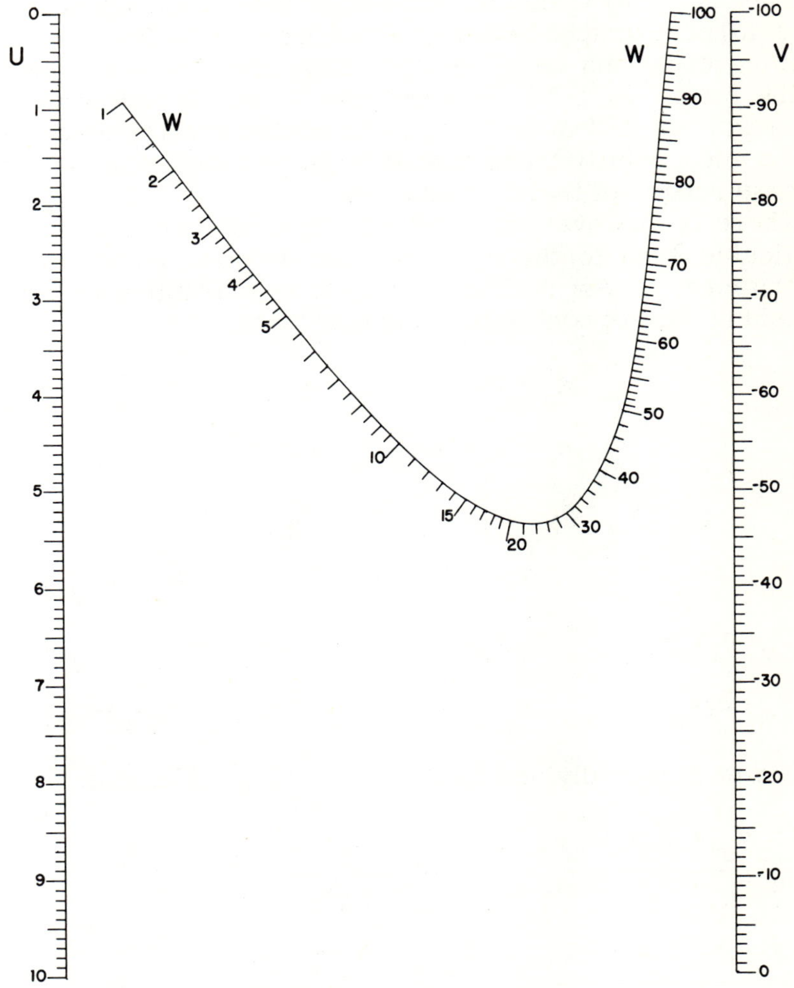

FIG. 9:5 CHART FOR $W^2 + VW + U = 0$, CORRECTED FOR UNEQUAL RANGES

MATRIX TRANSFORMATIONS

9.4 Transformation by Matrix

The foregoing transformation of the determinant is applicable to all nomographs having a pair of parallel scales, for their defining equations have one row of the form $| y_n \ 0 \ 1 |$ and one of the form $| y_v \ 1 \ 1 |$. Nomographs in which both ordinates and abscissas of two or three of the scales are variables may not be successfully transformed in this manner to accomodate diverse ranges of their parameters.

There is, however, an algebraic method of transformation applicable even to the most general defining equation. The determinant of the defining equation is multiplied by a nonvanishing matrix containing nine constants:

$$\begin{vmatrix} y_u & x_u & 1 \\ y_v & x_v & 1 \\ y_w & x_w & 1 \end{vmatrix} \times \begin{vmatrix} a_1 & a_2 & a_3 \\ b_1 & b_2 & b_3 \\ c_1 & c_2 & c_3 \end{vmatrix} =$$

$$\begin{vmatrix} a_1y_u+b_1x_u+c_1 & a_2y_u+b_2x_u+c_2 & a_3y_u+b_3x_u+c_3 \\ a_1y_v+b_1x_v+c_1 & a_2y_v+b_2x_v+c_2 & a_3y_v+b_3x_v+c_3 \\ a_1y_w+b_1x_w+c_1 & a_2y_w+b_2x_w+c_2 & a_3y_w+b_3x_w+c_3 \end{vmatrix} = 0$$

Each row is then divided by its third-column element:

$$\begin{vmatrix} \dfrac{a_1y_u + b_1x_u + c_1}{a_3y_u + b_3x_u + c_3} & \dfrac{a_2y_u + b_2x_u + c_2}{a_3y_u + b_3x_u + c_3} & 1 \\ \dfrac{a_1y_v + b_1x_v + c_1}{a_3y_v + b_3x_v + c_3} & \dfrac{a_2y_v + b_2x_v + c_2}{a_3y_v + b_3x_v + c_3} & 1 \\ \dfrac{a_1y_w + b_1x_w + c_1}{a_3y_w + b_3x_w + c_3} & \dfrac{a_2y_w + b_2x_w + c_2}{a_3y_w + b_3x_w + c_3} & 1 \end{vmatrix} = 0 \qquad (9.7)$$

This determinant meets the requirements for a defining equation; each variable, u, v, or w, is confined to a single row, and all the elements in the third column are 1's. The constants—the a's, b's, and c's—may take on any value, independently of one another. They may take on whatever values are necessary to provide the scales of the nomograph with suitable moduli.

Actually, as one may readily observe, only two of the scales may be independently modified. The third will acquire a position and modulus determined by positions and moduli of the other two scales. Generally, the scales chosen for modification are the exterior scales, it being desirable that their terminal points occupy positions corresponding to the corners of a rectangle. Thus, if the nomograph constructed by Equation 9.1 is of the form of Figure 9:4, it may, by modification of its exterior scales, be converted into the form of Figure 9:5.

The method is this: If the two exterior scales are u and v, the elements of columns 1 and 2, rows 1 and 2 of Equation 9.7 are employed to determine the values of the a's, b's, and c's. The value of the parameter u at the lower end of its scale is u_1, at the upper end, u_2. The required ordinate of the u scale at its lower end is Y_{u_1}, at its upper end, Y_{u_2}. The corresponding abscissas are X_{u_1} and X_{u_2}. These values are arbitrarily decided upon to give the nomograph the desired form. The corresponding coordinates in the unmodified nomograph are y_{u_1}, y_{u_2}, x_{u_1}, and x_{u_2}, whose values depend upon the values of u_1 and u_2 and the functional nature of y_u and x_u. These values, however, as well as the values of X_{u_1}, X_{u_2}, Y_{u_1}, and Y_{u_2} are known.

A similar set of values is known for the original and modified coordinates of the v scale. One may write eight equations,

$$a_1y_{u_1} + b_1x_{u_1} + c_1 = Y_{u_1}\left(a_3y_{u_1} + b_3x_{u_1} + c_3\right),$$

$$a_1y_{u_2} + b_1x_{u_2} + c_1 = Y_{u_2}\left(a_3y_{u_2} + b_3x_{u_2} + c_3\right),$$

$$a_2y_{u_1} + b_2x_{u_1} + c_2 = X_{u_1}\left(a_3y_{u_1} + b_3x_{u_1} + c_3\right),$$

$$a_2y_{u_2} + b_2x_{u_2} + c_2 = X_{u_2}\left(a_3y_{u_2} + b_3x_{u_2} + c_3\right),$$

$$a_1y_{v_1} + b_1x_{v_1} + c_1 = Y_{v_1}\left(a_3y_{v_1} + b_3x_{v_1} + c_3\right),$$

$$a_1y_{v_2} + b_1x_{v_2} + c_1 = Y_{v_2}\left(a_3y_{v_2} + b_3x_{v_2} + c_3\right),$$

$$a_2y_{v_1} + b_2x_{v_1} + c_2 = X_{v_1}\left(a_3y_{v_1} + b_3x_{v_1} + c_3\right),$$

$$a_2y_{v_2} + b_2x_{v_2} + c_2 = X_{v_2}\left(a_3y_{v_1} + b_3x_{v_1} + c_3\right),$$

in which the a's, b's and c's are the only unknown factors. There being nine unknowns and eight equations, one of the unknowns may be assigned an arbitrary value; c_3 seems a likely candidate for this assignment, since it occurs in every equation, and a proper value for c_3 is 1. Each equation then has one term free of unknowns, and the equations are conveniently written,

$$a_1y_{u_1} + b_1x_{u_1} + c_1 - a_3y_{u_1}Y_{u_1} - b_3x_{u_1}Y_{u_1} = Y_{u_1}$$

$$a_1y_{u_2} + b_1x_{u_2} + c_1 - a_3y_{u_2}Y_{u_2} - b_3x_{u_2}Y_{u_2} = Y_{u_2}, \text{ etc.}$$

From the equations in this form, the values of the a's, b's and c's may be written down directly; the value of a_1, for example, is

$$a_1 = \frac{1}{D} \times \begin{vmatrix} Y_{u_1} & x_{u_1} & 1 & 0 & 0 & 0 & -y_{u_1}Y_{u_1} & -x_{u_1}Y_{u_1} \\ Y_{u_2} & x_{u_2} & 1 & 0 & 0 & 0 & -y_{u_2}Y_{u_2} & -x_{u_2}Y_{u_2} \\ X_{u_1} & 0 & 0 & y_{u_1} & x_{u_1} & 1 & -y_{u_1}X_{u_1} & -x_{u_1}X_{u_1} \\ X_{u_2} & 0 & 0 & y_{u_2} & x_{u_2} & 1 & -y_{u_2}X_{u_2} & -x_{u_2}X_{u_2} \\ Y_{v_1} & x_{v_1} & 1 & 0 & 0 & 0 & -y_{v_1}Y_{v_1} & -x_{v_1}Y_{v_1} \\ Y_{v_2} & x_{v_2} & 1 & 0 & 0 & 0 & -y_{v_2}Y_{v_2} & -x_{v_2}Y_{v_2} \\ X_{v_1} & 0 & 0 & y_{v_1} & x_{v_1} & 1 & -y_{v_1}X_{v_1} & -x_{v_1}X_{v_1} \\ X_{v_2} & 0 & 0 & y_{v_2} & x_{v_2} & 1 & -y_{v_2}X_{v_2} & -x_{v_2}X_{v_2} \end{vmatrix}$$

$$\text{where } D = \begin{vmatrix} y_{u_1} & x_{u_1} & 1 & 0 & 0 & 0 & -y_{u_1}Y_{u_1} & -x_{u_1}Y_{u_1} \\ y_{u_2} & x_{u_2} & 1 & 0 & 0 & 0 & -y_{u_2}Y_{u_2} & -x_{u_2}Y_{u_2} \\ 0 & 0 & 0 & y_{u_1} & x_{u_1} & 1 & -y_{u_1}X_{u_1} & -x_{u_1}X_{u_1} \\ 0 & 0 & 0 & y_{u_2} & x_{u_2} & 1 & -y_{u_2}X_{u_2} & -x_{u_2}X_{u_2} \\ y_{v_1} & x_{v_1} & 1 & 0 & 0 & 0 & -y_{v_1}Y_{v_1} & -x_{v_1}Y_{v_1} \\ y_{v_2} & x_{v_2} & 1 & 0 & 0 & 0 & -y_{v_2}Y_{v_2} & -x_{v_2}Y_{v_2} \\ 0 & 0 & 0 & y_{v_1} & x_{v_1} & 1 & -y_{v_1}X_{v_1} & -x_{v_1}X_{v_1} \\ 0 & 0 & 0 & y_{v_2} & x_{v_2} & 1 & -y_{v_2}X_{v_2} & -x_{v_2}X_{v_2} \end{vmatrix}$$

D is the determinant of a matrix composed of the coefficients of a_1, b_1, c_1, a_2, b_2, c_2, a_3, and b_3 of the preceding equations. The determinant which is multiplied by 1/D to give the value of a_1 is obtained by substituting the terms from the right hand side of those equations for the coefficients of a_1 in the matrix of D. Similar substitutions for the coefficients of b_1, c_1, etc., result in determinants which, when divided by D, give the values of b_1, c_1, etc.

This is a perfectly general method that may be applied to any nomograph, regardless of its original shape. The final shape of the nomograph need not be rectangular; it may, if preferred, be trapezoidal, or of any other desired shape, depending entirely upon the choice of values for X_{u_1}, Y_{u_1}, X_{u_2}, etc. The values of these coordinates may, indeed, include the ultimate dimensions of the nomograph, although evaluation of the constants of the matrix of transformation is more difficult than with the normalized form of the defining equation. If this method is to be applied, it is recommended that the values chosen for X_{u_1}, Y_{u_1}, X_{u_2}, etc., be confined, if possible, to 0 and 1. Thus, the normalized nomograph might be considered as lying in a unit square having two of its sides on the two axes of the cartesian coordinate system:

$$X_{u_1} = 0 \qquad Y_{u_1} = 0 \qquad X_{u_2} = 0 \qquad Y_{u_2} = 1$$

$$X_{v_1} = 1 \qquad Y_{v_1} = 0 \qquad X_{v_2} = 1 \qquad Y_{v_2} = 1$$

Expansion of the determinants for evaluation of a_1, b_1, and so on is thereby greatly simplified. The abscissas and ordinates of the modified defining equation may then be multiplied by any desired factors to impart to the nomograph the desired dimensions.

9.5 Simplified Matrix Transformation

Cases so general as to require application of the full set of eight independent constants of the matrix of transformation are rare by comparison with those in which two of the scales, by the original defining equation, are already parallel. Where this is the case, transformation by matrix methods may be greatly simplified.

The original defining equation, which is to be transformed, is of the type

$$\begin{vmatrix} y_u & 0 & 1 \\ y_v & 1 & 1 \\ y_w & x_w & 1 \end{vmatrix} = 0$$

This is multiplied by the matrix of a's, b's and c's, as before, c_3 being given the value 1 at the outset:

$$\begin{vmatrix} y_u & 0 & 1 \\ y_v & 1 & 1 \\ y_w & x_w & 1 \end{vmatrix} \times \begin{vmatrix} a_1 & a_2 & a_3 \\ b_1 & b_2 & b_3 \\ c_1 & c_2 & 1 \end{vmatrix} =$$

$$\begin{vmatrix} a_1y_u + c_1 & a_2y_u + c_2 & a_3y_u + 1 \\ a_1y_v + b_1 + c_1 & a_2y_v + b_2 + c_2 & a_3y_v + b_3 + 1 \\ a_1y_w + b_1x_w + c_1 & a_2y_w + b_2x_w + c_2 & a_3y_w + b_3x_w + 1 \end{vmatrix} = 0,$$

or
$$\begin{vmatrix} \dfrac{a_1y_u + c_1}{a_3y_u + 1} & \dfrac{a_2y_u + c_2}{a_3y_u + 1} & 1 \\ \dfrac{a_1y_v + b_1 + c_1}{a_3y_v + b_3 + 1} & \dfrac{a_2y_v + b_2 + c_2}{a_3y_v + b_3 + 1} & 1 \\ \dfrac{a_1y_w + b_1x_w + c_1}{a_3y_w + b_3x_w + 1} & \dfrac{a_2y_w + b_2x_w + c_2}{a_3y_w + b_3x_w + 1} & 1 \end{vmatrix} = 0$$

From this equation, the values of the new coordinates are

$$X_u = (a_2y_u + c_2)/(a_3y_u + 1),$$

$$Y_u = (a_1y_u + c_1)/(a_3y_u + 1), \text{ etc.}$$

With values for x_{u_1}, y_{u_1}, x_{u_2}, etc., the constants of the matrix can be evaluated from a set of simultaneous equations for any values of X_{u_1}, X_{u_2}, Y_{u_1}, etc. It is usually desirable, however, to transform the nomograph to rectangular form, taking as the values of these coordinates,

$$X_{u_1} = 0, \quad Y_{u_1} = 0, \quad X_{u_2} = 0, \quad Y_{u_2} = 1,$$

$$X_{v_1} = 1, \quad Y_{v_1} = 0, \quad X_{v_1} = 1, \quad Y_{v_2} = 1.$$

With these values of the coordinates, the eight simultaneous equations become

$$a_2y_{u_1} \qquad + c_2 \qquad = 0$$

$$a_2y_{u_2} \qquad + c_2 \qquad = 0$$

$$
\begin{aligned}
a_2 y_{v_1} + b_2 + c_2 - a_3 y_{v_1} - b_3 &= 1 \\
a_2 y_{v_2} + b_2 + c_2 - a_3 y_{v_2} - b_3 &= 1 \\
a_1 y_{u_1} + c_1 &= 0 \\
a_1 y_{u_2} + c_1 - a_3 y_{u_2} &= 1 \\
a_1 y_{v_1} + b_1 + c_1 &= 0 \\
a_1 y_{v_2} + b_1 + c_1 - a_3 y_{v_2} - b_3 &= 1
\end{aligned}
$$

This is clearly a much simpler set of equations from which to evaluate the a's, b's, and c's; so much simpler, indeed, that their values can be readily stated without resorting to a determinant form of expression:

$$
\left.
\begin{aligned}
a_1 &= 1/\left(y_{u_2} - y_{u_1}\right), \\
a_2 &= 0, \\
a_3 &= 0, \\
b_1 &= a_1\left(y_{u_1} - y_{v_1}\right), \\
b_2 &= a_1\left(y_{v_1} - y_{v_1}\right), \\
b_3 &= a_1\left[\left(y_{u_1} - y_{u_2}\right) - \left(y_{v_1} - y_{v_2}\right)\right], \\
c_1 &= a_1 y_{u_1}, \\
c_2 &= 0, \\
(c_3 &= 1).
\end{aligned}
\right\} \quad (9.8)
$$

Three of these constants are zeros; hence, the defining equation of the transformed nomograph becomes

$$\begin{vmatrix} a_1 y_u + c_1 & 0 & 1 \\ \dfrac{a_1 y_v + b_1 + c_1}{b_3 + 1} & \dfrac{b_2}{b_3 + 1} & 1 \\ \dfrac{a_1 y_w + b_1 x_w + c_1}{b_3 x_w + 1} & \dfrac{b_2 x_w}{b_3 x_w + 1} & 1 \end{vmatrix} = 0 \qquad (9.9)$$

To obtain the transformed defining equation, it is necessary only to evaluate a_1, b_1, b_2, b_3, and c_1 from the values of the original and final ordinates, using Equations in 9.8, and substitute those values in Equation 9.9.

9.6 Application to a Quadratic Formula

The merits of the simplified matrix transformation may be compared with those of the method employed in Section 9.3 by using it to solve the same problem. The values of the parameters U and V at their termini are $u_1 = 0$, $u_2 = 1$, $v_1 = 0$, $v_2 = 100$. From the defining equation (Eq. 9.4) the original ordinates of the points on which these termini lie are $y_{u_1} = 0$, $y_{u_2} = 1$, $y_{v_1} = 0$, and $y_{v_2} = 100$. The values of the constants derived from these, by Equations 9.8, are $a_1 = 1$, $b_1 = 0$, $b_2 = 100$, $b_3 = 99$, and $c_1 = 0$. These values, substituted in Equation 9.9, together with the expressions for x_u, x_v, y_v, x_w and y_w from Equation 9.4

$$x_u = 0$$

$$y_u = -U$$

$$x_v = 1$$

$$y_v = -V$$

$$x_w = \frac{W}{(W + 1)}$$

$$y_w = \frac{W^2}{(W + 1)}$$

give the defining equation for the transformed nomograph:

$$\begin{vmatrix} -V & 0 & 1 \\ -V/100 & 1 & 1 \\ \frac{W^2}{100W + 1} & \frac{100W}{100W + 1} & 1 \end{vmatrix} = 0$$

This is identical with the form obtained by manipulating the determinant in Section 9.3. The untransformed nomograph is shown in Figure 9:4, the transformed nomograph in Figure 9.5.

EXERCISES

1. Express each of the following formulas in determinant form and manipulate the determinant into the form of a defining equation: (a) $B = 108\ F^{0.44}\ (Q^{-0.21} - 1.28)$* where B is the brightness of a source of light and F the brightness of the field; (b) the relation between the Bragg angle 0 and the interplanar spacing d in a crystal, as measured by X-ray of wave length $\lambda : \lambda = 2d \sin \theta$; (c) the pull on a glass furnace by a bottle making machine, L = 0.45 SW, where L is tons pull per 24 hr, W is bottle weight (oz), and S is number of bottles per minute produced by the machine.

2. Convert each of the expressions in Problem 1 into logarithmic form, write the determinant for each, and manipulate the determinant into a defining equation.

*Luckiesh, M., and Guth, S. K. Brightness in Visual Field, Illuminating Engineering, 44:650 (1949).

3. In Problem 1 (c), the ranges of the parameters are $L\Big|_0^{100}$ tons, $W\Big|_0^{80}$ oz, and $S\Big|_0^{100}$ min^{-1}. Select suitable scale moduli for the W and S scales, and insert these moduli into the defining equation after the method of Section 9.3.

4. Using the simplified matrix transformation of Section 9.5, modify the defining equation for Furnace Load to accommodate the ranges given in Problem 3.

5. From Equation 9.8 it is obvious that the matrix of transformation may be written

$$\begin{vmatrix} a_1 & 0 & 0 \\ b_1 & b_2 & b_3 \\ c_1 & 0 & 1 \end{vmatrix}$$

for the general nomograph in which the U and V scales are parallel. What constants in this matrix vanish (a) when y_{u_1} lies at the origin $\left(y_{u_1} = 0\right)$, (b) when the U and V scales are the same length $\left[\left(y_{u_1} - y_{u_2}\right) - \left(y_{v_1} - y_{v_2}\right)\right] = 0$? What particular form must the original defining equation have for others of these constants to vanish?

chapter 10

NOMOGRAPHS WITH HYPERBOLIC COORDINATES

The analytic approach to the construction of nomographs has, in the preceding chapter, shown the following advantages over the synthetic approach.

1. A higher degree of generalization is made possible; the same form of defining equation may with slight modifications, represent several nomographic forms which, by synthetic methods, are entirely different types.
2. The defining equation may be stated in determinant form, greatly facilitating the adaption of mathematical formulas to the defining equation and providing a far simpler form upon which to effect transformations for the accommodation of diverse ranges of the various parameters.
3. Nomographic forms that would be virtually impossible to approach through synthetic methods may, because of its generality, be treated by the matrix method of transformation.

The algebra of the transformation is, however, still intricate, for the cartesian coordinate system lacks the flexibility to adapt itself to transformations of the type required for the nomograph. The present chapter describes a second analytic approach based upon a hyperbolic coordinate system. This coordinate system may be readily altered in form by changing the values of one or two constants of the coordinate system. These alterations are sufficient to effect the same changes in the shape of the nomograph as are effected by the matrix of transformation used with cartesian coordinates. No change in the defining equation is required.

THE SEMIHYPERBOLIC COORDINATE SYSTEM

10.1 The Hyperbolic Scale

The form of hyperbolic coordinate system presented in this chapter differs from the cartesian system primarily in the fact that the length of the X axis is finite rather than infinite. Indeed, the length of the X axis, which may be any length convenient for the purpose at hand, is defined as unit length. As this definition is purely arbitrary it has no effect upon the generality of the coordinate system, but it greatly simplifies use of the system.

Just as the infinite length of the cartesian X axis is subdivided linearly into convenient units of length (the modulus), the point $x = 0$ lying at the origin and the point $x = \infty$ at the terminus (also ∞) of that axis, so is the X axis of the hyperbolic coordinate system subdivided by a scale whose zero point is its origin and whose infinity point is its terminus. This scale is not linearly divided; it is divided into segments of length s, where $s = p/(p + r)$. The symbol p represents the numbers from $-\infty$ to $+\infty$, inclusive, and r is a constant functionally related to the modulus of a linear scale.

The hyperbolic scale is not easily visualized; however, this is not the reader's first encounter with this scale. The construction of such a scale is described in Section 1.9 and shown in Figure 1:8. This scale was also employed in the construction of nomographs in Chapter 9. A study of Figure 10:1 which shows scales corresponding to several values of r, should prove helpful in visualizing the effect of a change in the constant r. Some relationships of this family of hyperbolic scales are easily apparent:

1. The length of scale from the origin to any given scale point has the value s, where $s = p/(p + r)$. The scale point, however is identified with the value of p, not s. This is in keeping with the policy mentioned in Chapter I of marking scale points for successive values of the independent variable in the scale function.

2. The point $p = 0$ lies at the origin and the point $p = \infty$ at the terminus of each scale. With the exception of these two points, no two hyperbolic scales with different values of r have any scale points in common.

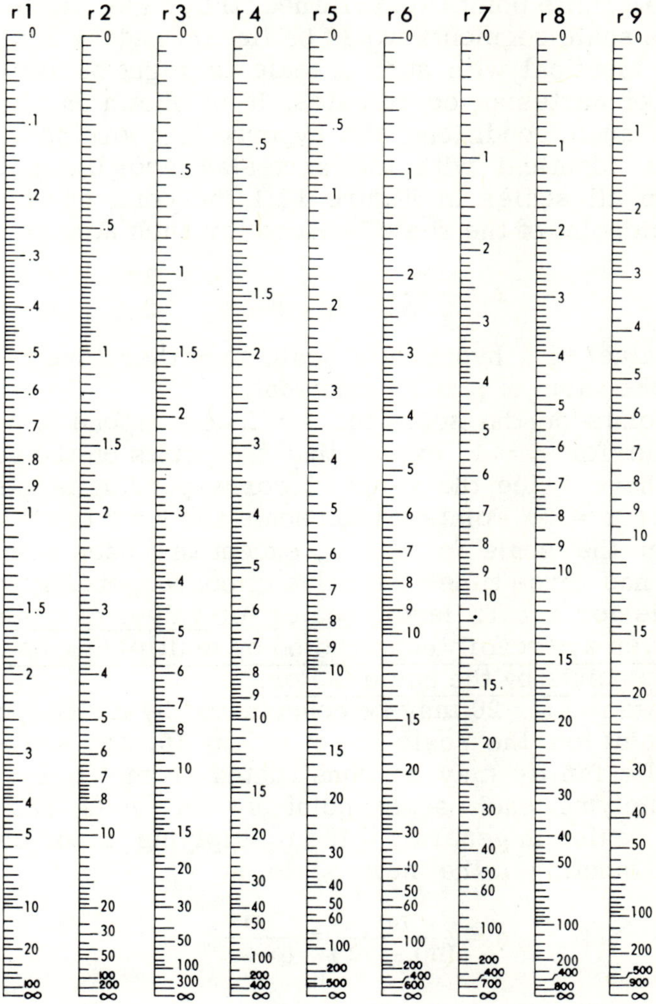

FIG. 10:1 HYPERBOLIC SCALES

3. The subdivisions of these scales are almost linear near the origin, but become less linear as their distance from the origin increases. On a hyperbolic scale of infinite length, all finite points would lie near the origin; consequently, all finite scale segments would be linear, and the scale would become identical with such a scale as might be used for the X axis of cartesian coordinates. It is shown in Chapter 14 that, in such conditions the hyperbolic coordinate system becomes identical with the cartesian coordinate system.

4. On all scales in Figure 10:1 the point at which p = r is the midpoint of the scale because for such a point,

$$s = \frac{p}{p + r} = \frac{p}{p + p} = \frac{p}{2p} = \frac{1}{2}$$

The r-value of a hyperbolic scale may therefore be identified as the value of p at its midpoint.

5. Points on the scale for r = 2 correspond to points on the scale for r = 1, except that the points on the scale for r = 2 have twice the value of corresponding points on the scale for r = 1. Points on the scale for r = 3 correspond to those on the scale for r = 1, except that each point on the former has three times the value of corresponding points on the scale for r = 1. Hence, scales for values of r other than those given are readily constructed by multiplying both r and the scale points by the same factor.

A scale for r = 20 may be constructed by multiplying each scale point on the scale for r = 2 by 10. Any scale in this hyperbolic family may be constructed from the r = 1 scale by multiplying each scale point by the value of r for the desired scale. In general, if the multiplying factor employed is n, the function of the new scale is

$$s = \frac{np}{(np + nr)} = \frac{p}{(p + r)}$$

It is often unnecessary to locate more than a limited number of points on such a scale, and it is easier to use the r = 1 scale with this method of multiplication than to construct an entire scale for the desired r-value. This method also has the virtue of greater accuracy, since the scale for r = 1 given in the supplement has sufficient accuracy for virtually all nomographic work.

Some of the more general properties of hyperbolic scales and coordinates are discussed in Chapter 14.

From this point on a simpler notation will be employed in designating hyperbolic scales, namely, the letter r followed by the r-value of the scale. Thus, r2 will designate a hyperbolic scale for which r has the value of 2, and so on.

10.2 Semihyperbolic Coordinates

The hyperbolic coordinate system described in this chapter should, in reality, be called semihyperbolic, as only one of its axes is hyperbolic, and the others are linear. Chapter 12 discusses the general form of hyperbolic coordinates, in which all axes are hyperbolic. Semihyperbolic coordinates are by far the most useful form of hyperbolic coordinates; hence, unless otherwise specified, it will be understood that this is the form intended by the term "hyperbolic coordinates."

It has been mentioned already that this coordinate system differs from cartesian coordinates in that the X axis is of finite length. The hyperbolic subdivision of this axis permits representation within this finite length of all values of p from zero to infinity. This range is sufficient for representation of all points commonly represented on the X axis of cartesian coordinates, because negative abscissas may be brought into this range either by changing their sign or by adding to them a sufficiently large number.

The abscissa of a point is measured in terms of p. All points having the abscissa 2 lie on the vertical (parallel to the Y axis) through the point $p = 2$ on the X axis. This is similar to the location of abscissas in cartesian coordinates.

The location of ordinates is somewhat different. In cartesian coordinates, ordinates are located by lines parallel to the X axis through the value of y on the Y axis. Such lines, of course, converge at infinity. In hyperbolic coordinates, the lines for locating ordinates are drawn through the value of q on the Y axis, they are straight, and they converge at the point $p = \infty$. But they are not parallel to the X axis or to each other, because the length of the X axis is finite.

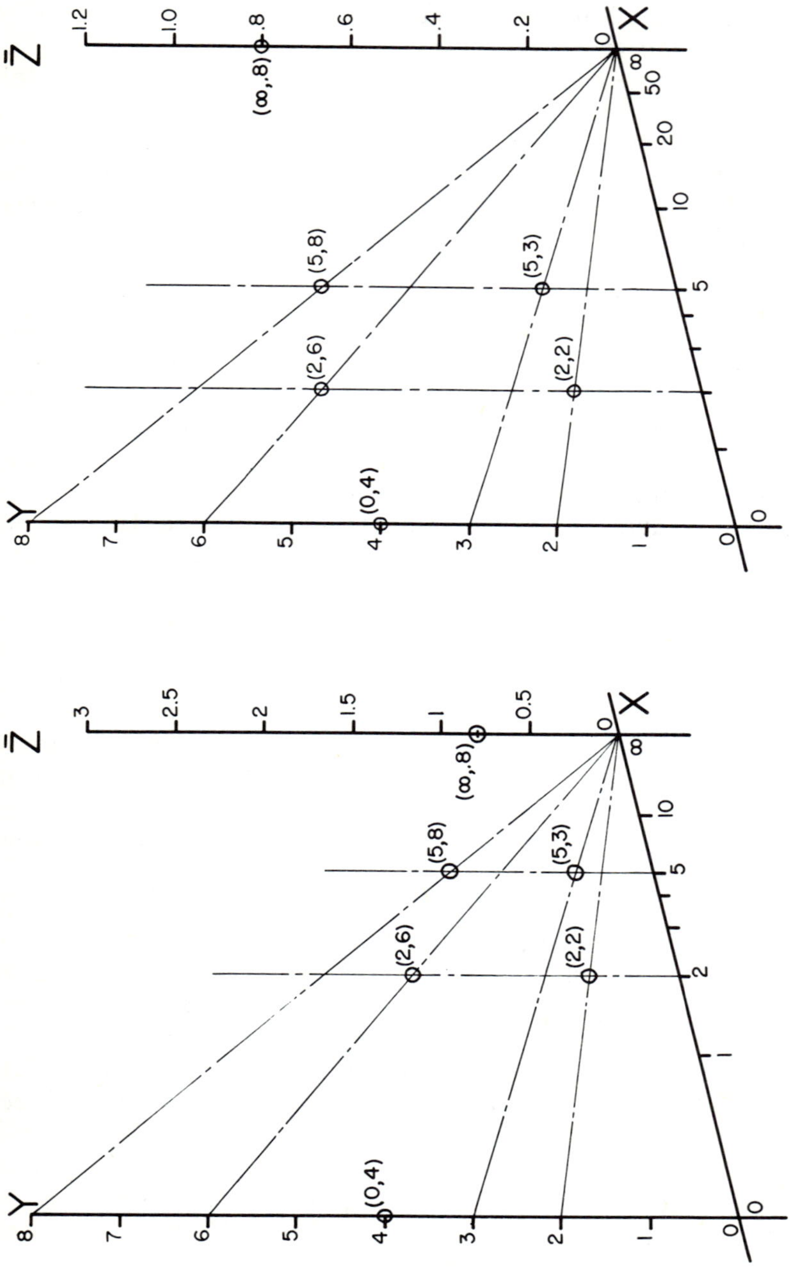

FIG. 10:2 LOCATION OF POINTS ON HYPERBOLIC COORDINATE SYSTEM
a) r = 2; b) r = 5

Figure 10:2(a) shows the location of several points on the r2 hyperbolic system. It is obvious from this diagram that both the ordinate and abscissa (cartesian) of each of these points would be entirely different for a different value of r; Figure 10:2(b) shows the same points located on an r5 hyperbolic system. Note that points that are colinear in Figure 10:2(a)—e.g., (0,4), (5,8) and ∞ , 0.8—are also colinear in Figure 10:2(b). Figure 10:3 shows the same points located on r 0.5, r1, r3, and r6 hyperbolic coordinates.

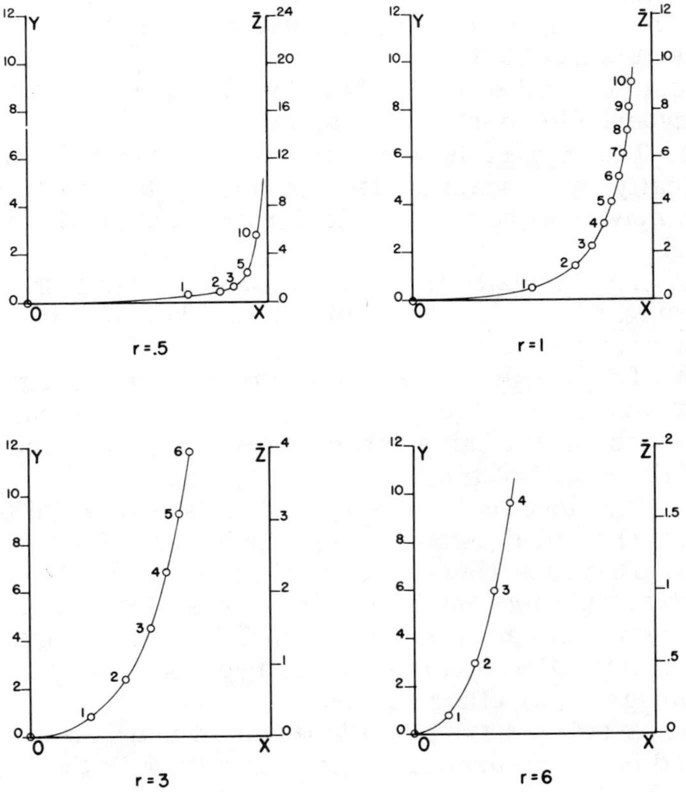

FIG. 10:3 PLOT OF THE FUNCTION $q = p^2$ ON FOUR COORDINATE SYSTEMS
(Showing Relationship of r value to Moduli and Positions of Scales

It would appear that all points whose abscissas are $p = \infty$ would lie on the X axis, but this is not so. Points whose abscissas place them at the terminus of the X axis have very real ordinates; they just cannot be located with respect to the Y axis. Consequently, a third axis is required; namely, the $\overline{Z}$ axis, parallel to the Y axis through the terminus of the X axis. In developing the equation of the straight line in Chapter 14, it is shown to be desirable that the modulus of the $\overline{Z}$ axis be r times as great as that of the Y axis. This would make points on the $\overline{Z}$ axis lie on a straight line with corresponding points that have finite abscissas, regardless of the value of r. It will be noted that this is indicated in Figures 10:2 and 10:3.

It would be well to remember the following characteristics of the hyperbolic coordinate system:

1. The X axis is a line segment subdivided hyperbolically by a scale as described in Section 10.1. This axis may be either perpendicular or oblique to the other axes.

2. The Y axis is a vertical line passing through the origin ($p = 0$) of the X axis. It is subdivided linearly with modulus m_y.

3. The $\overline{Z}$ axis is a vertical line passing through the terminus ($p = \infty$) of the X axis. Its subdivisions are linear and in the same sense as the Y axis. Its modulus is r times that of the Y axis, or $m_{\overline{Z}} = rm_y$.

4. The lengths of the Y or $\overline{Z}$ axes may be infinite.

5. The subdivisions (linear or hyperbolic) of these axes are properties of the axes, not necessarily of the functions plotted on these axes. Thus, logarithmic or other functions may be plotted upon these axes, producing scales other than linear or hyperbolic as parts of a nomograph or other device.

6. Abscissas are given in terms of p and are determined by lines parallel to the Y axis through the points p on the X axis.

7. Ordinates for all points with finite abscissas are given in terms of q and are determined by lines through the points q on the Y axis and the terminus of the X axis.

8. Ordinates of points with infinite abscissas are given in terms of $q_{\bar{Z}}$ and are located at the points $q_{\bar{Z}}$ on the $\bar{Z}$ axis.

10.3 Relationship to Cartesian Coordinates

Section 10.4 shows the derivation of the equation of the straight line in hyperbolic coordinates, and Chapter 11 presents the construction of nomographs using hyperbolic coordinates. For these purposes it is desirable to establish

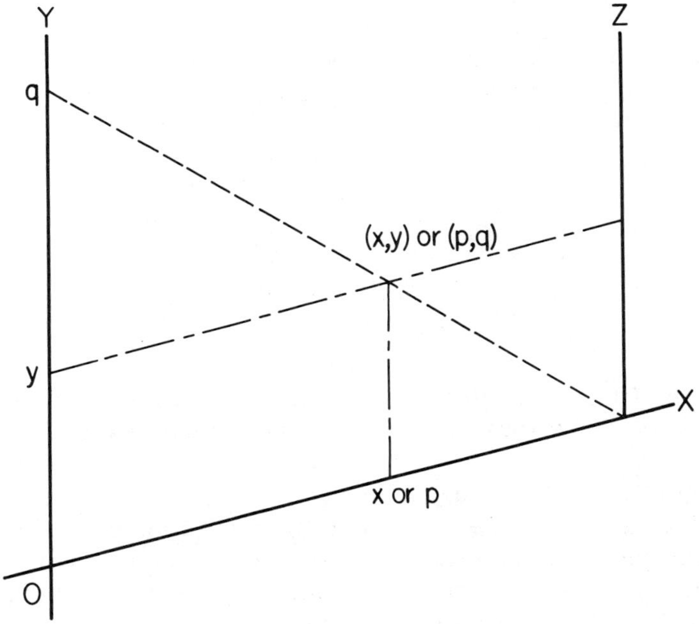

FIG. 10:4 RELATIONSHIP OF CARTESIAN AND HYPERBOLIC COORDINATES

first the relationship between the cartesian and hyperbolic coordinates of a point.

Figure 10:4 shows a set of cartesian coordinates superimposed upon hyperbolic coordinates so that the two X axes coincide, and the two Y axes coincide. As the length of the hyperbolic X axis is unit length (Sec. 10.1), its terminus

($p = \infty$) lies at the point $x = 1$ on the cartesian X axis. A point (x,y) on the cartesian system is then defined as the point (p,q) on the hyperbolic system.

From the foregoing description of the method of locating ordinates on hyperbolic coordinates, it is apparent that this point lies on a straight line whose hyperbolic intercepts are (∞,0) and (0,q). The cartesian intercepts of this line are the points (1,0) and (0,q). From the intercept form of the equation of the straight line in cartesian coordinates,

$$y = q(1 - x). \quad (10.1)$$

But x is a point on the hyperbolic scale whose distance from the origin is $s = p/(p + r)$; hence

$$x = p/(p + r). \quad (10.2)$$

Therefore,

$$\begin{aligned} y &= q[1 - p/(p + r)] \\ &= qr/(p + r). \end{aligned} \quad (10.3)$$

Equation 10.2 and 10.3 define the cartesian coordinates, x and y of any point in terms of its hyperbolic coordinates.

10.4 Equations of the Straight Line

Equation 9.1 is the equation in cartesian coordinates of the straight line passing through the points (x_u, y_u), (x_v, y_v) and (x_w, y_w). The cooresponding equation in hyperbolic coordinates is obtained by substitution in Equation 9.1 of the values of x and y given in Equations 10.2 and 10.3.

As the equation of the straight line will eventually be used as defining equation for the construction of nomographs, it is desirable to use the determinant form rather than the linear form. Consequently, it is preferable to make the substitutions in Equation 9.3, which is the determinant form of Equation 9.1. The resulting determinant is

$$\begin{vmatrix} q_u r/(p_u + r) & p_u/(p_u + r) & 1 \\ q_v r/(p_v + r) & p_v/(p_v + r) & 1 \\ q_w r/(p_w + r) & p_w/(p_w + r) & 1 \end{vmatrix} = 0. \qquad (10.4)$$

Equation 10.4 is simplified by making the following transformations:

1. Subtracting the second column from the third to obtain a new third column,
2. Factoring r from the first and third columns, and
3. Factoring $1/(p_u + r)$ from the first row, $1/(p_v + r)$ from the second row and $1/(p_w + r)$ from the third row. This latter step is valid only for finite values of p. In the case of infinite values of p, this step would involve division by zero; consequently, the case for infinite abscissas must be dealt with separately.

As a result of these transformations, Equation 10.4 becomes

$$\frac{r^2}{(p_u + r)(p_v + r)(p_w + r)} \begin{vmatrix} q_u & p_u & 1 \\ q_v & p_v & 1 \\ q_w & p_w & 1 \end{vmatrix} = 0 \qquad (10.5)$$

As $r^2 \neq 0$, Equation 10.5 may be reduced to

$$\begin{vmatrix} q_u & p_u & 1 \\ q_v & p_v & 1 \\ q_w & p_w & 1 \end{vmatrix} = 0 \qquad (10.6)$$

which is the general equation of the straight line in hyperbolic coordinates for the example in which the three defining points of the straight line have finite abscissas.

If one of the points has an infinite abscissa in hyperbolic coordinates, its cartesian abscissa is 1. Its ordinate is measured on the $\overline{Z}$ axis, whose modulus is r times that of the Y axis; consequently, this ordinate has the value

$$y = q_{\overline{z}} m_{\overline{z}} = r q_{\overline{z}} m_y .$$

The modulus of the Y axis, however, may be taken as unity in all subsequent applications of this theory, as no loss of generality is incurred by so doing. Hence, the cartesian ordinate of a point with infinite hyperbolic abscissa is

$$y = r q_{\overline{z}}$$

and its cartesian abscissa is

$$x = 1$$

These substitutions are made in Equation 9.3 for the point, say (x_w, y_w), which has an infinite abscissa. The substitutions of Equations 10.2 and 10.3 are made for those points with finite abscissas, and the equation is simplified by the transformations outlined previously. The only change in this transformation is in the third step, where the factoring is applied only to those rows that contain finite abscissas. The resulting equation is

$$\begin{vmatrix} q_u & p_u & 1 \\ q_v & p_v & 1 \\ q_{\overline{z}} & 1 & 0 \end{vmatrix} = 0 \qquad (10.7)$$

The appearance of the cipher in the third column of this determinant need cause no concern, because together with the "1" in the second column, it denotes a point with infinite abscissa. On the hyperbolic coordinate system, such a point can be readily located. In cartesian coordinates the location of a point with an infinite abscissa is impossible; hence, it is necessary to effect a transformation that produces a third column of ones when dealing with cartesian coordinates.

As Equations 10.6 and 10.7 are the basic equations of the straight line in hyperbolic coordinates these will be applied in the following chapter to the problem of constructing a nomograph. It may be recalled that the position of points relative to one another and to the axes of hyperbolic coordinates may be varied greatly by varying the value of the constant r. This constant does not appear as a factor in either Equation 10.6 or Equation 10.7; hence, its value may be changed without involving any change in the equation. If these equations are employed as defining equations for nomographs, it is then possible to effect radical changes in the moduli and positions of the nomograph scales without changing the defining equations.

APPLICATION TO CONSTRUCTION OF NOMOGRAPHS

10.5 Defining Equation

As has been previously explained the defining equation of a nomograph that comprises scales for three variables intersected by a straight index line is the equation of the straight line through three points. In cartesian coordinates this is Equation 9.1, or its somewhat more useful counterpart, 9.3. The latter, being in determinant form, has the advantage of showing in its first column the ordinates and in its second column the abscissas of the points on the three scales that are intersected by the index line.

In hyperbolic coordinates the defining equation is 10.6 for all nomographs in which all three scales have finite abscissas. Most nomographs contain one scale with infinite abscissa, and for these the defining equation is 10.7. These, however, are general equations; several modifications of these equations occur with such frequency that they merit separate consideration at this point. These are the modifications in which one or more scales lie on the X, Y or $\overline{Z}$ axes.

If one scale, say, the U scale, lies on the Y axis, the abscissa of all of its scale points is zero; hence, $p_u = 0$.

The resulting equation of the nomograph is

$$\begin{vmatrix} q_u & 0 & 1 \\ q_v & p_v & 1 \\ q_w & p_w & 1 \end{vmatrix} = 0 \qquad (10.8)$$

If one scale—say, the V scale—lies on the X axis, the ordinate of all of its scale points is zero; hence $q_V = 0$. The resulting equation of the nomograph is

$$\begin{vmatrix} q_u & p_u & 1 \\ 0 & p_v & 1 \\ q_w & p_w & 1 \end{vmatrix} = 0 \qquad (10.9)$$

If one scale, say, the W scale, lies on the $\overline{Z}$ axis, the abscissa of all of its scale points is infinite and is denoted in the W row of the determinant by a second-element "1" and a third-element "0". The resulting equation of the nomograph is

$$\begin{vmatrix} q_u & p_u & 1 \\ q_v & p_v & 1 \\ q_w & 1 & 0 \end{vmatrix} = 0 \qquad (10.10)$$

In Equations 10.8 through 10.10, the position of the cipher has the following significance: Occurring in column 1, it indicates a scale lying on the X axis; occurring in column 2, it indicates a scale lying on the Y axis; occurring in column 3, it indicates a scale lying on the $\overline{Z}$ axis, provided the second element in the same row is 1.

Three somewhat more specific examples employ combinations of these modifications. The first shows two parallel scales accompanied by one curved scale. This may be expressed as

a) Parallel scales on the Y and $\overline{Z}$ axes,

$$\begin{vmatrix} q_u & 0 & 1 \\ q_v & p_v & 1 \\ q_w & 1 & 0 \end{vmatrix} = 0 \qquad (10.11)$$

b) one scale on the Y-axis and another parallel to it at abscissa k_V,

$$\begin{vmatrix} q_u & 0 & 1 \\ q_v & k_v & 1 \\ q_w & p_w & 1 \end{vmatrix} = 0 \qquad (10.12)$$

c) one scale on the $\overline{Z}$-axis and another parallel to it at abscissa k_V,

$$\begin{vmatrix} q_u & p_u & 1 \\ q_v & k_v & 1 \\ q_w & 1 & 0 \end{vmatrix} = 0 \qquad (10.13)$$

In (b) and (c) the curved scale may or may not be between the parallel scales. It might conceivably even cross one or both of them.

The second example is that of three parallel scales, which may always be expressed as scales lying on the Y and $\overline{Z}$ axes with a third scale between;

$$\begin{vmatrix} q_u & k_u & 1 \\ q_v & 0 & 1 \\ q_w & 1 & 0 \end{vmatrix} = 0$$

The third example is that of two parallel scales and one oblique scale. This type may always be expressed in the form of a nomograph with one scale lying on each of the three axes;

$$\begin{vmatrix} 0 & p_u & 1 \\ q_v & 0 & 1 \\ q_w & 1 & 0 \end{vmatrix} = 0$$

This type is generally constructed on oblique axes to bring the two parallel scales into symmetrical alignment, as was shown in the nomograph for the hyperbola T×R = S (Fig. 9:3).

10.6 Construction of Scales Lying on the Axes

The value of the factor r determines the relative values of the moduli of the Y and $\overline{Z}$ axes, because $r = m_{\overline{z}}/m_y$. It also determines the subdivision of the X axis, that axis being subdivided by an hyperbolic scale, $x = p/(p + r)$. The selection of a suitable value of r for the construction of a given nomograph must, therefore, take into consideration the range and degree of subdivision desired in each of the scales that lie on the X, Y, and $\overline{Z}$ axes. For scales lying on the Y and $\overline{Z}$ axes the range and degree of subdivision determine the moduli as in this example:

> Suppose a range of 0 to 500 is to be represented by a linear scale subdivided into integral units not less than 1/50 inch apart. The length of the scale will be 500 × 1/50 inches = 10 inches. The modulus is then 10/500 = .02 inch.

The moduli of the Y and $\overline{Z}$ axes usually are such that the scales lying on them are of equal length, since such an arrangement gives the nomograph greater balance, while the scales have equal relative accuracy of subdivision.

The value of r best suited to the range and degree of subdivision required of the X axis is determined by similar considerations:

> Suppose a range of 0 to 50 is to be represented by hyperbolic scale on which values up to 10 are subdivided into 0.1 units and values from 10 to 50 into units not exceeding 0.5 unit. In order to meet the first requirement, the value of r should be 8 to 10. Such a value, however, crowds the upper range so that the second condition does not easily apply. To meet both conditions, a value of r in the vicinity of 40 proves most satisfactory. The length of the X axis may be increased, if necessary, to give the desired degree of subdivision.

Consideration of these several factors provides one value of r from the range and degree of subdivision desired in the scales which lie on the Y and $\overline{Z}$ axes, while a second value of r is determined from the range and degree of subsivision desired in the scale that lies on the X axis. The nomographer always hopes that these values will agree, but they sometimes do not. In some of the examples encountered in Chapter 11 it is necessary to effect a compromise, as is shown in the discussion of each one.

By definition, the Y and $\overline{Z}$ axes are linear; consequently, the scales they bear are linear-linear, logarithmic-linear, exponential-linear, and so forth, depending upon the functions that define the ordinates of their scale points. Thus, if the ordinates of scale points are defined by $q = kv$, where k is a constant term and v is a variable, q is a linear function of v, and the v-scale is linear. Plotted on the Y of $\overline{Z}$ axis, this scale is linear-linear. Likewise, if $q = k \log v$, q is a logarithmic function of v, and the v scale is logarithmic. Plotted on the Y or $\overline{Z}$ axis, this scale is logarithmic-linear.

On the other hand, the X axis is hyperbolic, as defined in Chapter 3. The scales it bears are linear-hyperbolic, logarithmic-hyperbolic, etc., depending upon the functions defining the abscissas of their scale points. Thus, if $p = kv$, the v scale, plotted on the X axis is linear-hyperbolic. If p =

k log v, the v scale, plotted on the X axis is logarithmic-hyperbolic.

10.7 Construction of Vertical Scales

A vertical scale is a scale all of whose scale points have the same abscissa. If this abscissa is zero, the scale lies on the Y axis; if it is infinity, the scale lies on the $\bar{Z}$ axis. These cases were discussed in a previous section. If the abscissa has any value k between zero and infinity, the scale lies parallel to the Y and Z axes and at the point p = k on the X axis. The modulus of such a scale is not defined in the defining equation, but is determined by the modulus of the Y axis and the position of the scale with respect to that axis, as shown in the definition of the hyperbolic ordinate. Both of the methods described below for the construction of vertical scales are derived from the relationships involved in that definition.

> First Method: This method is a form of graphical projection. A scale of the type desired, having a modulus equal to that of the Y axis is superimposed on the Y axis with its origin at the origin of the Y axis. A vertical line is drawn at the point p = k on the X axis. The points of the scale on the Y axis are then projected onto the vertical line by means of a family of lines which converge at the terminus of the X axis.

This method is just another form of copying. It may be varied by the use of a linear scale instead of some other standard scale as a reference scale on the Y axis, as, indeed, it must be if no standard scale is available for the function to be plotted. Even in this form, however, it is subject to all of the inaccuracies and eyestrain associated with copying methods. Furthermore, if the vertical scale lies an appreciable distance from the Y axis, the reference scale from which the projection is made must frequently extend some distance beyond the end of the Y axis, or auxiliary vertical axes with fractional moduli must be used to supplement the scale on the Y axis as reference scales.

Perhaps a more serious handicap to this method lies in the fact that the lines of projection intersect the vertical obliquely, making it impossible to draw the graduations perpendicular to the scale line during the process of projection. It becomes necessary to locate the scale points by projection, then draw them in with a parallel edge. This double operation entails additional opportunities for inaccuracies to enter the construction process.

Second Method: In this method the scale is plotted from a set of calculated values. The modulus of the vertical scale is calculated from the modulus of the Y axis (m_y) and the abscissa of the vertical scale; the scale is then constructed with that modulus by methods discussed in Chapter 1.

The modulus of the vertical scale is determined by the relationship,

$$m_{y'} = m_y(1 - x),$$ [1]

in which $m_{y'}$ is the required modulus and (1 - x) is a factor whose value is determined by the values of p and r.

The actual calculations are quite simple. Values of the ordinate as defined by the defining equation are calculated for the desired range of the variable. These values are then multiplied by the constant factor (1 - x) given in the Appendix. The resulting products are plotted to form the desired scale.

This method might be clarified by an example. Suppose the Y axis has a modulus of 10 inches. A vertical scale is to be constructed at the point p = 2.35. The value of r for the desired nomograph is 5. The value of p/r is 0.47, for which the table in the Appendix gives the value of 1 - x as 0.68027. The modulus of the vertical scale is then 10 × 0.68027 = 6.8027 inches.

---

[1] No proof of this relationship is necessary, because it is obvious from the method of locating ordinates. As all points lying on a given ordinate line have the same numerical value, although their distances from the X axis decrease linearly as they approach the point x = 1, it is necessary that the moduli of vertical axes decrease similarly.

The preceding discussion is a general description of the hyperbolic coordinate system, its application to nomography, and the methods employed in the location and construction of scale elements on this system. Application to specific problems is deferred to Chapter 11.

chapter 11

PRACTICAL ASPECTS OF CONSTRUCTING NOMOGRAPHS

Preceding chapters have presented several theoretical approaches to the design of nomographs, with some mention of constructional aspects. Together with the discussions of mechanical phases of the preparation of scales and graph papers in Chapters 1 and 3, the reader should at this point be fairly well prepared to construct nomographs for most engineering formulas without undue tedium. Each formula, however, presents new demands upon one's art, and one's experience is greatly increased with the solution of each new problem. This is fortunate, for only a background of broad experience can enable one to produce nomographs of real value with dispatch. Toward that end, this chapter is devoted to the design of some practical nomographs and a discussion of certain mechanical phases peculiar to the art of nomography.

The theoretical approach employed in most of these problems is that of the hyperbolic coordinate system, which from the author's experience is the most expeditious method of designing nomographs. The thoughtful student, however, will apply the synthetic method and the method of cartesian coordinates to these same problems as a means of obtaining greater understanding of those methods.

NOMOGRAPHS WITH PARALLEL SCALES

$$\begin{vmatrix} q_u & k & 1 \\ q_v & 0 & 1 \\ q_w & 1 & 0 \end{vmatrix} = 0 \qquad (11.1a)$$

$$q_u = q_v + kq_w$$

11.1 Defining Equation

The equations shown are the determinant and linear forms of the defining equations of nomographs consisting of three parallel scales. Characteristics of these scales revealed by the determinant are:

1. The V-scale has the abscissa 0 and lies on the Y-axis.
2. The W-scale lies on the $\bar{Z}$-axis, as indicated by the second column "1" and third column "0" in the third row of the determinant.
3. The U-scale has a constant abscissa k and lies on a vertical axis Y' passing through the selected p-scale at the point p = k.

The modulus of the Y'-axis on which the U-scale lies is determined by the relationship $m_{y'} = m_y(1 - x)$, the value of the factor (1 - x) for the point p = k can be determined from the Appendix after a suitable value for r is selected.

The value of r is determined by the values of m_y and $m_{\bar{z}}$, which are, in turn, determined by the range and degree of subdivision of the scales that lie on the Y- and Z-axes.

Nomographs for the simple addition of linear terms are seldom constructed, as such addition is more easily handled by mental or mechanical processes. However, this form is widely applied to the addition of non-linear functions, as in the following examples.

11.2 Area of a Torus

The surface area of a doughnut, an innertube, or any other torus of circular cross-section (Fig. 11:1) is

$$A = \frac{\pi^2}{4} (R^2 - r^2).$$

This rather simple formula is typical of the addition of non-linear functions. Others are included in the exercises at the end of this chapter. The determinant form of this expression is

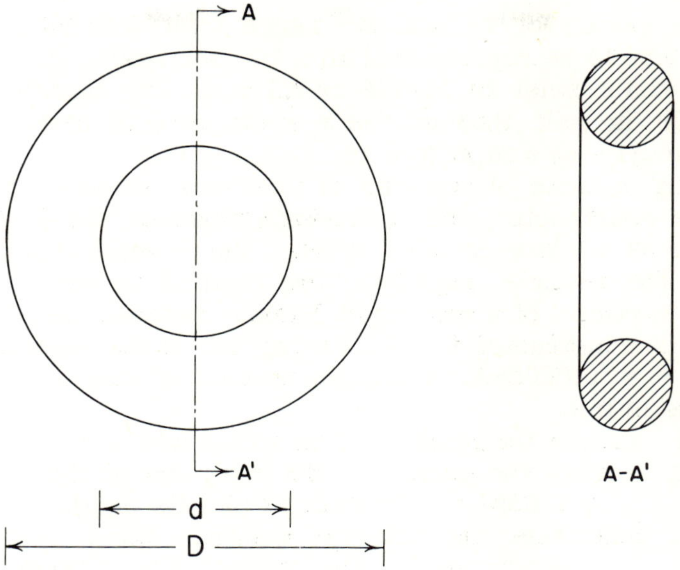

FIG. 11:1 TORUS OF CIRCULAR CROSS-SECTION

$$\begin{vmatrix} 4A/\pi^2 & 1 & 1 \\ R^2 & 1 & 0 \\ -r^2 & 0 & 1 \end{vmatrix} = 0 \qquad (11.2)$$

R lies on the $\overline{Z}$ axis, r on the Y axis, and A on a vertical axis through the point p = 1.

The ranges to be included in the nomograph are

$$r\Big|_0^{10} \text{ inch} \quad R\Big|_0^{10} \text{ inch} \quad \text{and } A\Big|_0^{300} \text{ sq inch}$$

The ranges covered by the scales on the Y and $\overline{Z}$ axes are equal; hence the moduli of those axes are equal. The value of the hyperbolic constant, therefore, is 1, placing the A scale at the midpoint of the X axis.

The values of r^2 and R^2 range from 0 to 100. If these terms are to be represented on a 10-inch scale, the modulus of the scale must be 10/100 or 0.1 inch. The modulus of the A scale is half that of the r scale, or 0.05 inch, because $m_{y'} = m_y(1 - x) = m_y r/(p + r)$.

Using a table of squares of numbers, scales for r^2 and R^2 are constructed with a 0.1-inch modulus, parallel to one another at a distance of 8 inches, the R scale in a positive sense, the r-scale, negative. The termini of the scales lie at the vertices of a rectangle. Midway between these a third scale is constructed for A, having the scale equation, $y' = 0.2A/\pi^2 = 0.02028A$. A simple method of constructing the A-scale is this:

1. Locate the point a = 0 on a line with r = R = 0;
2. Locate the point A = 300 at a point on the scale 300 x 0.0203 = 6.09 inches from its origin;
3. Subdivide the interval between these two points linearly, using the method for altering scale moduli described in Section 1.11.

The completed nomograph is shown in Figure 11:2. For other ranges, D and d may be simultaneously multiplied by any convenient factor (e.g., 0.1), provided that A is multiplied by the square of that factor. D, d, and A may be expressed in any corresponding units (e.g., 0.01) of length and area.

Cartesian Method

To meet the requirements of cartesian geometry, it is necessary to convert Equation 11.2 to a determinant whose third column elements are 1's. The second column is added to the third, and the first row is then divided by two to produce

$$\begin{vmatrix} 2A/\pi^2 & 1/2 & 1 \\ R^2 & 1 & 1 \\ -r^2 & 0 & 1 \end{vmatrix} = 0$$

which is the normalized form of the defining equation.

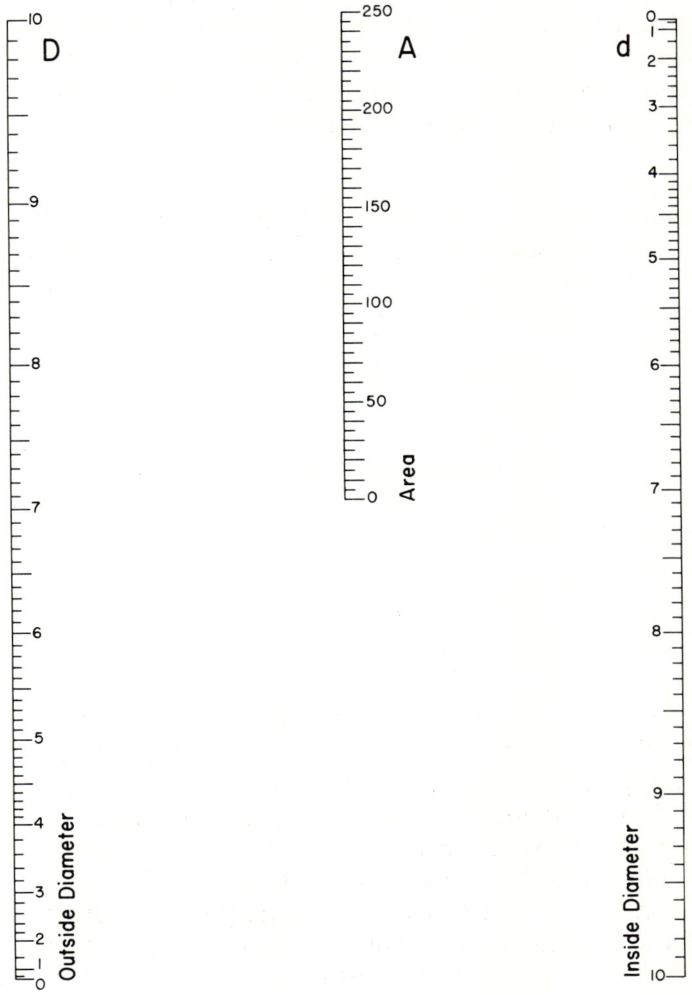

FIG. 11:2 NOMOGRAPH FOR SURFACE AREA OF A TORUS

Abscissas of the r, R, and A scales are, respectively 0, 1, and 1/2; the modulus for the scale $4A/\pi^2$ is 1/2 that of the other two scales. These results are clearly identical with those obtained above, and the construction of the nomograph is the same. If desired, the actual dimensions of the nomograph may be inserted by multiplying the first column by the height of the nomograph, and the second column by its width.

11.3 Products

Although the q's of Equation 11.1a may be defined in terms of various functions, linear or non-linear, by far the most common application of this form is that in which these functions are logarithmic:

$$\begin{vmatrix} \log U & k & 1 \\ \log V & 0 & 1 \\ \log W & 1 & 0 \end{vmatrix} = 0 \qquad (11.3)$$

Upon expansion this becomes

$$\log U = \log V + k \log W,$$

or

$$U = VW^k. \qquad (11.4)$$

Nomographs of this type may often be constructed wholly or partially of standard printed scales of the type included in the Supplement. In many instances, by proper selection of moduli and hyperbolic constant some of these scales may be used without modification. If the modulus requires modification, photographic methods are generally satisfactory.

The scales are cut to the proper range and width of stroke, and a base line is drawn. The values of the scale points may be applied with a lettering set or hand stamp, or the values printed on the standard scales may be used with addition of ciphers, decimal points, or cut-outs of the factors "$x10^{-2}$", "$x10^4$", and so on.

A non-aqueous cement must be used, as water destroys the dimensional stability of paper. A plastic cement, used for model airplanes, is very satisfactory. The scales must be quickly placed in position once the cement is applied, as it dries rapidly. Clear rubber cement may also be used, this is to be applied to both the scale and the background and allowed to dry. The dry surfaces of rubber cement form a strong bond when brought into contact.

11.4 Shafting Horsepower

For First Shaft from the Prime Mover under heavy loading (the shaft being of cold-finished steel), the horsepower is[1]

$$\text{H.P.} = D^3R/125,$$

which is identical with Equation 11.4 with the substitutions U = H.P., V = R/125, W = D, and k = 3. The determinant form is, therefore

$$\begin{vmatrix} \log \text{H.P.} & 3 & 1 \\ \log(R/125) & 0 & 1 \\ \log D. & 1 & 0 \end{vmatrix} = 0.$$

The elements of the second and third columns place the R scale on the Y axis, the D scale on the $\bar{Z}$ axis and the H.P. scale at p = 3, the three scales being parallel. The parameters, their scale functions, and corresponding ranges are:

Parameter	Range	Unit	Scale Position	Scale Function	Range
R	2500 – 125	min^{-1}	Y-axis	log (R/125)	1.301 – 0
H.P.	4000 – 1	H.P.	p = 3	log H.P.	3.602 – 0
D	6 – 1	inch	$\bar{Z}$-axis	log D	0.778 – 0

[1] Perry, John H., Chemical Engineers' Handbook, 3rd Ed., p. 1667, McGraw-Hill Book Company, New York, 1950.

The range of the function to be plotted on the Y-axis is approximately 1.67 times that of the function to be plotted on the $\overline{Z}$-axis; consequently, the modulus of the $\overline{Z}$ axis must be approximately 1.67 that of the Y axis. The value of the hyperbolic constant r is $m_{\overline{z}}/m_y$, or approximately 1.67. If an exact rectangle is desired, the ratio of moduli (and the

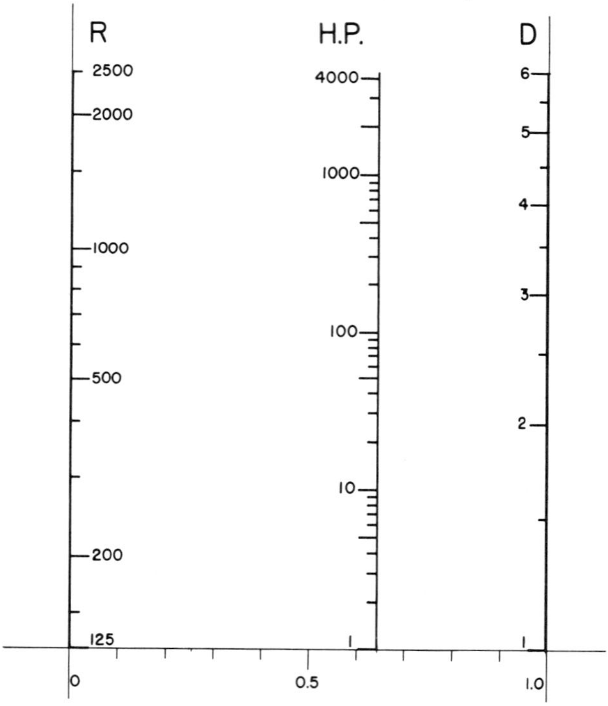

FIG. 11:3 LAYOUT FOR NOMOGRAPH FOR SHAFTING HORSEPOWER

value of the hyperbolic constant r) may be taken as 1.301/.778 = 1.67. The position of the D scale then becomes 3/4.67 = 0.64 times the length of the X axis, and its modulus, (1 - 0.64), or 0.36 times the modulus of the Y axis. A diagram for the construction of this nomograph is shown in Figure 11.3.

11.5 Products of More than Two Numbers

The defining equation of the nomograph is limited to three variables, as is the nomograph itself. This basic form may be extended to include more variables by subdividing the complete equation in such a manner as to include only three variables in each sub-equation. Such subdivision necessitates the introduction of additional parameters, common to the sub-equations, which serve to carry over the computation of the sub-equation into that of the next. Unless any of the intermediate parameters has some real significance, it is common practice to leave unscaled the lines, that represent those parameters, and to mark them simply as "pivot axes."

11.6 Scobey's Formula for Steel Pipes

In nomographs constructed for specific formulas, it is seldom desirable to have more than one variable located on a given scale, because of the confusion in notation that results. Separate nomographs are, instead, constructed for sub-equations of the original equation, with corresponding pivot-axes so superimposed that the calculation carries over automatically from one nomograph to the other. This is illustrated in the nomograph for Scobey's formula for Steel Pipes[2] (Fig. 11:4),

$$V = \frac{D^{0.58}H^{0.526}}{K_s^{\ 0.526}}$$

where K_s is a coefficient for steel, and the other parameters have the designations given in the table below.

If converted to the logarithmic form and divided by 0.526, this equation becomes

$$1.90 \log V = 1.10 \log D + \log H - \log K.$$

This equation may then be subdivided into the two equations:

---

[2] Davis, C. V., Handbook of Applied Hydraulics, 2nd Ed., p. 1213; McGraw-Hill Book Company, New York, 1952.

$$\left.\begin{aligned} P &= 1.10 \log D + \log H, \\ \text{and} \qquad 1.90 \log V &= P - \log K \end{aligned}\right\}$$

in which P is the parameter introduced to relate the sub-equations.

The determinant forms of these equations are

$$\left.\begin{aligned} \text{I} \quad \begin{vmatrix} P & 1.10 & 1 \\ \log D & 1 & 0 \\ \log H & 0 & 1 \end{vmatrix} = 0 \\ \text{II} \quad \begin{vmatrix} P & 0 & 1 \\ 1.90 \log V & 1 & 1 \\ -\log K & 1 & 0 \end{vmatrix} = 0 \end{aligned}\right\} \qquad (11.5)$$

The basic factors are summarized as follows:

Parameter		Designation	Range	q	p	Position of Scale
I	D	Diameter, ft	1.00 to 10.0	log D	∞	On $\overline{Z}$ axis
	H	Loss of Head, ft/1000 ft	0.05 to 50.0	log H	0	On Y axis
	P	Pivot parameter	---	---	1.10	Vertical at p = 1.10
II	P	Pivot parameter	---	---	1	On Y axis
	V	Velocity, fpm	0.10 to 20.0	1.90 log V	1	Vertical at P = 1
	K	Coefficient	0.30 to 0.60	- log K	∞	On $\overline{Z}$ axis

Although the P axis is not scaled, it nevertheless has a modulus, a sense (i.e., direction of increasing scale values) and a phantom zero point which are determined by the moduli of the scales and its position with respect to them. Obviously, the modulus, sense and zero point of the P axis must be the same in both parts of the nomograph, because each point on that axis represents a value that enters the computation in the first step and drops out in the second. This restriction makes the selection of positions and moduli of the nomographic scales somewhat more complex. For this reason, it is usually best to plan the nomograph from the standpoint of the most desirable positions and senses of the scales, based upon such factors as readability, ease of manipulation of index lines, appearance, etc. A rough sketch will help at this stage. Next, a set of approximate moduli may be selected. If it appears that the restrictions that these moduli place upon one another are incompatible with the rough sketch, it may be necessary to rearrange the scales to obtain a compromise design. All this may be done with very little computation. When a design has been found that meets all requirements, exact moduli and positions of the various scales may then be computed.

Construction: Parallel scales for H and D, spaced 7 inches apart, are constructed as a first step in preparing the nomograph shown in Figure 11.4. The modulus of the H scale is 3 inches, and that of the D scale is 9 inches. The resulting value of r is $m_{\bar{z}}/m_y = m_D/m_H = 3$. The pivot axis lies at the point p = 1.10 on the r3 hyperbolic X axis between these two scales. Its linear distance from the H scale is given by

$$s = l_x p/(p+r) = 7\,[1.1/(1.1+3)] = 7.7/4.1 = 1.878 \text{ inches.}$$

The modulus of the P axis is then

$$m_p = m_y r/(p + r) = 3[3/(1.1 + 3)] = 9/4.1 = 2.193$$

A convenient modulus for the V scale is one that will provide values for log V on a 3 inch cycle, similar to the H scale. As the function to be scaled is 1.9 log V, the modulus would be 3/1.9 inches. The range of K is narrow; hence, the problem of adapting the position and modulus of the K scale to the preferred modulus of the V scale is a minor one. For the sake of compactness, it is desirable to have the K scale coaxial with the D scale. If this is decided upon, there remains one additional variable whose value may be fixed arbitrarily. That variable may be selected from the following list: (a) the value of r; (b) the position of the V-scale; or (c) the modulus of the K-scale. Fixing a value for one of these automatically fixes values for the other two.

A preliminary sketch indicates a convenient position for the V scale, such that the resulting modulus of the K scale will provide adequate accuracy without too much spread. One may then select a definite numerical value for this position and calculate the modulus of the K scale. The position of the K scale on the $\overline{Z}$ axis is best located by use of a known solution of the formula. By an alternate approach, the nomograph is completed as follows: The V scale is constructed at the position designated in the preliminary sketch, using a 3-inch logarithmic standard scale. Positions of the points K = 0.3 and K = 0.6 (the extremes of the K scale) are located by using known solutions of the Scobey formula, and remaining points on the K scale are plotted by graphical interpolation. Inasmuch as the final nomograph must be checked by the use of known solutions, regardless of the method of construction, it is often possible to save time and computation through the use of known solutions in the construction, as is done here. The nomograph is shown in Figure 11:4. The form of determinant chosen in Equation 11.5 is by no means unique for this formula. Other arrangements may be used in setting up the scales, and it would be advantageous to the reader at this point to experiment with some other arrangements.

11.7 Nomograph With Two or More Pivot Axes

The device of the pivot axis employed in this nomograph for four parameters (Fig. 11:4) may be extended to formulas involving any number of parameters, an additional pivot axis being required for each additional parameter. These constructions often appear quite simple in theory, but prove difficult in practice because of the idiosyncracies of the formula to be represented, the ranges of the variables, and the limitations of chart dimensions. Figure 7:2 shows a nomograph having six variables, the scales being interrelated by three pivot axes.

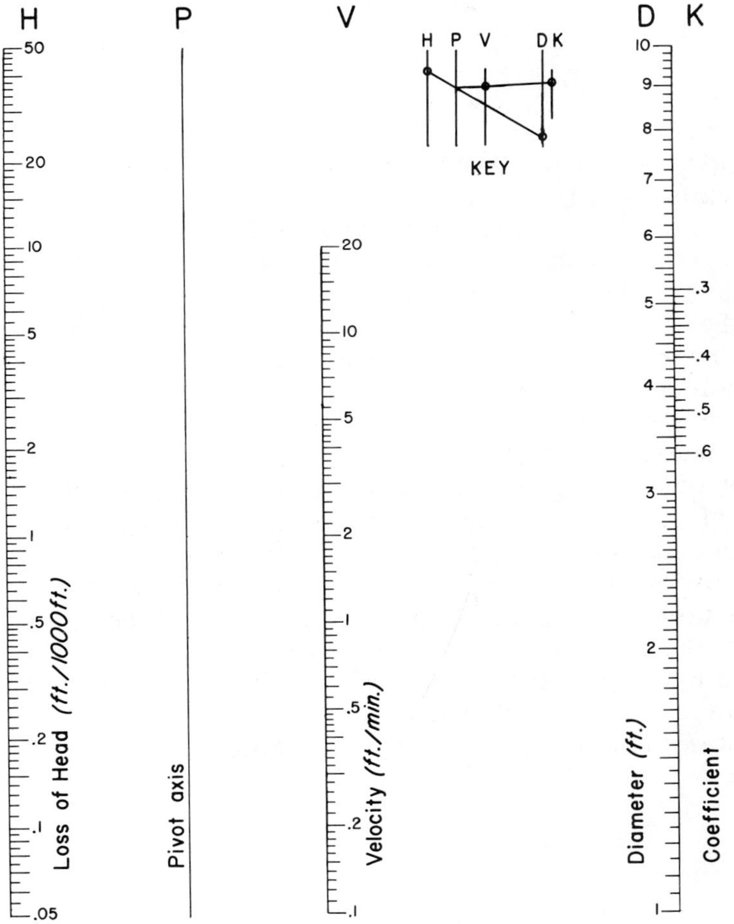

FIG. 11:4 NOMOGRAPH FOR SCOBEY'S FORMULA FOR STEEL PIPES

NOMOGRAPHS WITH OBLIQUE SCALES

$$\begin{vmatrix} 0 & p_u & 1 \\ q_v & 0 & 1 \\ q_w & 1 & 0 \end{vmatrix} = 0$$

This is the familiar Z or N type nomograph. From the determinant it is apparent that:

a) The U scale has zero ordinate and lies on the X axis. Its scale is the p scale for selected value of r.

b) The V scale lies on the Y axis; the W scale, on the $\bar{Z}$ axis. These two are parallel. Their scales are linear with moduli in proportion to the value of r selected for the X axis.

For the three scales to be intersected by a straight line, it is necessary that the two parallel scales lie on opposite sides of the diagonal scale. Nomographs represented by this type of equation will, therefore, consist of two parallel scales of opposite sense with a third scale diagonal to these. For greater ease of reading, the diagonal scale is usually set at an acute angle to the other scales, the two parallel scales being as nearly opposite as possible. This amounts to the use of oblique axes for the coordinate system and involves no change in the defining equation.

Expansion of the preceding determinant gives the equation,

$$q_v + p_u q_w = 0$$

or

$$p_u q_w = -q_v$$

An application that is obvious from this expansion is the calculation of products. This is, indeed, the most common application of this type, although other calculations, such as evolution and involution, are sometimes made on charts of this type.

11.8 Theoretical Horsepower of a Pump

The formula for the theoretical horsepower developed by a pump at maximum head and rate of discharge is:[3]

$$H.P. = 0.1136\ h\ Q$$

in which h is the dynamic head in feet and Q is the maximum discharge in cubic feet per second.

A determinant form of this equation is

$$\begin{vmatrix} 0 & h & 1 \\ H.P. & 0 & 1 \\ -0.1136Q & 1 & 0 \end{vmatrix} = 0, \qquad (11.6)$$

from which the following table may be compiled:

Var.	Designation	Range	q	p	Position of Scale
h	Dynamic Head (ft)	0 to 40	0	h	On X axis
H.P.	Horse power (theoretical)	0 to 200	H.P.	0	On Y axis
Q	Discharge (cfs)	0 to 50	0.1136Q	∞	On $\overline{Z}$ axis

The range of 0.1136Q is (0 to 50) x 0.1136 = 5.68. If the length of the Q scale is to be 10 inches, the modulus of the $\overline{Z}$ axis must be approximately 10/5.68 = 1.75 inches. The range of H.P. (0 to 200) can be represented on a 10-inch scale with a modulus of 10/200 = 0.05 inches. If these moduli are used, the value of r will be 1.75/.05 = 35, although the more convenient values 30 or 40 would probably prove equally satisfactory.

The range of h (0 to 40) has its optimum distribution on the r20 scale, higher values of r tending to crowd the scale points too much for accurate reading. The compromise value of r30 would seem to be most satisfactory from every standpoint so it is used in the following construction. The value of the modulus of the Y axis remains 0.05 inches, while that of the Z axis becomes 0.05 x 30 = 1.50.

[3] Davis, C. V., Handbook of Applied Hydraulics, 2nd Ed., p. 610; McGraw-Hill Book Company, New York, 1952.

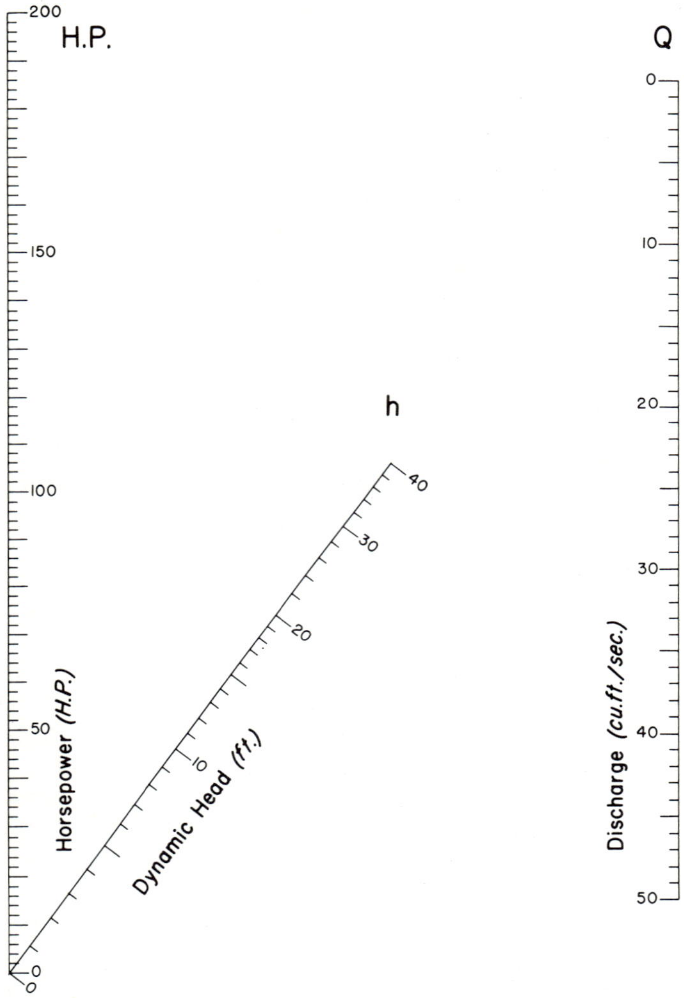

FIG. 11:5 NOMOGRAPH FOR THEORETICAL HORSE-POWER OF A PUMP

The two linear scales are constructed first. The H.P. scale is laid off with a modulus of 0.05 and a range of 200 (length 10 inches). The scale for -0.1136Q is laid off on a line parallel to the H.P. scale at a distance of eight or ten inches. These scales are centered with respect to each other (Fig. 11.5).

A diagonal is drawn between the origins of the Y and $\bar{Z}$ axes, and the r30 scale is laid off with its origin at the origin of the Y axis and its terminus at the origin of the $\bar{Z}$ axis. This scale may be constructed either from the values given in the Appendix or by projection from the r3 scale. In the latter case, the values of p are multiplied by 10 to form the r30 scale and are marked for values of h.

11.9 Combined Addition and Multiplication

Many formulas involving simple products, such as the formula for horsepower of pumps discussed in the preceding section, are as well represented by nomographs constructed of parallel logarithmic scales as by the Z type of nomograph. However, this Z type has two linear scales and it thereby can be combined with other nomographic forms involving linear scales. The following example illustrates the combination of a Z type of nomograph for products with a parallel-scale nomograph for sums.

11.10 Volume of Earthwork

The formula for calculating the volume of earthwork included between two parallel cross-sections is[4]

$$V = \frac{L}{27}\left[\frac{A_1 + A_2}{2}\right]$$

where V is the volume in cubic yards between sections, L is the perpendicular distance between sections in feet, and A_1 and A_2 are the areas of the two sections in square feet. This formula may be expressed in parametric form as

[4] Ives, H. C., <u>Highway Curves</u>, p. 113; John Wiley and Sons, Inc., New York, 1929.

$$\left.\begin{aligned} P &= A_1 + A_2 \\ \text{and} \quad P &= 54V/L. \end{aligned}\right\}$$

The determinant forms of this are

$$\left.\begin{aligned} &\begin{vmatrix} P & 1 & 1 \\ A_1 & 1 & 1 \\ A_2 & 1 & 0 \end{vmatrix} = 0 \\ \text{and}\ldots \quad &\begin{vmatrix} 0 & L & 1 \\ -54V & 0 & 1 \\ P & 1 & 0 \end{vmatrix} = 0 \end{aligned}\right\} \qquad (11.7)$$

The basic specifications for the nomograph are

Var.	Designation	Range	q	p	Position of Scale
P	(parameter)	0 to 4000	P	1	Vertical at p = 1.
A_1	Area sect (sq ft)	0 to 2000	A_1	0	On Y axis
A_2	Same	0 to 2000	A_2	∞	On $\overline{Z}$ axis
P	(parameter)	0 to 4000	P	∞	On $\overline{Z}$ axis
V	Volume (cu yd)	0 to 15,000	-54V	0	On Y' axis
L	Length Sect (ft)	0 to 200	0	L	On X' axis

For the first sub-equation a value of $r = 1$ is used, as A_1 and A_2 are of equal range. This places the pivot axis for P halfway between these two scales.

For the second sub-equation, if the V and P scales are to be 10 inches in length, the modulus of the P axis is 10/4000, and that of the V scale is $10/(54 \times 15{,}000) = 10/810{,}000$. The use of these moduli requires a value of $r' = 810{,}000/4000 = 200$ (approx.). This value of r' is also satisfactory for the

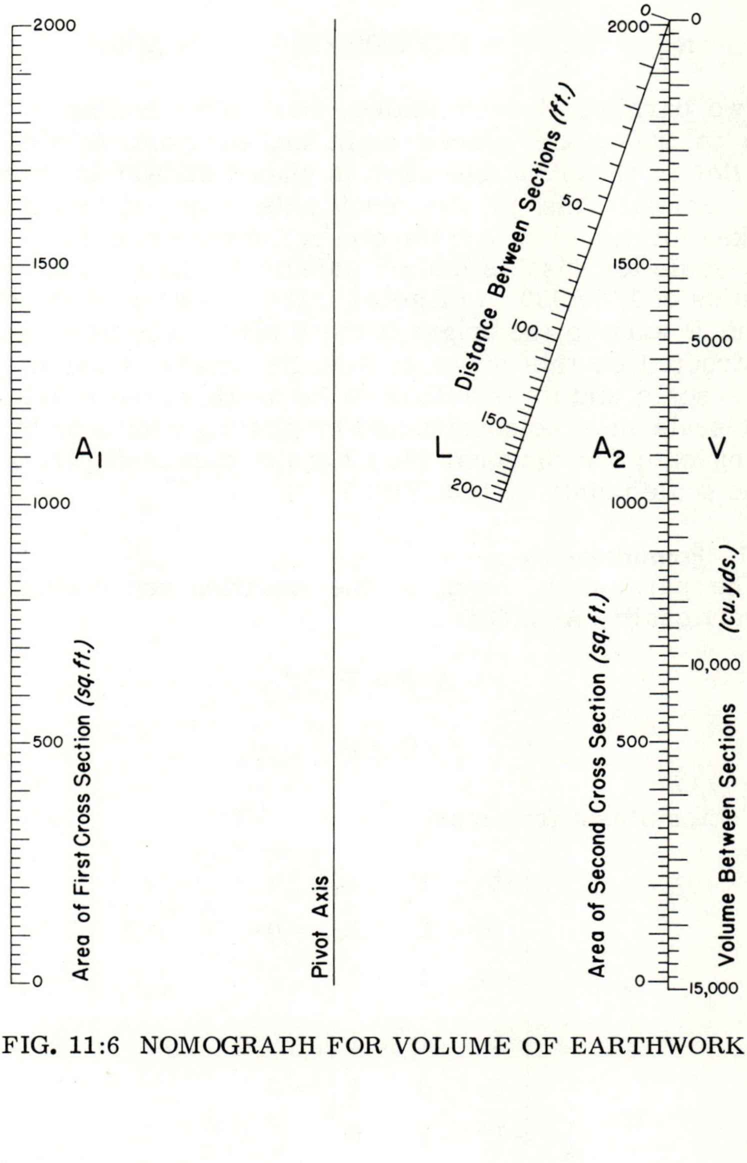

FIG. 11:6 NOMOGRAPH FOR VOLUME OF EARTHWORK

range of L and it is used in the construction. The modulus of the V scale is slightly modified to correspond with the value of 200 for r, whence

$$m_V = m_P/r = (10/4000)/200 = 1/80{,}000.$$

Two parallel 10-inch scales, each representing a range of 0 to 2000, are placed eight inches apart. A third line, parallel to these but unscaled, is placed midway between the two scales. This is the pivot axis. One of the scales is marked for values of A_1; the other, for values of A_2.

A scale for -54V is laid off parallel to the P axis, using a modulus of 1/80,000. A diagonal is then drawn from the origin of the P axis to the origin of the V scale. The r200 scale is constructed on this diagonal with its origin at the origin of the V scale and its terminus at the origin of the P axis. The r200 scale may be constructed by plotting values or by projecting onto the diagonal the r2 scale, then multiplying each of the scale points by 100 (Fig. 11:6).

11.11 Proportionality

The parametric form of the equation representing the proportionality A:B::C:D is

$$A/B = P$$

and

$$C/D = P$$

The determinant forms are

$$\begin{vmatrix} 0 & P & 1 \\ A & 0 & 1 \\ -B & 1 & 0 \end{vmatrix} = 0$$

and

$$\begin{vmatrix} 0 & P & 1 \\ C & 0 & 1 \\ -D & 1 & 0 \end{vmatrix} = 0$$

In both parts of the nomograph, P lies on the X axis, which, consequently, need not be scaled. Both A and C lie on the Y axis; B and D, on the $\overline{Z}$ axis.

It is unnecessary to determine any set values for r as the X axis is not to be scaled. It must be remembered, however, that the value of r determines the relative moduli of the Y and $\overline{Z}$ axes and it must be the same for both parts of the nomograph, if the two parts are to be superimposed. As $r = m_{\overline{Z}}/m_Y$, it follows that

$$m_B/m_A = r = m_D/m_C,$$

or

$$m_A : m_B = m_C : m_D$$

This is the only restriction upon the moduli of the four scales.

11.12 Weight of Gas Flowing through an Orifice

The formula for determining the weight of gas flowing through an orifice is

$$w = \pi d^2 v/576\ V$$

where W = weight of gas flowing (lb/sec) Range 0 to 0.4
d = diameter of orifice (inch) Range 0 to 2.0
(scaled as πd^2) Range 0 to 12.7
V = specific volume of the gas (cu ft) Range 0 to 200
(scaled as 576V) Range 0 to 115,000
and v = velocity of the gas (ft/sec) Range 0 to 3000.

For construction of the nomograph, the equation is written as a proportionality:

$$w/\pi d^2 = v/576\ V \qquad (11.8)$$

From the above ranges, the following ratio of moduli appears to be most satisfactory:

$$m_w/m_d = m_v/m_V = 30$$

If $m_W = 10/0.4 = 25$, then

$$m_d = 25/30 = 5/6,$$

giving the w scale a length of 10 inches, and the d scale a length of 12.7 x 5/6 = approx. 10.5 inches.

Also, if $m_V = 10/115{,}200 = 8.65 \times 10^{-5}$, then

$$m_v = 30 \times 8.65 \times 10^{-5} = 2.595 \times 10^{-3},$$

giving the V scale a length of 10 inches, and the v scale a length of 3000 x 2.595×10^{-3} = approx. 7.8 inches.

The four scales are constructed about a common diagonal. The w and v scales have a common origin at one end of the diagonal; the d and V scales originate at the other end. The w and d scales are parallel, as are the v and V scales. The actual length of the diagonal is of no consequence and may be any length which seems to contribute balance and ease of reading to the nomograph. The completed nomograph is shown in Figure 11:7.

NOMOGRAPHS WITH CURVED SCALES

11.13 Defining Equation

The occurrence of curved scales in nomographs is indicated by rows containing two functions of the same variable in the determinant of the defining equation. For instance, the row

$$\begin{vmatrix} q_u & p_u & 1 \end{vmatrix}$$

occurring in a determinant indicates that both the abscissas and ordinates of the points on the U scale are variable functions of u, and that the U scale will therefore be curved.

Quadratic and Cubic Equations: Among the simpler nomographs falling within this class are the polynomials, such as the quadratic,

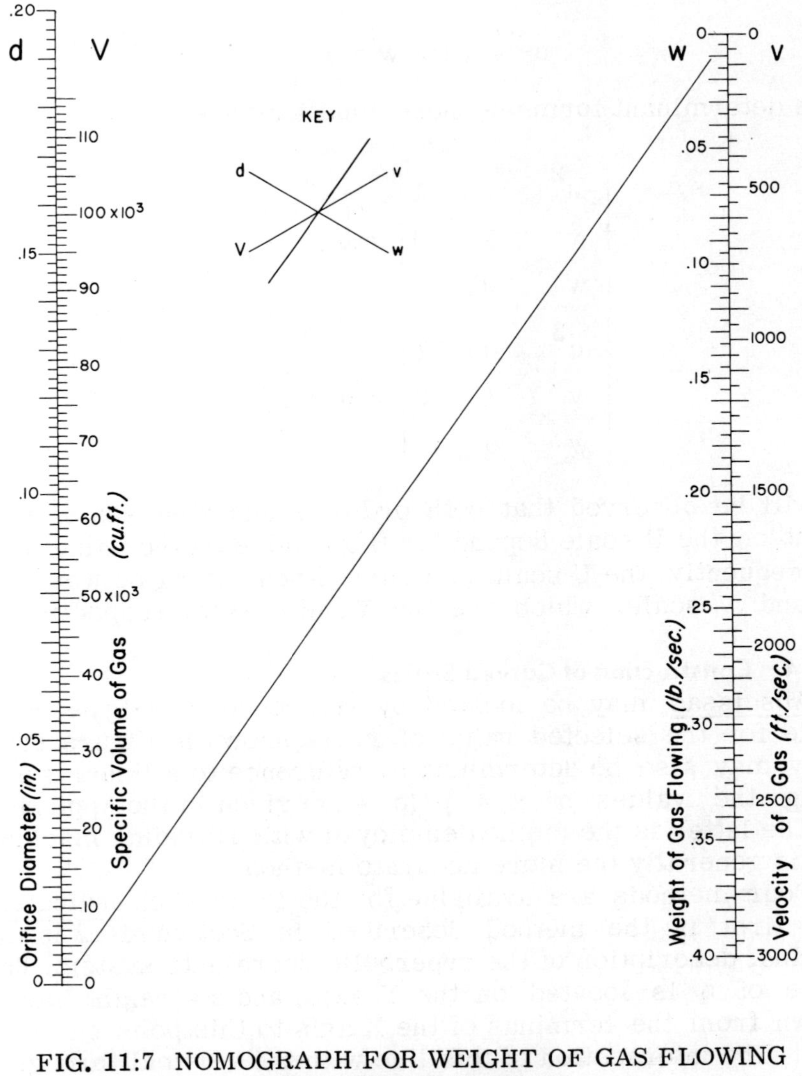

FIG. 11:7 NOMOGRAPH FOR WEIGHT OF GAS FLOWING THROUGH AN ORIFICE

$$u^2 + vu + w = 0,$$

and the reduced cubic,

$$u^3 + vu + w = 0.$$

The determinant forms of these equations are

$$\begin{vmatrix} -u^2 & u & 1 \\ v & 1 & 0 \\ w & 0 & 1 \end{vmatrix} = 0$$

and

$$\begin{vmatrix} -u^3 & u & 1 \\ v & 1 & 0 \\ w & 0 & 1 \end{vmatrix} = 0.$$

It will be observed that both ordinate and abscissa of each point on the U scale depend for their value on the value of u. Consequently, the U scale is a curved scale lying between the W and V scales, which are the Y and $\overline{Z}$ axes, respectively.

11.14 Construction of Curved Scales

Abscissas may be located by reference to a hyperbolic scale for the selected value of r, as shown in Figure 11:8. They may also be determined by reference to a linear scale using the values of $x = p/(p + r)$ given in the Appendix.

The latter is the method employed with a dividing machine and is generally the more accurate method.

Four methods are available for the location of ordinates. The first is the method described in Section 10.2 in the original description of the hyperbolic coordinate system. The value of q is located on the Y axis, and a straight line is drawn from the terminus of the X axis to this point on the Y axis. The intersection of this line with the vertical through p on the X axis is the desired point. This method is shown in Figure 11:9(a).

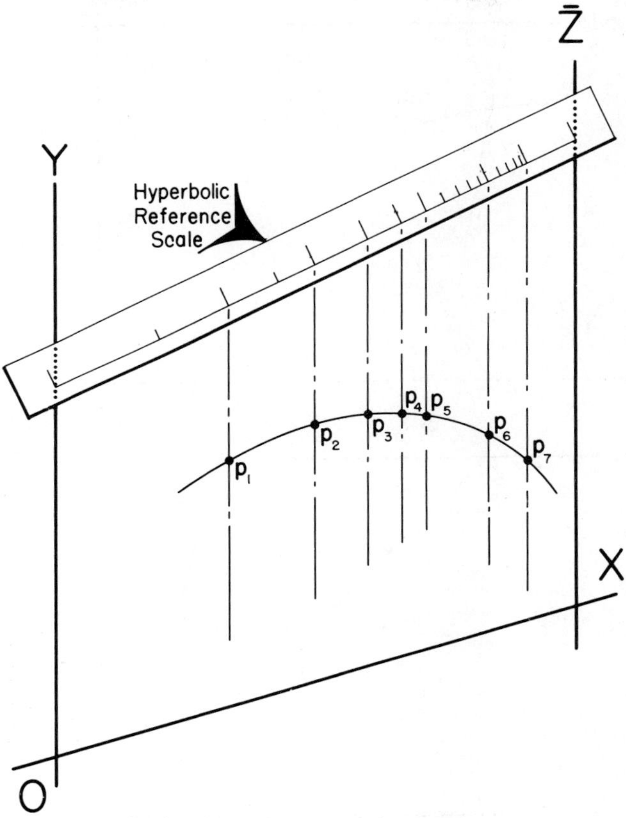

FIG. 11:8 HYPERBOLIC SCALES FOR LOCATION OF ABSCISSAS FOR CURVED SCALES

The second method is similar to the first, except that the value of $q_{\bar{Z}} = q/p$ is located on the $\bar{Z}$ axis, and a straight line is drawn to this point from the origin of the X axis. This method is shown in Figure 11:9(b). Examination of similar triangles in Figure 11:9(b) reveals the simple relationship, $q_{\bar{Z}} \times m_{\bar{Z}} = y/x$. But $m_{\bar{Z}} = rm_y$, $y = qm_y r/(p + r)$, and $x = p/(p + r)$. Hence, $q_{\bar{Z}} = q/p$.

The principal advantage of the second method lies in the plotting of curved scales in polynomials, such as the quadratic and reduced cubic. For the quadratic, $q_{\bar{Z}} = -u^2/u = -u$. For

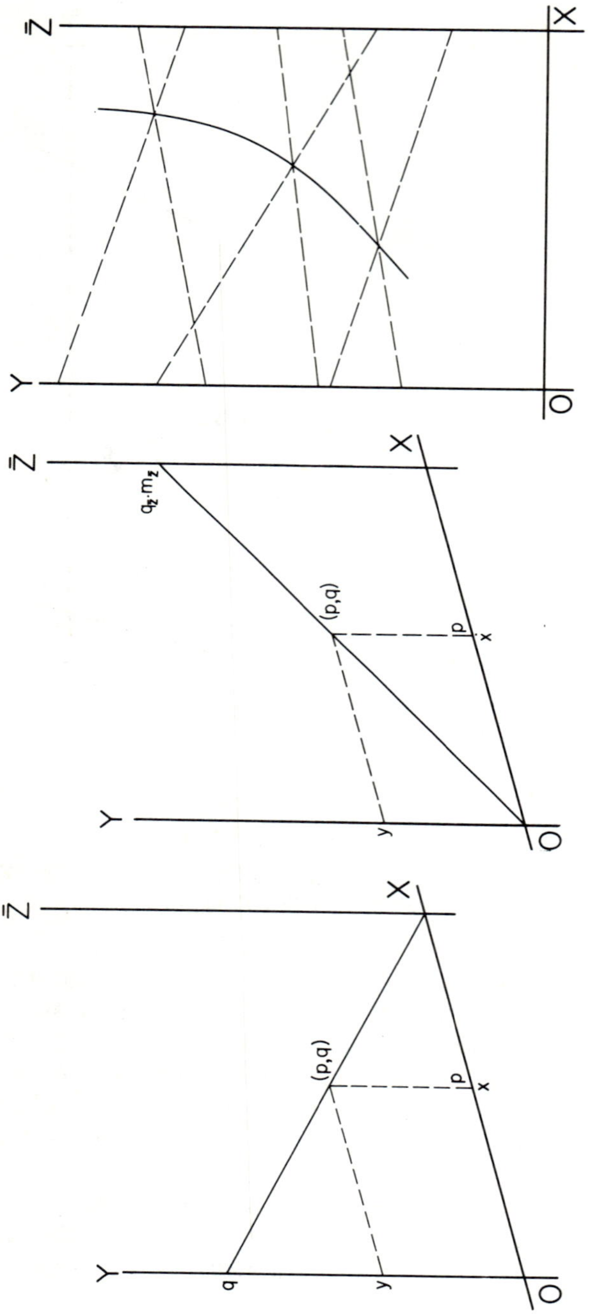

FIG. 11:9 LOCATION OF ORDINATES
a) First Method; b) Second Method; c) By Known Solutions

the reduced cubic, $q_z = -u^3/u = -u^2$. Since q_z is in each case of a lower power than q, the calculations necessary for plotting by this method are somewhat simpler.

The third method is the plotting of the cartesian ordinate, $y = qr/(p + r)$. The calculation of the values of y is greatly simplified by the use of the table of values of the factor $(1 - x)$ given in the Appendix. The modulus of a vertical scale, $m_{y'}$, is related to the modulus of the Y axis, m_y, by the factor $(1 - x)$, thus: $m_{y'} = m_y(1 - x)$.

By this method, the function which determines the ordinate of the point is $y = q.m_{y'} = q.m_y(1 - x)$. An example of the calculations necessary for plotting both x and y by this method will serve to illustrate the simplification resulting from the use of the table in the Appendix.

Suppose the curve to be plotted is the U scale for the reduced cubic equation. From the determinant, $p = u$, and $q = -u^3$. Suppose, further, that u has a range of 5 to 15, that the value of r chosen for the construction is 10, and that the value of m_y is 2.0 in. The following table shows the calculation of x and y:

u	p	p/r	x*	q	$q.m_y$	(1 - x)*	y**
5	5	0.5	0.33333	-125	-250	0.66667	-166.7
6	6	0.6	.37500	-216	-432	.62500	-270.0
7	7	0.7	.41176	-343	-686	.58824	-403.5
8	8	0.8	.44444	-512	-1024	.55556	-568.9
9	9	0.9	.47368	-729	-1458	.52632	-767.4
10	10	1.0	.50000	-1000	-2000	.50000	-1000
11	11	1.1	.52381	-1331	-2662	.47619	-1267.6
12	12	1.2	.54545	-1728	-3456	.45455	-1570.9
13	13	1.3	.56522	-2197	-4394	.43478	-1910.6
14	14	1.4	.59333	-2734	-5468	.41667	-2276.4
15	15	1.5	0.60000	-3375	-6750	0.40000	-2700.0

* From Table ** Product of two preceding columns

This method provides values of x and y which are sufficiently accurate for the most meticulous construction. On the other hand, it is obvious that the only calculations that are necessary are:

a) divide the values of p by the constant r
b) Multiply the values of q by the constant m_y;
c) Multiply the product $q.m_y$ by the factor $(1-x)$.

The only one of these calculations that is directly affected by a change in the ranges of the variables is the first, because for different ranges a different value of r would be necessary.

The fourth method for locating ordinates makes use of known solutions. The equation is solved to obtain a pair of values of the parameters plotted on the Y and $\overline{Z}$ axes for each value of the third parameter. A line drawn between these points on the extreme scales intersects the vertical through the corresponding abscissa at its ordinate value (Figure 11:9(c).

The last two methods are decidedly more tedious, in that they require more preliminary calculation than either of the first two methods for locating ordinates: consequently, they are seldom used except for constructions requiring a high degree of accuracy. Most engineering formulas leading to nomographs with curved scales are simplified approximations, applicable over limited ranges of the variables. To demand accuracy in the degree of $\pm 0.1\%$ in the construction of a nomograph for a formula that deviates as much as $\pm 1.0\%$ from experimental data is hardly reasonable. Instead, the following suggestions are made as a means of obtaining a substantial degree of accuracy without monumental effort.

1. With the use of the table there is little difference in the two methods of locating abscissas, but the second method is more accurate, especially with the assistance of a vernier or a dividing machine.
2. The first two methods of locating ordinates are substantially equivalent. But neither should be used to the complete exclusion of the other. Instead, the q or $q_{\overline{Z}}$ ordinate line should be used that most nearly approaches a right angle in its contact with the vertical abscissa line. If desired, both may be used as a check on the location of the point. Occasionally, as in the location of very high or low points on the curve, it may be desirable to abandon both methods for ordinates and use either a calculated cartesian ordinate or a known solution.
3. Although all abscissas may generally be located

with little effort, it is seldom necessary to locate ordinates for all points on the curve. Enough points are first determined to provide a smooth and accurate curve and the curve is drawn with the aid of a carefully selected guide. Intermediate points on the curve thus drawn are then indicated by the intersection of the curve with the abscissa lines previously drawn.

4. Interpolation between points on the curve must follow the same principles as those outlined in Chapter 3.

Evaluation of r: Figure 11:10 shows the nomograph for quadratic and reduced cubic equations for ranges of v and w from -10 to +10. As these scales have equal moduli, a value of r = 1 was employed for the p scale. This is not the only factor to consider, however; the range of u is equally important. Had the range of u been 5 to 15, a more satisfactory value of r would have been 10, the value used in Figure 11:11. In the latter nomograph, however, the ranges of v and w are such as to indicate this higher value of r, also.

In engineering formulas it is generally found that these two factors work together in the selection of the value of r; the ranges of v and w usually indicate a value of r that leads to satisfactory placing of the U scale. Where this is not true, a compromise value of r must be selected.

11.15 The Complete Cubic; Network Scales

The complete cubic equation is

$$u^3 + tu^2 + vu + w = 0$$

This equation contains four variables and can be represented in the usual determinant form only by designating one of the variables as a variable constant. For example, this equation might be represented in determinant form as

$$\begin{vmatrix} -\left(u^3 + tu^2\right) & u & 1 \\ v & 1 & 0 \\ w & 0 & 1 \end{vmatrix} = 0.$$

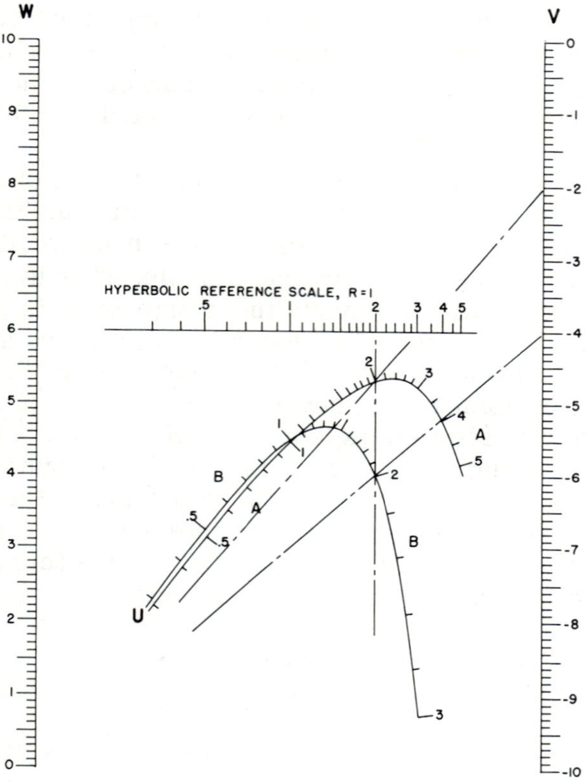

FIG. 11:10 CONSTRUCTION OF NOMOGRAPH FOR QUADRATIC AND REDUCED CUBIC EQUATIONS
Second Method Used to Locate Ordinates
$q_{\bar{z}} = q/p = -u$ (quadratic) or $-u^2$
(reduced cubic scale)

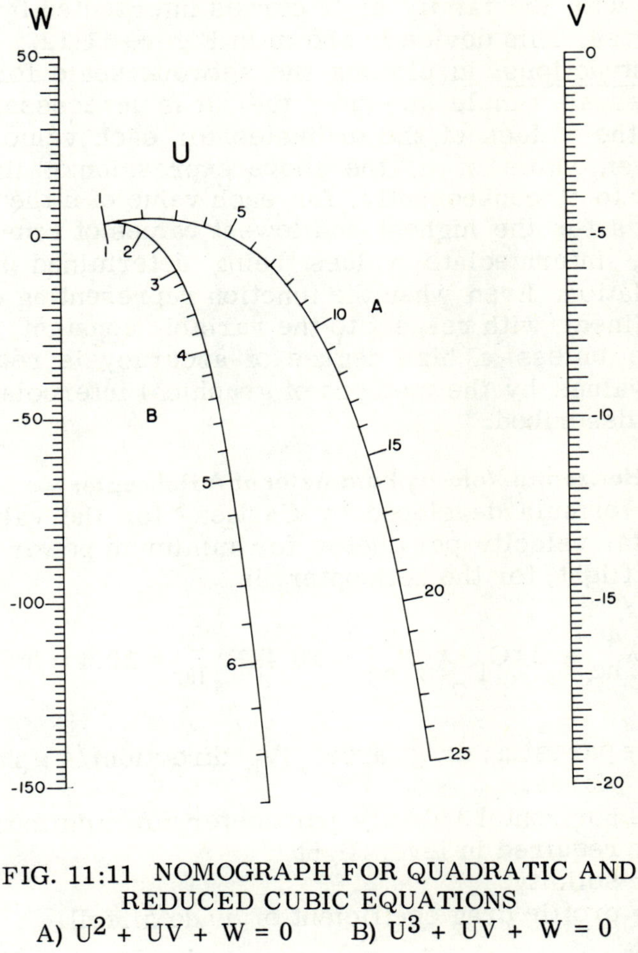

FIG. 11:11 NOMOGRAPH FOR QUADRATIC AND REDUCED CUBIC EQUATIONS

A) $U^2 + UV + W = 0$ B) $U^3 + UV + W = 0$

For each value of t a nomograph can be constructed as described in the previous sections showing the relationship between values of u, v and w. The position of the U scale will depend on the value of t. If nomographs for several values of t are superimposed, the resulting U scale becomes a "network scale," with the family of U curves intersected by a family of t curves. This device is shown in Figure 11:12.

Constructions: In plotting the network scale for a set of ordinates as simple as $-(u^3 + tu^2)$, it is unnecessary to calculate the values of the ordinates for each value of t. For any given value of u, the above expression is linear with respect to t; consequently, for each value of u the values of ordinates for the highest and lowest values of t may be calculated, intermediate values being determined by linear interpolation. Even when the function representing ordinates is not linear with respect to the variable constant, it is still possible unless a high degree of accuracy is required, to obtain values by the methods of graphical interpolation previously described.

11.16 Horizontal Velocity Parameter of A Helicopter

The formula developed by Castles [5] for the value of the horizontal velocity parameter for minimum power required in level flight, for the helicopter, is

$$24F\lambda_{hc}^{4} + 3\sigma C_{D_o}\lambda_t\lambda_{hc}^{3} + 10.4kF\lambda_{hc}^{2} - 10.4 = 0$$

where F = parasite drag area (V_h direction)/4 × rotor disk area,

λ_{hc} = horizontal velocity parameter for minimum power required in level flight,

σ = solidity,

C_{D_o} = profile drag coefficient of blade airfoil,

k = a constant, whose value for this case may be taken as 7, and

[5] Castles, Walter, Jr., A Direct Method of Estimating the Performance of a Helicopter in Powered Flight, J. Aero. Sci. 12:477 (1945).

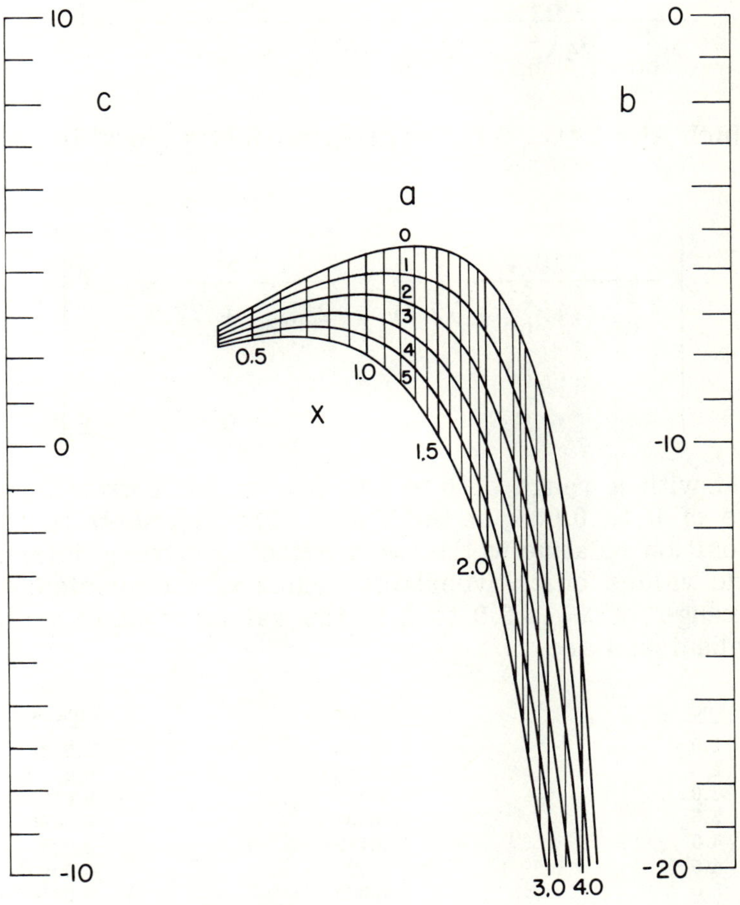

FIG. 11:12 NOMOGRAPH FOR COMPLETE CUBIC EQUATION

$$\lambda_t = \frac{\text{tangential velocity of blade tips}}{\text{vertical component of velocity along flight path.}}$$

This equation may be written

$$\frac{10.4}{\lambda_{hc}^2 \left(24\lambda_{hc}^2 + 72.8\right)} - \frac{A' \lambda_{hc}}{24\lambda_{hc}^2 + 72.8} - F = 0,$$

in which $A' = 3\sigma C_{D_o} \lambda t$. The determinant form of this equation is

$$\begin{vmatrix} \dfrac{10.4}{\lambda_{hc}^2 (24\lambda_{hc}^2 + 72.8)} & \dfrac{\lambda_{hc}}{24\lambda_{hc}^2 + 72.8} & 1 \\ A' & 1 & 0 \\ F & 0 & 1 \end{vmatrix} = 0$$

A', with a range of 0 to 1.0, lies on the $\bar{Z}$ axis; F, with a range of 0 to 0.005, on the Y axis. The λ_{hc} scale is curved, its position relative to the two vertical axes being determined by the values of its hyperbolic ordinates and abscissas. For the range of λ_{hc}, 2.0 to 5.0, the values of these ordinates and abscissas are

λhc	Ordinate	Abscissa
2.0	0.01540	0.01185
2.5	0.00782	0.01120
3.0	0.00400	0.01040
3.5	0.00231	0.00955
4.0	0.00142	0.00878
4.5	0.00092	0.00806
5.0	0.00062	0.00744

If the scale for A' is to have a range of 0 to 1.0 with subdivisions for every interval of 0.01 spaced 0.1 inch apart, the length of the scale will be 10 inches. Its modulus will be 10 inches. The F scale, over the range 0 to 0.005, with 0.2 inch subdivisions for every interval of 0.0002, will have a length of 5 inches. Its modulus will be 5/.005 = 1000. These moduli indicate a value of r = 10/1000 = 0.01. Abscissas fall in the neighborhood of this same value; consequently, the curve will lie approximately halfway between the A' and F scales, if the r .01 scale is used.

Were the modulus of the F scale doubled to increase the accuracy of its subdivision, the resulting value of r would be 0.005, and the λ_{hc} scale would lie nearer the A scale. The resulting equal lengths of the A and F scale might enhance the appearance of the nomograph; a shift in the other direction is objectionable, because it decreases the accuracy of subdivision of the λ_{hc} scale in its upper range.

Not all of the values on the λ_{hc} scale were included in the calculation of ordinates and abscissas. In this case, intermediate points were determined by graphical interpolation. The nomograph appears in Figure 11:13.

11.17 Horizontal Velocity of Helicopter in Level Flight

Castles' formula for horizontal velocity parameter for top speed in level flight is

$$16\,F\lambda_{hm}^{4} + 3\sigma C_{D_o}\lambda_t\lambda_{hm}^{3} + 20.8\,kF\lambda_{hm}^{2} +$$

$$(\sigma C_{D_o}\lambda_t^{3} - 16\lambda_p)\lambda_{hm} + 20.8 = 0$$

in which λ_{hm} = the horizontal velocity parameter for top speed in level flight,
λt = tangential blade tip velocity parameter
λ_p = the power parameter,

and the other terms carry the meanings defined in the preceding section. This equation is written

$$\frac{16\left(F\lambda_{hm}^{4} + 9.1\,F\lambda_{hm}^{2} + 1.3\right)}{\lambda_{hm}} + A\lambda_{hm}^{2} + C = 0$$

or in the determinant form,

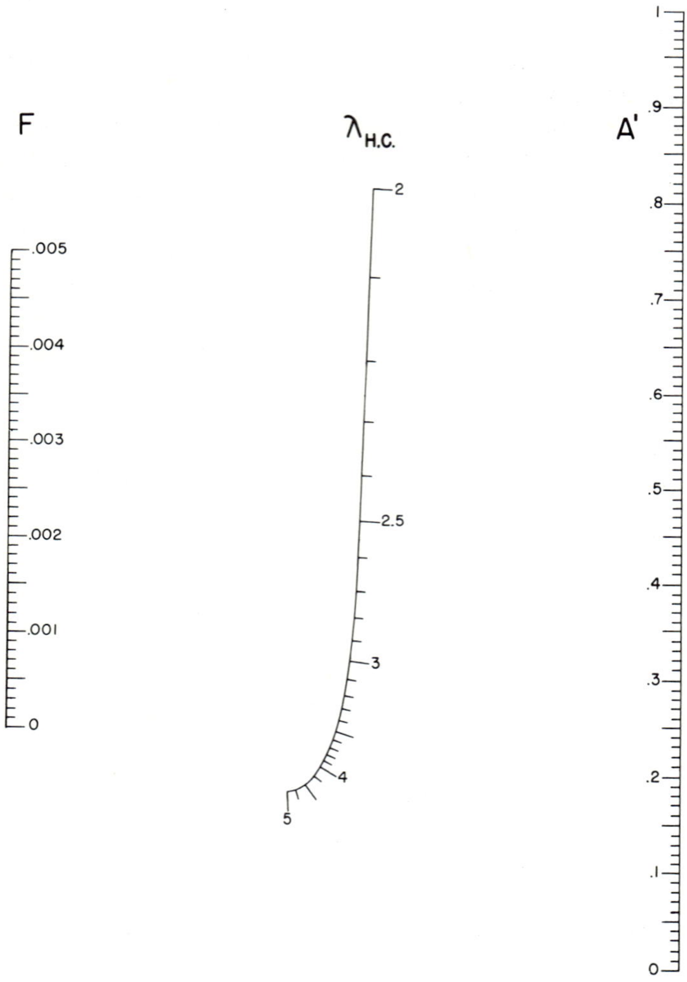

FIG. 11:13 NOMOGRAPH FOR HORIZONTAL VELOCITY PARAMETER OF HELICOPTER FOR MINIMUM POWER REQUIRED IN LEVEL FLIGHT

$$\begin{vmatrix} \dfrac{16\left(F\lambda_{hm}^{4} + 9.1\ F\lambda_{hm}^{2} + 1.3\right)}{\lambda_{hm}} & \dfrac{\lambda_{hm}^{2}}{} & 1 \\ A' & 1 & 0 \\ C' & 0 & 1 \end{vmatrix} = 0.$$

In these equations A' has the meaning assigned in the preceding section, and

$$C' = \sigma C_{D_o} \lambda_t^3 - 16\lambda_p.$$

In this nomograph, A' lies on the $\overline{Z}$ axis, and C' on the Y axis. A' with a range of 0 to 0.1 may be represented with sufficient accuracy on a two-inch scale, the modulus being 2/0.1 = 20 inches. The range of C is 0 to -50 and requires a ten-inch scale, the modulus being 10/50 = 0.2 inch. These two values would indicate a value of r = 20/0.2 = 100.

The range of λ_{hm} is 5 to 15, with a consequent range of the abscissas of the λ_{hm} scale of 25 to 225. These abscissas are well separated on the r100 scale, but would be too crowded on scales of higher r values. Here again, the value of r indicated by the moduli of the Y and $\overline{Z}$ axes is in remarkable agreement with the value required by the values of the abscissas of the curved scale.

Ordinates of the λ_{hm} scale are functions of both λ_{hm} and F. As, for any fixed value of λ_{hm}, the value of the ordinate is a linear function of F, it is unnecessary to calculate ordinates for all values of F. Ordinates for the higher and lower values of F may be calculated, and intermediate points may be determined by interpolation. The completed nomograph is in Figure 11:14.

11.18 Vertical Velocity of Helicopter

In the same paper with the two preceding formulas, Castles gives as the value of the vertical velocity parameter for maximum rate of climb,

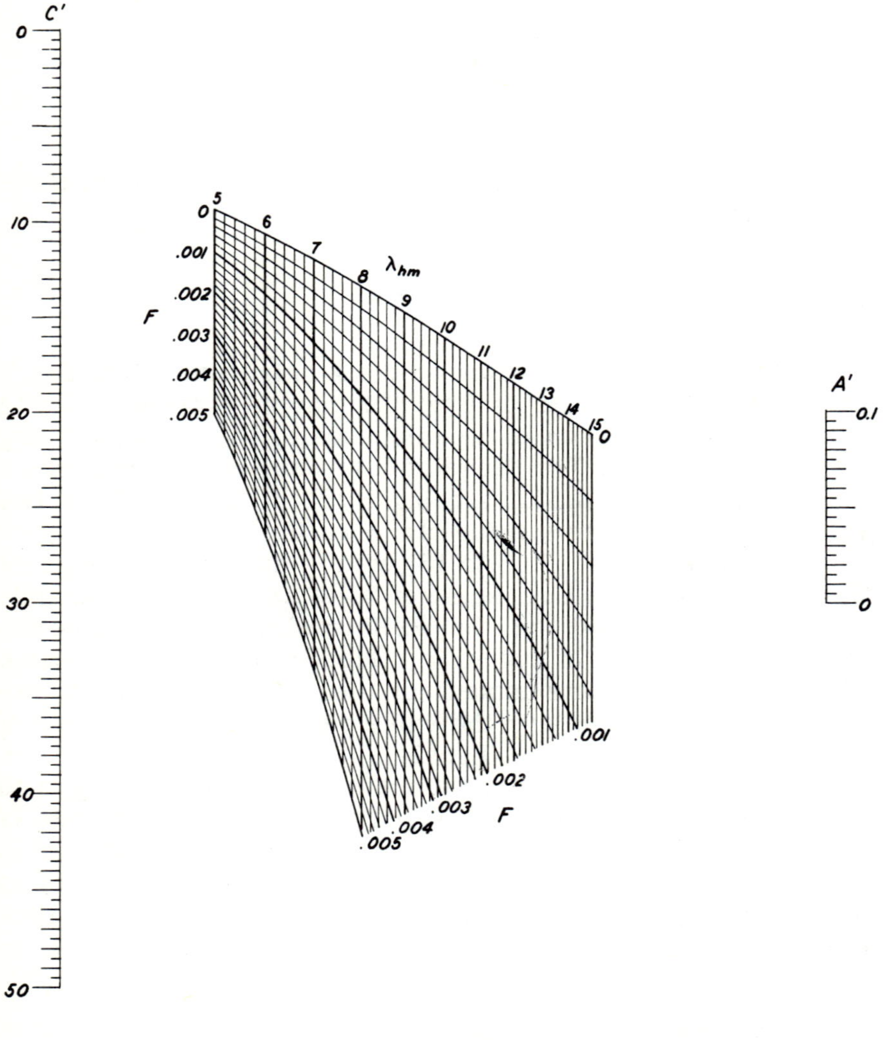

FIG. 11:14 NOMOGRAPH FOR VELOCITY PARAMETER OF HELICOPTER FOR TOP SPEED IN LEVEL FLIGHT

$$\lambda_{v_{max}} = Y - \frac{1.3(1 + kFY^2)}{\left(Y^2 + \lambda_{hc}^{\;2}\right)^{1/2}} -$$

$$kFY^3 - FY^2\lambda_{hc} - kFY\lambda_{hc}^{\;2} \qquad (11.9a)$$

where

$$Y = \lambda_{p_{max}} - \left(\frac{\sigma C_{D_o}}{16}\right)\lambda_t^{\;3} - \left(\frac{3\sigma C_{D_o}}{16}\right)\lambda_t\lambda_{hc}^{\;2} - F\lambda_{hc}^{\;3}$$

Equation 11.9a may be written,

$$f_1\left(Y,\lambda_{hc}\right) + F\, f_2\left(Y,\lambda_{hc}\right) + \lambda_{v\,max} = 0 \qquad (11.9b)$$

where

$$f_1\left(Y,\lambda_{hc}\right) = -\,Y + \frac{1.3}{\left(Y^2 + \lambda_{hc}^{\;2}\right)^{1/2}}$$

and

$$f_2\left(Y,\lambda_{hc}\right) = \frac{1.3\,kY^2}{\left(Y^2 + \lambda_{hc}^{\;2}\right)^{1/2}} + kY^3 + Y^2\lambda_{hc} + kY\lambda_{hc}^{\;2}$$

The determinant form of Equation 11.9b is

$$\begin{vmatrix} -f_1(Y,\lambda_{hc}) & f_2(Y,\lambda_{hc}) & 1 \\ F & 1 & 0 \\ \lambda_{v_{max}} & 0 & 1 \end{vmatrix} = 0$$

This nomograph is interesting from the standpoint that both ordinates and abscissas of points on the network scale are functions of two variables. The details of selecting a value of r and of plotting the scales are no different from those of the two preceding nomographs and are obvious from an examination of the nomograph itself (Figure 11:15).

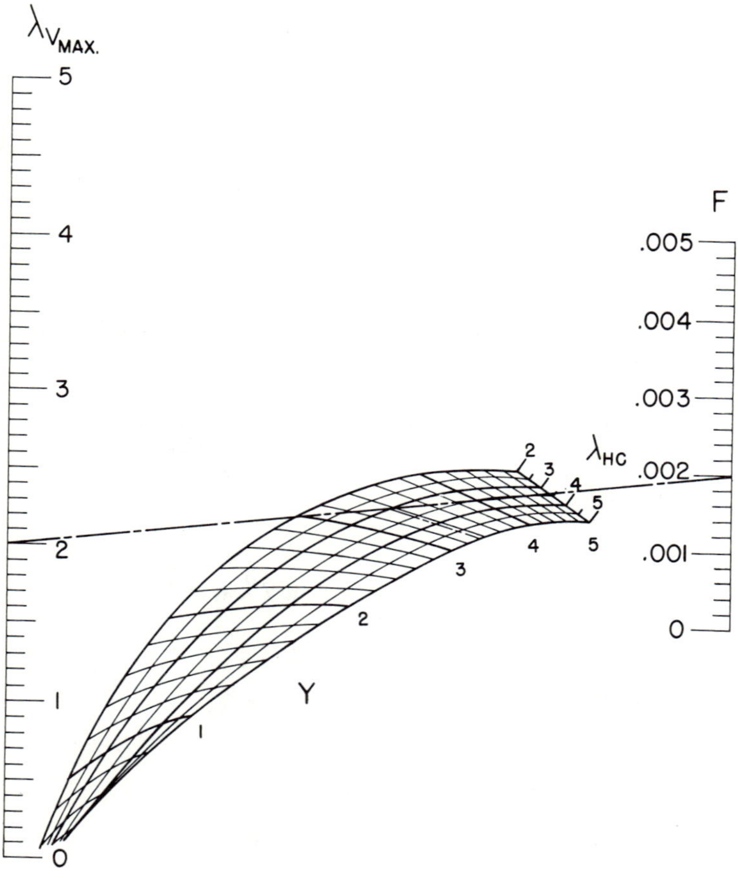

FIG. 11:15 NOMOGRAPH FOR VERTICAL VELOCITY PARAMETER OF HELICOPTER FOR MAXIMUM RATE OF CLIMB

11.19 The Complete Quartic

In the treatment of the foregoing examples, the original polynomial in four variables was put into the determinant form of three variables by incorporating one of the coefficients as a variable constant in the ordinates of the power term scale. The resulting nomograph contains a network scale, in which the abscissas represent the power term, and the ordinates, the variable constant.

Occasionally it might be feasible to apply this principle to an equation in five variables, assigning two of the coefficients as variable constants in the expression for the ordinates of the power term scale. It would then be necessary to construct a series of nomographs, one for each value of one of the coefficients, the other coefficient being represented by means of a network scale on each of the nomographs. This would be practical only when one of the coefficients had a limited number of possible values.

The following methods of representing the complete quartic are based rather upon the division of the original equation into parametric form, which involves the introduction of a sixth parameter P. Each of the two parametric equations will contain the power term, the parameter P and two of the coefficients. The nomographs for the two equations will be so adjoined that simultaneous solution will eliminate the parameter P and yield the desired root of the equation. Both methods will be applied to the general quartic,

$$x^4 + ax^3 + bx^2 + cx + d = 0, \tag{11.10}$$

but they may also be applied to other polynomials in five variables.

First Method: The general equation for the quartic is written in parametric form as

$$\left.\begin{aligned} (x^4 + d) + bx^2 + P &= 0 \\ -ax^3 - cx + P &= 0 \end{aligned}\right\}$$

and

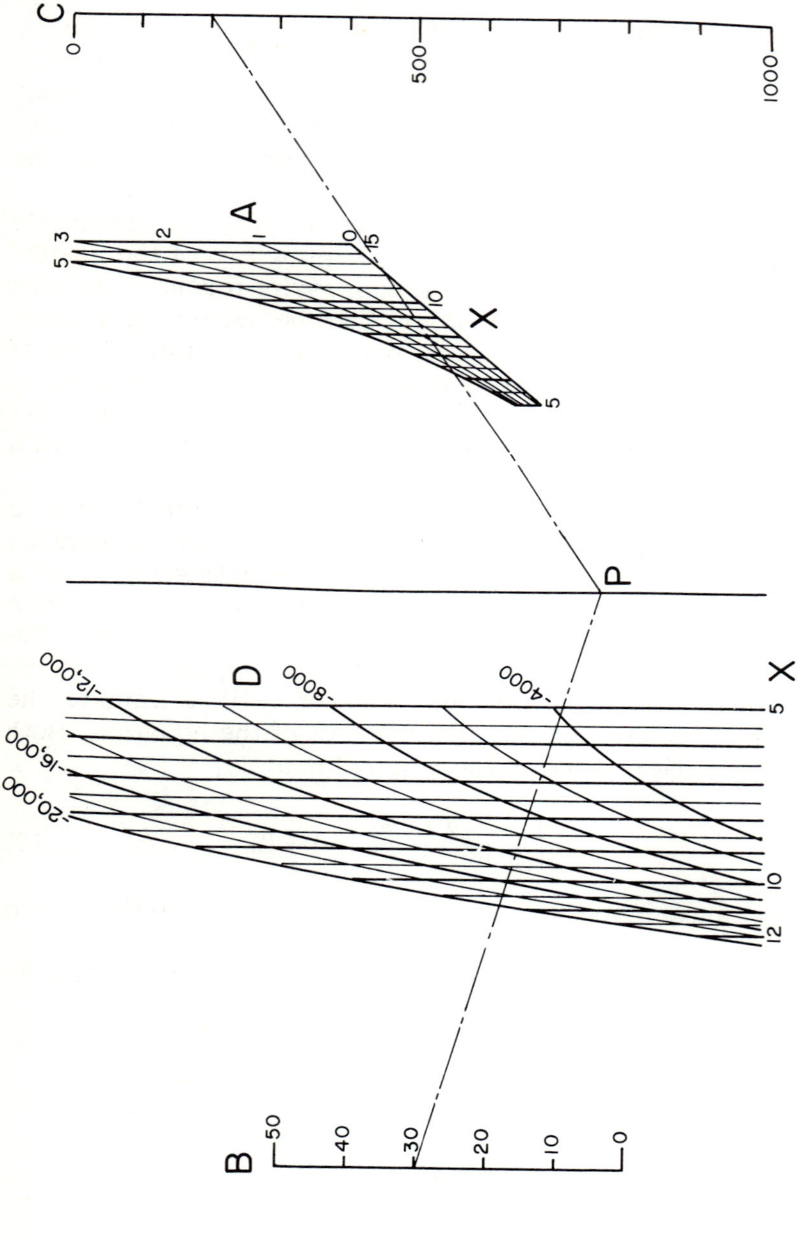

FIG. 11:16 NOMOGRAPH FOR COMPLETE QUARTIC EQUATION, FIRST FORM

the determinant forms of these are

$$\left.\begin{aligned} &\begin{vmatrix} -(x^4 + d) & x^2 & 1 \\ b & 1 & 0 \\ P & 0 & 1 \end{vmatrix} = 0 \\ \text{and}\quad &\begin{vmatrix} ax^3 & x & 1 \\ -c & 1 & 0 \\ P & 0 & 1 \end{vmatrix} = 0 \end{aligned}\right\}$$

In both parts of the nomograph, P lies on the Y axis. The values of r for the two parts are so selected that the modulus of the P-scale is the same for both. In the example shown (Figure 11:16) the direction of the X axis for one part of the nomograph is reversed so that the P scale will lie in the center of the combined figure. The P scale is not ruled in the final drawing, although a temporary ruling is necessary for plotting the x scales during construction.

Problems are solved in the following manner: Place pivot pins at the values of b and c on their respective scales. Lay straight edges from these pins to a common point on the P scale until the straight edges simultaneously intercept the a and d lines on the network scales at identical values of x. The value of x so obtained is a root of the equation.

Second Method: The general equation is factored into parametric form, with P as the variable constant:

$$\left.\begin{aligned} &(x^2 - P) + ax + b = 0 \\ \text{and}\quad &Px^2 + cx + d = 0 \end{aligned}\right\}$$

or, in determinant form,

$$\left.\begin{array}{l} \begin{vmatrix} -(x^2 - P) & x & 1 \\ a & 1 & 0 \\ b & 0 & 1 \end{vmatrix} = 0 \\ \text{and} \\ \begin{vmatrix} -Px^2 & x & 1 \\ c & 1 & 0 \\ d & 0 & 1 \end{vmatrix} = 0 \end{array}\right\}$$

In the two parts of the nomograph, b and d lie on the Y axis; a and c, on the Z axis. Both a and c are coefficients of the same power of x; hence, if both parts of the nomograph are constructed on vertically colined coordinates with the same value of r, the abscissas of the x scales will be identical. The ordinates of the x scale will be linear functions of P, and will give rise to network scales in x and P. (Figure 11:17).

Problems are solved as follows: Lay straight edges from a to b and from c to d. The value of x at which the straight edges simultaneously intercept identical values of x and P is a root of the equation.

11.20 Heterogeneous Formulas

The methods given here for polynomials—the quadratic, cubic and quartic—are not confined to the polynomials at all, but may be applied equally well to any formula whose determinant is of the form that gives rise to curved scales. The formula discussed next, although it is not a polynomial in the strict sense of the word, is treated by exactly analogous methods.

The Rehbock formula[6] for discharge over a sharp-crested weir is:

[6] Rehbock, Theodore, Discussion of Precise Weir Measurements, Trans. Am. Soc. C. E., 93:1148 (1929).

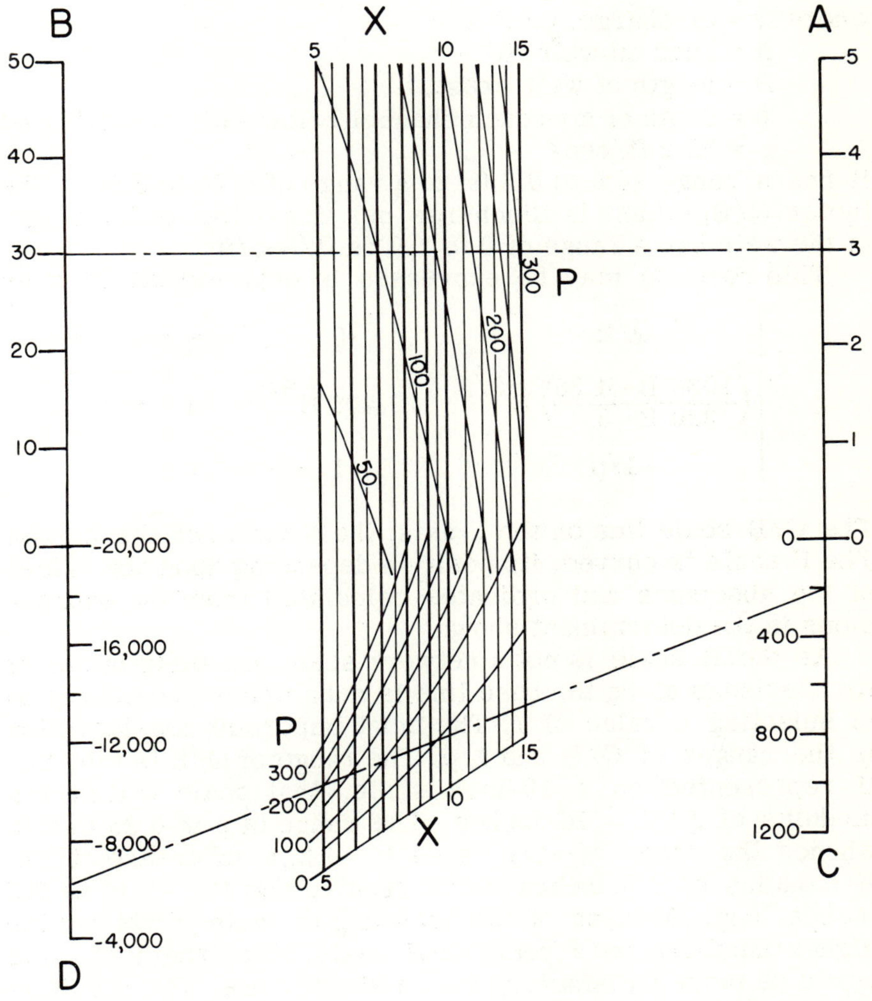

FIG. 11:17 NOMOGRAPH FOR COMPLETE QUARTIC EQUATION, SECOND FORM

$$Q = \frac{2}{3}\sqrt{2g}\left(0.605 + \frac{1}{320H - 3} + 0.08\frac{H}{p}\right)BH^{3/2}$$

where Q = discharge, cu ft/sec
H = head on weir, ft
B = length of weir crest, ft
p = depth of approach channel below weir crest, ft, and
g = 32.2 ft/sec^2

H has a range of 0 to 0.5 ft; p, a range of 0.25 to 5.0 ft. The factor Q/B, which is discharge per linear foot of the length of the weir has a range of 0 to 1.0 cu ft/sec/ft.

This formula may be expressed in determinant form as

$$\begin{vmatrix} Q/B & 0 & 1 \\ \left(\frac{1036\ H - 4.36}{320\ H - 3}\right) H^{3/2} & 0.428\ H^{5/2} & 1 \\ -1/p & 1 & 0 \end{vmatrix} = 0$$

The Q/B scale lies on the Y axis; the p scale, on the Z axis. The H scale is curved, its position depending upon the values of the abscissas and ordinates calculated from the expressions in the determinant shown.

As the H scale is not a network scale, the distribution of its abscissas along the base line is not a prime consideration in selecting a value of r. The most important consideration is the ranges of Q/B and 1/p. The range of Q/B is 0 to 1.0; if represented on a 10-inch scale, that scale will have a modulus of 10/1 = 10 inches. The range of p is 0.25 to 5.0, whence the range of -1/p is -4 to -0.2, a difference of 3.8. A modulus of 2.5 inches would require that the scale be 9.5 inches long. Because of the crowding of scale points for the higher numbers on a reciprocal scale, this length of scale would be more satisfactory than a shorter one. The resulting value of r is 2.5/10 = 0.25.

For the most accurate construction, all scale points are plotted from calculated values. An examination of the background of this formula, however, reveals discrepancies between observed and calculated values of an order which are

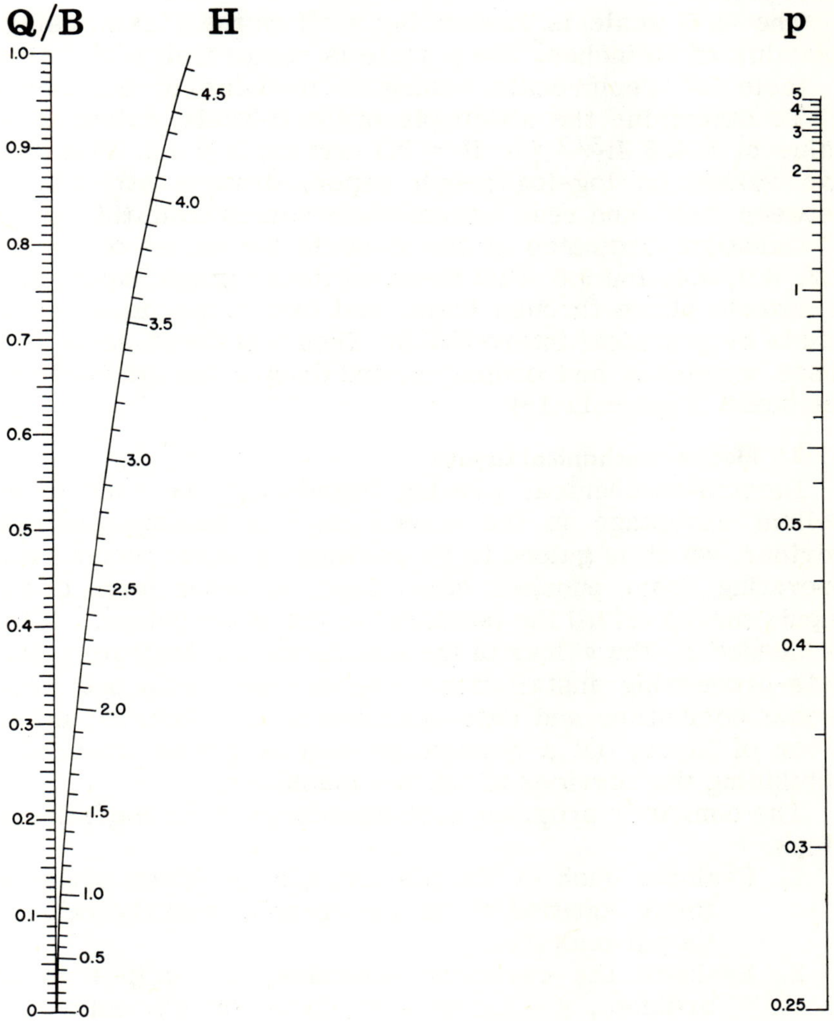

FIG. 11:18 NOMOGRAPH FOR REHBOCK FORMULA

too great to warrant a high degree of accuracy in the nomograph. A nomograph of sufficient accuracy can be produced as follows:

The Q/B scale is linear; lay it off on the Y axis, using a modulus of 10 inches. The p scale is reciprocal; plot it from a table of reciprocals, using a modulus of 2.5 inches.

To determine the abscissas of the H-scale, calculate the value of 0.428 $H^{5/2}$ for H = 0.1 and for H = 0.5. Plot these two values on log-log graph paper, draw a straight line between them, and read intermediate values from this graph.

Calculate ordinates of the H-scale for values of H = 0.1, 0.2, 0.3, 0.4, and 0.5. Plot these on linear graph paper, draw a smooth curve through them, and determine intermediate points by graphical interpolation. Then plot the H-scale from these abscissas and ordinates, and draw a smooth curve for the scale (Figure 11:18)

11.21 Electro-mechanical Layout

Electro-mechanical plotting boards may be used to excellent advantage in the construction of nomographs. The scriber, which is guided to its position by servo mechanisms operating from punched card, tape, or other form of data input, marks off all the positions of the scale points. All that is needed is the values of the abscissas and ordinates. Most data-processing installations are equipped with both electronic computers and data-operated plotting boards, and the labor of laying out a nomograph may be greatly reduced by combining the services of the two machines.

The computer program is designed to perform the following steps:

1. Evaluate each of the p's and q's, as given in the defining equation of the nomograph, from the values of its parameters.
2. Evaluate the cartesian abscissa, $x = p/(p + r)$, and ordinate, $y = qr/(p + r)$, from the p's and q's, for the selected value of r. These are the normalized abscissas and ordinates of the scale points.
3. Multiply the values of the x's and y's by the desired width and height, respectively, of the nomograph.

This gives the plotting values for rectangular coordinates.

4. If oblique coordinates are to be used for the sake of symmetry, multiply each abscissa by the slope of the X-axis and add this factor to the value of the corresponding ordinate obtained in Step 3.
5. Print out on data input medium, the abscissas of Step 3 and the ordinates of Step 4 against the values of the parameters for each scale point. This print-out is then fed to the plotting board for laying out the nomograph.

The resulting layout is converted into a finished nomograph by drawing the scale strokes and lines on the plotted points, adding the numerical designations, and inserting symbols and legends.

EXERCISES

1. Write defining equations for the following examples for both hyperbolic and cartesian coordinates:

1) $\cos 2x = \cos^2 x - \sin^2 x$

2) $\sin 3x = 3 \sin x - 4 \sin^3 x$

3) $\sin 3x = 3 \sin x \cos^2 x - \sin^3 x$

4) $\cos 3x = 4 \cos^3 x - 3 \cos x.$

2. Examples in this chapter are worked with hyperbolic coordinates. For each one, convert the defining equation to the form required for cartesian coordinates and modify the resulting determinant (using the method of Section 0.5) to accommodate the ranges given for the parameters.

3. The weight of rubber in an innertube is W = Asd', where A is the average of the inside and outside areas of the tube, s is the wall thickness and d' is the density of the rubber. A may be computed directly from the inside diameter d, outside

diameter D and wall thickness s, using the formula for the area of a torus (Section 11.2). In this formula D will be outside diameter minus s ($D_o - s$), and d will be inside diameter plus s ($D_i + s$).

Make a composite nomograph for W in terms of $D\Big|_{15}^{45}$ inch, $d\Big|_{10}^{40}$ inch, $s\Big|_{1/16}^{1/4}$ inch (by 16ths), $d'\Big|_{.032}^{.050}$ lb/cu inch.

4. For aperture illumination (by microwave radiation of constant amplitude and phase) of a parabolic reflector surface the various properties, diameter (D), depth (d), focal length (F), half-power angle (θ), and gain (G) are related by

$$Fd = \left(\frac{D}{4}\right)^2$$

$$G_{db} = 20 \log \left(\frac{\pi d}{\lambda}\right)$$

and

$$\theta = \frac{50.5\lambda}{D},$$

where λ is the wave length of the microwave radiation. Prepare a composite nomograph for these three formulas covering the ranges $D\Big|_{0.5}^{10}$ ft, $d\Big|_{1}^{40}$ inch, $G\Big|_{15}^{65}$ db (decibels), $F\Big|_{1}^{100}$ inch, $\lambda\Big|_{0.5}^{10}$ cm, and $\theta\Big|_{0.1}^{25}$ degrees.

5. The range of critical temperatures for the hydrocarbons up to M.W. 125 is 190 to 640°K. The corresponding values of the critical pressures range from 24 to 64 atmospheres. For a temperature range of 200 to 500°K, construct a nomograph for the Berthelot equation of state, $\Delta c_p = 5.03\, T_c^3/P_c T^3$. Consult the tables in <u>Industrial and Engineering Chemistry</u> 35:1303, and mark the positions of the following

gases on the T_c and P_c scales: methane, ethane, propane, butane, pentane, hexane, heptane, and octane.

6. In the design of countercurrent absorbers, , the efficiency of the operation is taken as

$$E = \frac{\left(A^{N+1} - A\right)}{\left(A^{N+1} - 1\right)}$$

where A is the absorption factor and N, the number of theoretical plates in the system. Construct a nomograph using the determinant form

$$\begin{vmatrix} A^{N+1} & -A+1 & 1 \\ 1 & E-1 & 1 \\ 1 & 0 & 1 \end{vmatrix} = 0$$

Ordinates of the A-N scale are functions of both A and N; consequently, a network scale is the result. The bottom row of the determinant defines a single point.

Ranges are $A\left|\begin{matrix}10\\ \\1\end{matrix}\right.$, $N\left|\begin{matrix}40\\ \\1\end{matrix}\right.$, and $E\left|\begin{matrix}99.9999\\ \\0\end{matrix}\right.$ %.

Compare the resulting nomograph with that given by Mapstone. (see Ind. Eng. Chem. 37:1222, 1945).

7. The intrinsic strength of a cord or fiber is given by $D^2I/1{,}273{,}000 = P_k$, where P_k is the knot pull, D the diameter, and I the intrinsic strength. Ranges are $D\left|\begin{matrix}40\\ \\1\end{matrix}\right.$ mils (1 mil = 0.001 inch), $P_k\left|\begin{matrix}10\\ \\1\end{matrix}\right.$ pounds, and $I\left|\begin{matrix}100{,}000\\ \\10{,}000\end{matrix}\right.$ lb/sq. inch.

Convert to logarithmic form and construct the nomograph.

8. The Castles formula (see J. Aero. Sci., 12-4:477, 1945) for vertical velocity parameter, λV_v, for vertical

climb of a helicopter is

$$\lambda V_v = C'(1 + 7F) - 1.3/C'.$$

Construct the nomograph showing ranges of

$$\lambda V_v \Big|_{0}^{3}, \quad C' \Big|_{-50}^{-18}, \text{ and } F \Big|_{0}^{.005}.$$

9. Castles (Ibid) also gives the formula for Y used to obtain the maximum rate of climb of the helicopter:

$$Y = C' + (A'/16)\,\lambda_{hc}^{2} - F\lambda_{hc}^{3}.$$

Design a nomograph incorporating the ranges $Y \Big|_{0}^{3}$, $C' \Big|_{-50}^{0}$, $\lambda_{hc} \Big|_{1}^{5}$, $F \Big|_{0}^{.005}$, and $A' \Big|_{0}^{0.1}$. The nomograph may be constructed with a network scale in F and λ_{hc}, and a scale for A' on one side of a pivot line; scales for C and Y on the other side. Index lines for the two sides meet at the pivot line. Other designs, however, are also possible. Compare your nomographs for Exercises 8 and 9 with those shown in Castles' article.

10. Various statistical measures require the averaging, on a weighted basis, of the items in a population; for instance,

$$A^2 = \frac{\underline{w}_1\underline{x}_1^{2} + \underline{w}_2\underline{x}_2^{2} + \underline{w}_3\underline{x}_3^{2} + \cdots}{w_1 + w_2 + w_3 + \cdots\cdots}.$$

Using the principle of Section 4.3, construct a nomograph for the accumulation of such averages.

11. Numerous engineering formulas involve products of several terms, each raised to a fractional power: $P = u^a v^b w^c x^d \ldots$ Express this product in logarithmic form. Using the method of Section 4.4, construct a nomograph for obtaining products of this nature. Compare the nomograph with those of McMillen (Ind. Eng. Chem. 30:71, 1938) and Burrows (Ibid, 38:586, 1946).

12. The Spitzglass formula for rate of flow of low pressure gas through pipes is

$$q_h = 3550 \sqrt{\frac{h_w d^5}{GL(1 + 3.6/d + 0.03d)}}$$

where h_w is pressure drop, inches of water (0 to 25), d is diameter, inches (0 to 12), G is specific gravity of gas relative to air (0.6 to 3.0), L is length of pipe, feet (10 to 1000), and q_h is rate of flow, cubic feet per hour at standard conditions (14.7 psi abs, and 60 F). Construct a nomograph for the solution of this formula. The factor $d^5/(1 + 3.6/d + 0.03d)$ may, be represented on a single scale.

13. The nomograph for the complete cubic equation in Figure 11:12 includes a range of 0 to 5 for the parameter a. Increase the range of a in this nomograph to include -5 and +15.

chapter 12

GENERAL HYPERBOLIC COORDINATES IN CONSTRUCTION OF NOMOGRAPHS

The preceding chapters have described the construction of nomographs by the use of a form of coordinates termed semi-hyperbolic, consisting of one principal axis (the X axis) subdivided hyperbolically and another principal axis (the Y axis) subdivided linearly, as in cartesian coordinates. A more general form of plane hyperbolic coordinates, consisting of two principal axes both of which are subdivided hyperbolically, will now be described.[1] The properties of this coordinate system are derived in Chapter 14 and will be freely quoted here without derivation. Two of the problems of nomography that are simplified by application of this geometrical device, are the constructing of the V-type nomograph, and the problem of fitting nomographs to a rectangle.

HYPERBOLIC PLANE COORDINATES

12.1 Location of Points

Both the X and Y axes of the general hyperbolic coordinate system are finite in length and divided hyperbolically into parts whose lengths are $x = p_x/(p_x + r_x)$ and $y = p_y/(p_y + r_y)$, the lengths of the axes being taken as unity, although these lengths may be different for the two axes.

Ordinates are located, as previously, by ordinate lines passing through the terminus of the X axis and the points p_y on the Y axis, where the value of p_y is the value of the ordinate (Figure 12:1). Abscissas are located similarly by

[1] Adapted from article published in Ind. Eng. Chem., 43:1193 (1951).

abscissa lines passing through the terminus of the Y axis and the points p_x on the X axis, the value of p_x being the value of the abscissa. The values of r_x and r_y are independent of one another.

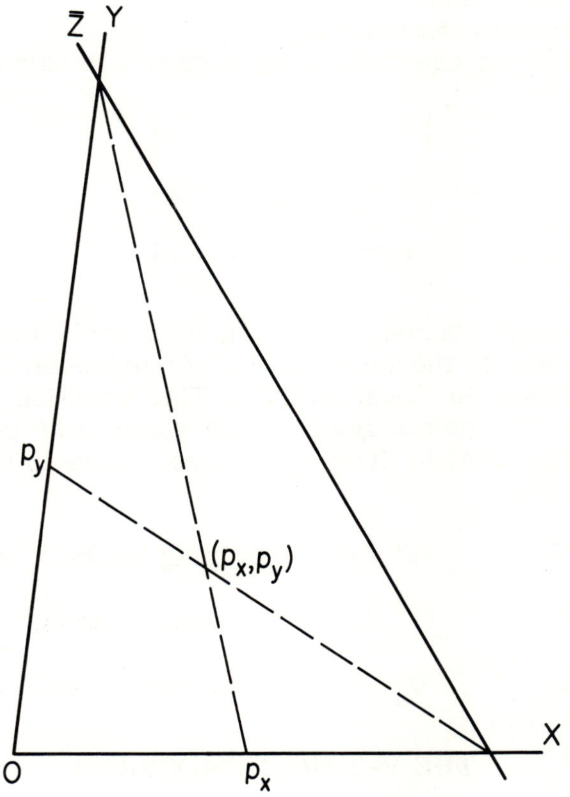

FIG. 12:1 LOCATION OF THE POINT p_x, p_y ON HYPERBOLIC COORDINATE SYSTEM

There is an auxiliary axis, the $\bar{Z}$ axis, for the location of points with infinite abscissas and/or ordinates. This axis passes through the termini of the X and Y axes, having its origin at its x intercept and its terminus at its y intercept. It is thus finite in length, that length also being taken as unity. This axis is subdivided into segments such that $y = p_{\bar{z}}/(p_{\bar{z}} + r_{\bar{z}})$. In order to simplify relations involving the coordinates on the $\bar{Z}$ axis and the coordinates of other points in the plane, it is desirable to define the hyperbolic constant of this axis, $r_{\bar{z}}$, in terms of the hyperbolic constants of the principal axes, thus $r_{\bar{z}} = r_y/r_x$.

12.2 Application to Nomography

The defining equation of any nomograph having a straight index line is

$$\begin{vmatrix} p_y(u) & p_x(u) & 1 \\ p_y(v) & p_x(v) & 1 \\ p_y(w) & p_x(w) & 1 \end{vmatrix} \qquad (12.1)$$

the p_x's being abscissas and the p_y's ordinates of the u, v, and w scales of the nomographs, as indicated by the letter in parentheses in Equation 12.1. This equation is a general expression for points lying in the plane. For points (or, in nomography, scales) lying on the axes, rows in the determinant becoming

$\begin{vmatrix} 0 & p_x & 1 \end{vmatrix}$ for points lying on the X axis,

$\begin{vmatrix} p_y & 0 & 1 \end{vmatrix}$ for points on the Y axis,

$\begin{vmatrix} p_{\bar{z}} & 1 & 0 \end{vmatrix}$ for points on the $\bar{Z}$ axis.

THE V-TYPE NOMOGRAPH

12.3 Lens Formula

If the substitutions, $p_y(u) = s_i$, $p_x(u) = 0$, $p_y(v) = p_x(v) = s_f$, $p_y(w) = 0$, and $p_x(w) = s_o$, are made in Equation 12.1, the resulting determinant

$$\begin{vmatrix} s_i & 0 & 1 \\ s_f & s_f & 1 \\ 0 & s_o & 1 \end{vmatrix} = 0$$

may be taken as the defining equation of a nomograph for the familiar lens formula, $1/s_f = 1/s_o + 1/s_i$, in which s_i, s_o, and s_f are respectively the distance of the image, of the object, and the focal length, respectively, of the lens.

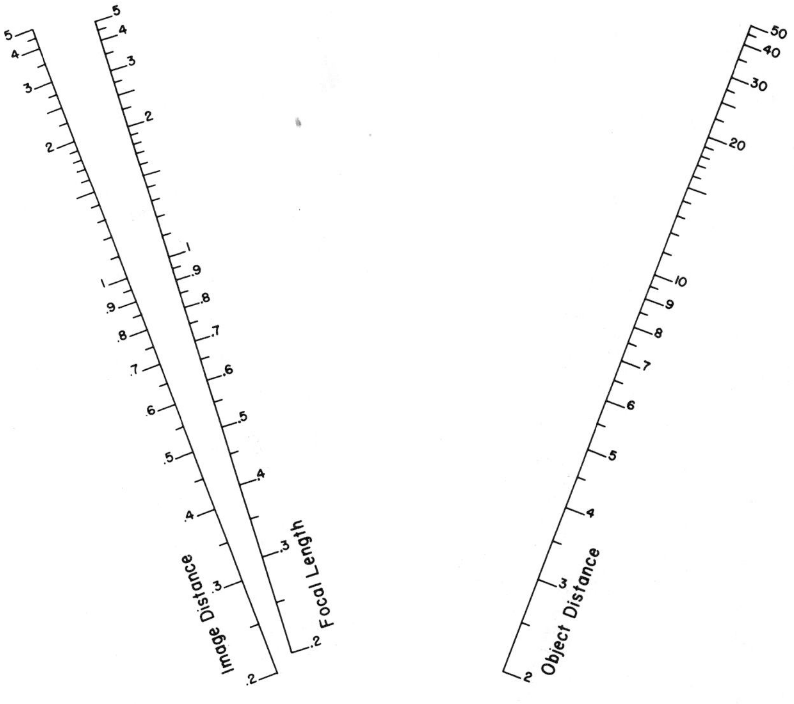

FIG. 12:2 NOMOGRAPH FOR LENS FORMULA

$$\frac{1}{s_f} = \frac{1}{s_o} + \frac{1}{s_i}$$

In Figure 12:2 this nomograph is shown constructed with hyperbolic coordinates employing the values $r_x = 0.5$ and $r_y = 5.0$ to emphasize ranges (in feet) encountered in certain fields of commercial photography. The same nomograph is shown in Figure 12:3 constructed with cartesian coordinates, using the same defining equation, but having the modulus of the Y axis ten times as great as that of the X axis. Figure 12:2 has the rather obvious advantage of greater detail in the lower range, while the flexibility of the hyperbolic coordinate system

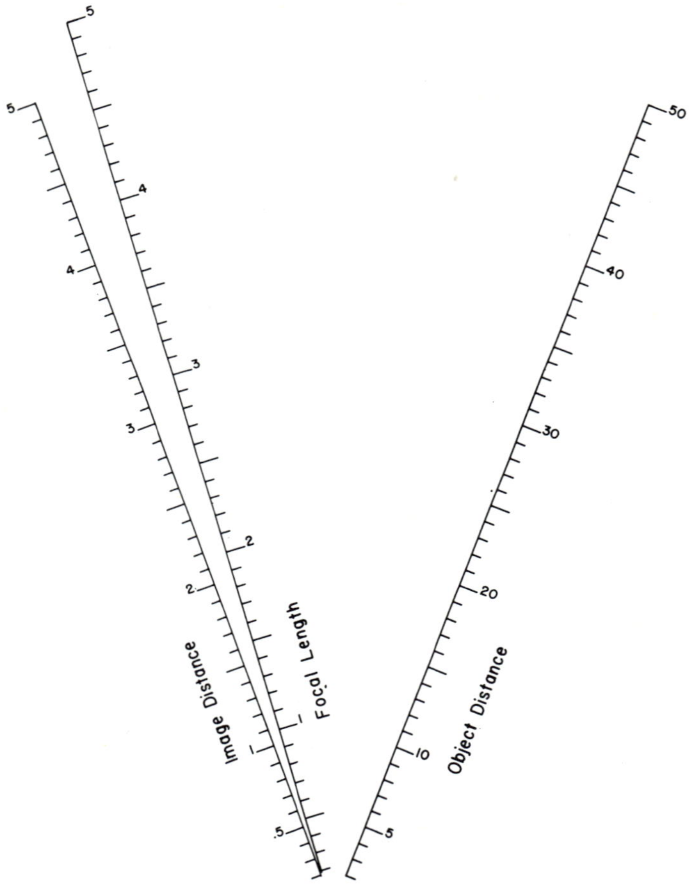

FIG. 12:3 NOMOGRAPH FOR LENS FORMULA CONSTRUCTED ON CARTESIAN COORDINATES

makes it possible to emphasize any portion of the ranges involved without increasing the length of the scales.

A similar effect may be achieved with cartesian coordinates only by extensive manipulation of the defining equation.

On the other hand, a nomograph of the form of Figure 12:3 is easily constructed with hyperbolic coordinates if the defining equation is written in the form

$$\begin{vmatrix} \frac{1}{s_f} & 1 & 1 \\ \frac{1}{s_i} & 1 & 0 \\ \frac{1}{s_o} & 0 & 1 \end{vmatrix} = 0,$$

the three scales being linear and having their origins at the termini of the Y and $\bar{Z}$ axes. The s_o scale lies on the Y axis, and the s_i scale on the $\bar{Z}$ axis. For either form of the defining equation the s_i scale is easily located, as it always passes through the point $p = 1$ on the third axis, regardless of the values of the hyperbolic constants.

FITTING THE NOMOGRAPH TO A RECTANGLE

12.4 Comparison of Methods

One of the most interesting problems in nomography is that of modifying the original form of a nomograph so that its two exterior scales terminate at the corners of a rectangle. This form of the nomograph has the advantages of greater accuracy, increased legibility, and more pleasing appearance. The treatment of this problem when the two exterior scales are already parallel is discussed in preceding chapters. In the more usual treatment, in which the termini of the exterior scales fall on any four points, the hyperbolic coordinate system provides an interesting mode of approach.

There are several phases to the problem:

1) Adjustment of the hyperbolic constants of both vertical and horizontal components of the reference frame may convert any quadrilateral to a parallelogram;
2) A change in the angle that those components make with one another—e.g., the angle between the X and Y

axes of oblique coordinates—may convert the parallelogram into a rectangle;

3) Adjustment of the moduli of the axes may provide the desired length-to-width ratio for the rectangle; and

4) Rotation of the reference frame may bring the sides of the rectangle into vertical and horizontal positions.

The requirements for steps 2, 3 and 4 are obvious, once step 1 has been accomplished, as may be seen in the example given in Section 12.6. In many problems, step 1 may be achieved simply by trial-and-error adjustment of the values of r_x and r_y, the resulting nomograph being approximately, if not exactly, a rectangle.

For a treatment of step 1 which is at once accurate and direct, it is necessary to resort to analytical methods. In the classical method, using cartesian coordinates (see Chapter 9), all four steps are usually merged into one, and the entire problem is solved by the introduction of additional variables into the defining equation. The values of those variables are then determined by solving a series of simultaneous equations involving the desired dimensions of the nomograph. By this method one arrives at a new defining equation which gives the desired results, but which is often very tedious to employ.

12.5 Values of r_x and r_y

The method that follows avoids any change in the defining equation and eliminates the solution of simultaneous equations. By this method one proceeds from given coefficients directly to the required values of r_x and r_y, which values give an exact rectangle. An outline of the derivation of this method follows the discussion and application of it and may be omitted for most practical purposes.

If the defining equation for a nomograph relating the variables u, v, and w is

$$\begin{vmatrix} p_y(u) & p_x(u) & 1 \\ p_y(v) & p_x(v) & 1 \\ p_y(w) & p_x(w) & 1 \end{vmatrix} = 0$$

and u and v are the exterior scales with lower termini u_1 and v_1 and upper termini u_2 and v_2, respectively, the numerical values of p_x and p_y corresponding to these termini may be taken as

$$p_x(u_1) = p_{x_1}, \; p_x(u_2) = p_{x_2}, \; p_x(v_1) = p_{x_3}, \; p_x(v_3) = p_{x_4},$$

$$p_y(u_1) = p_{y_1}, \; p_y(u_2) = p_{y_2}, \; p_y(v_1) = p_{y_3} \text{, and } p_y(v_2) = p_{y_4}.$$

These values are employed to evaluate r_x and r_y by the following steps:

$$a_1 = p_{x_1} - p_{x_2}, \; a_2 = p_{x_3} - p_{x_4},$$

$$a_3 = p_{x_1} - p_{x_3}, \; a_4 = p_{x_2} - p_{x_4},$$

$$b_1 = p_{y_1} - p_{y_2}, \; b_2 = p_{y_3} - p_{y_4},$$

$$b_3 = p_{y_1} - p_{y_3}, \; b_4 = p_{y_2} - p_{y_4},$$

$$c_1 = p_{x_1}p_{y_2} - p_{x_2}p_{y_1}, \; c_2 = p_{x_3}p_{y_4} - p_{x_4}p_{y_3},$$

$$c_3 = p_{x_1}p_{y_3} - p_{x_3}p_{y_1}, \; c_4 = p_{x_2}p_{y_4} - p_{x_4}p_{y_2}.$$

Then $r_y = \frac{1}{D}\Big[\left(b_1c_2 - b_2c_1\right)\left(a_4c_3 - a_3c_4\right)$

$$- \left(b_3c_4 - b_4c_3\right)\left(a_2c_1 - a_1c_2\right)\Big]$$

and $r_z = \frac{1}{D}\Big[\left(a_1b_2 - a_2b_1\right)\left(b_3c_4 - b_4c_3\right)$

$$- \left(a_3b_4 - a_4b_3\right)\left(b_1c_2 - b_2c_1\right)\Big].$$

where $D = \left(a_1 b_2 - a_2 b_1\right)\left(a_4 c_3 - a_3 c_4\right)$

$$- \left(a_3 b_4 - a_4 b_3\right)\left(a_2 c_1 - a_1 c_2\right).$$

12.6 Formula for Retaining Wall

By way of illustration, this method may be applied to the formula[2] for a retaining wall

$$(1 + L)h^2 - Lh(1 + p) - \frac{1}{3}(1 - L)(1 + 2p) = 0.$$

In determinant form this is

$$\begin{vmatrix} \frac{L-1}{3(L+1)} & \frac{2(L-1)}{3L} & 1 \\ \frac{2p+1}{3} & \frac{2(2p+1)}{3(p+1)} & 1 \\ h^2 & h & 1 \end{vmatrix} = 0$$

For both p and L, the lower limit is 0.50 and the upper limit, 1.00. From a preliminary plot (Figure 12:4) of the corresponding limits of the scale functions in the above defining equation, it appears that p and L are the exterior scales. The p scale lies between the limits (8/9, 2/3) and (1,1), and the L scale lies between (-2/3, -1/9) and (0,0). If u is then taken to represent the present function p and v to represent the function L, the values of these coordinates are inserted in the above formulas to determine the values of the a's, b's and c's, as follows:

$$a_1 = -1/9, \quad a_2 = 2/3, \quad a_3 = 8/9, \quad a_4 = 5/3,$$

$$b_1 = -1/3, \quad b_2 = 1/9, \quad b_3 = 2/3, \quad b_4 = 10/9,$$

$$c_1 = 2/9, \quad c_2 = 0, \quad c_3 = 0, \quad c_4 = 5/9.$$

[2] Allcock, H. J., and Jones, J. R., The Nomogram, p. 107 ff; Sir Isaac Pitman and Sons, London, 1932.

The values of the hyperbolic constants are, then, $r_y = 1/2$, $r_{\bar{z}} = -7/8$, and $r_x = (1/2)(-7/8) = -4/7$. The negative sign of r_x indicates merely that positive values of p are to be located in the negative region of the scale, and vice versa, since $p/(p + r) = -p/(-p-r)$. The nomograph is constructed using these values of r_x and r_y. From a preliminary plot it is apparent that an angle of 30° between the X and Y axes will convert the

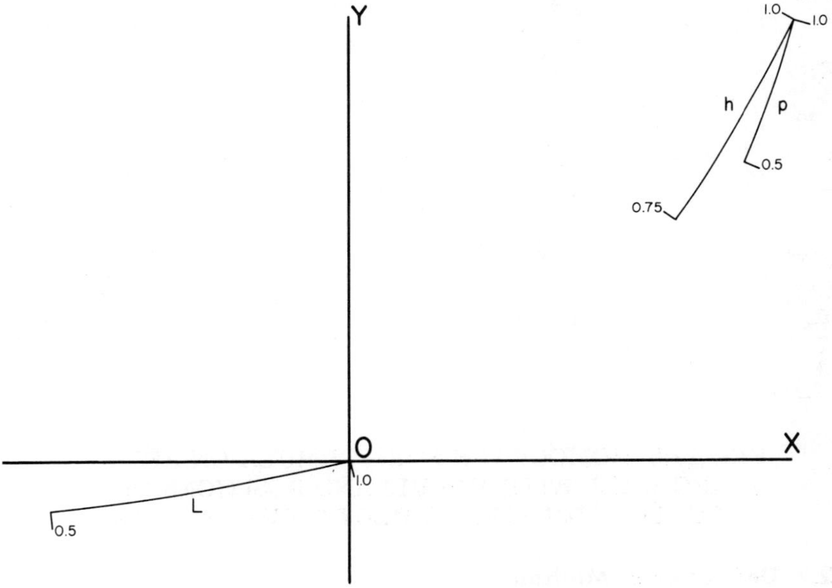

FIG. 12:4 POSITIONS OF SCALES OF NOMOGRAPH FOR RETAINING WALL BEFORE RECTIFICATIONS

parallelogram to a rectangle and that the length-to-width ratio desired may be achieved by setting the length (or modulus) of the X axis approximately twice as great as that of the Y axis. The nomograph thus constructed and rotated to place the scales in vertical position is shown in Figure 12:5.

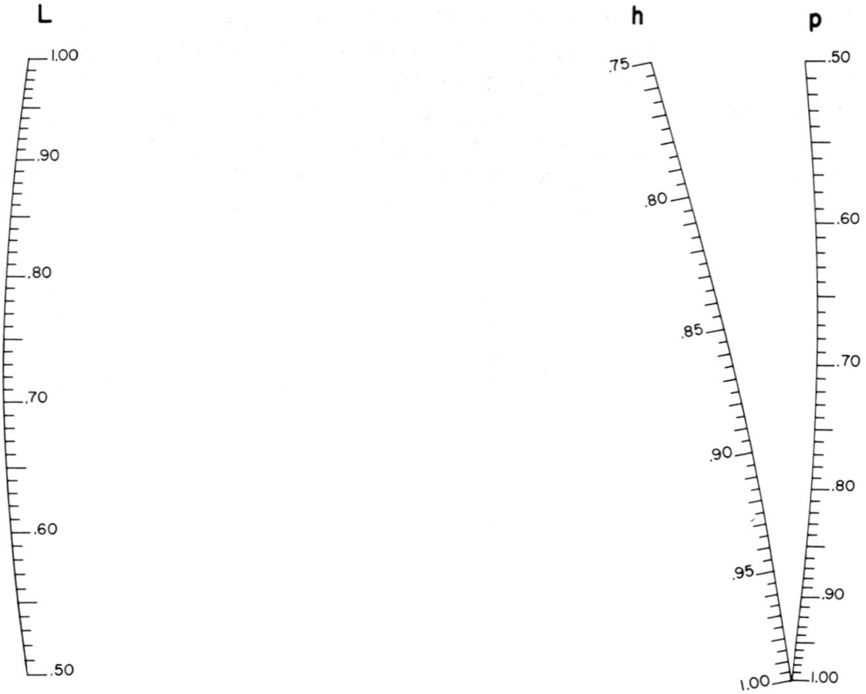

FIG. 12:5 NOMOGRAPH FOR FORMULA ON RETAINING WALL WITH MODULI AND POSITIONS OF SCALES ADJUSTED TO RECTANGULAR FORM

12.7 Derivation of Method

Let $\left(p_{x_1}, p_{y_1}\right)$ and $\left(p_{x_2}, p_{y_2}\right)$ be any two points in the general hyperbolic plane. A line passed through these two points meets the Y axis at a point $(0, p_y')$ whose value is given by

$$\begin{vmatrix} p_y' & 0 & 1 \\ p_{y_1} & p_{x_1} & 1 \\ p_{y_2} & p_{x_2} & 1 \end{vmatrix} = 0$$

or

$$p'_y = \frac{\left(p_{x_1} p_{y_2} - p_{x_2} p_{y_1}\right)}{\left(p_{x_1} - p_{x_2}\right)} \tag{12.2}$$

This line also meets the $\bar{Z}$ axis at a point $(p'_{\bar{z}}, \infty)$ given by

$$\begin{vmatrix} p'_{\bar{z}} & 1 & 0 \\ p_{y_1} & p_{x_1} & 1 \\ p_{y_2} & p_{x_2} & 1 \end{vmatrix} = 0,$$

or

$$p'_{\bar{z}} = \frac{\left(p_{y_1} - p_{y_2}\right)}{\left(p_{x_1} - p_{x_2}\right)} \tag{12.3}$$

Every line in the hyperbolic plane is characterized by an hyperbolic constant, $r_{x'}$, which is related to the hyperbolic constants of the axes (see Chapter 14) and its Y and $\bar{Z}$ intercepts by

$$r_{x'} = (p'_y + r_y)/(p'_z + r_z) \tag{12.4}$$

Substituting the values of p'_y and $p'_{\bar{z}}$ (Eqs. 12.2, 12.3) in Equation 12.4, the values of $r_{x'}$ in terms of the two general points is obtained as

$$r_{x'_1} = \left(\frac{p_{x_1} p_{y_2} - p_{x_2} p_{y_1}}{p_{x_1} - p_{x_2}} + r_y\right) \Bigg/ \left(\frac{p_{y_1} - p_{y_2}}{p_{x_1} - p_{x_2}} + r_{\bar{z}}\right).$$

The hyperbolic constant for a line passing through $\left(p_{x_3}, p_{y_3}\right)$

and $\left(p_{x_4}, p_{y_4}\right)$ is, similarly,

$$r_{x_2'} = \left(\frac{p_{x_3}p_{y_4} - p_{x_4}p_{y_3}}{p_{x_3} - p_{x_4}} + r_y\right) \Bigg/ \left(\frac{p_{y_3} - p_{y_4}}{p_{x_3} - p_{x_4}} + r_{\overline{z}}\right).$$

It is shown in Chapter 14 that the necessary and sufficient condition for these two lines to be parallel is that $r_{x_1'}$, and $r_{x_2'}$ be equal. In this case, one may write

$$\frac{p_{x_1}p_{y_2} - p_{x_2}p_{y_1} + r_y\left(p_{x_1} - p_{x_2}\right)}{p_{y_1} - p_{y_2} + r_{\overline{z}}\left(p_{x_1} - p_{x_2}\right)}$$

$$= \frac{p_{x_3}p_{y_4} - p_{x_4}p_{y_3} + r_y\left(p_{x_3} - p_{x_4}\right)}{p_{y_3} - p_{y_4} + r_{\overline{z}}\left(p_{x_3} - p_{x_4}\right)}$$

or

$$r_y(a_1b_2 - a_2b_1) + r_{\overline{z}}\,(a_2c_1 - a_1c_2) = b_1c_2 - b_2c_1,$$

where the a's, b's, and c's have the values given in Section 12.5.

If $\left(p_{x_1}, p_{y_1}\right)$, $\left(p_{x_2}, p_{y_2}\right)$, $\left(p_{x_3}, p_{y_3}\right)$ and $\left(p_{x_4}, p_{y_4}\right)$ are the vertices of a quadrilateral, Equation 12.5 expresses the condition that two opposite sides be parallel. If the figure is to be a parallelogram, the other two sides must also be parallel; whence

$$r_y\,(a_3b_4 - a_4b_3) + r_{\overline{z}}\,(a_4c_3 - a_3c_4) = b_3c_4 - b_4c_3 \quad (12.6)$$

The values of r_y and $r_{\overline{z}}$ are obtained by simultaneous solution of Equations 12.5 and 12.6:

$$r_y = \frac{\begin{vmatrix} b_1c_2 - b_2c_1 & a_2c_1 - a_1c_2 \\ b_3c_4 - b_4c_3 & a_4c_3 - a_3c_4 \end{vmatrix}}{\begin{vmatrix} a_1b_2 - a_2b_1 & a_2c_1 - a_1c_2 \\ a_3b_4 - a_4b_3 & a_4c_3 - a_3c_4 \end{vmatrix}} \tag{12.7}$$

and

$$r_{\bar{z}} = \frac{\begin{vmatrix} a_1b_2 - a_2b_1 & b_1c_2 - b_2c_1 \\ a_3b_4 - a_4b_3 & b_3c_4 - b_4c_3 \end{vmatrix}}{\begin{vmatrix} a_1b_2 - a_2b_1 & a_2c_1 - a_1c_2 \\ a_3b_4 - a_4b_3 & a_4c_3 - a_3c_4 \end{vmatrix}} \tag{12.8}$$

The value of r_x is $r_x = r_y/r_{\bar{z}}$.

chapter 13

THREE-DIMENSIONED NOMOGRAPHS

Construction of three-dimensional nomographs by synthetic methods or by analytic methods based upon cartesian coordinates is complicated and tedious, not only in effecting necessary modifications of scale moduli, but also in deriving suitable defining equations. This chapter presents a simple theoretical approach[1] to the construction of the three-dimensional nomographs for all formulas capable of representation in a nomograph with a planar index surface. The hyperbolic coordinate system on which the construction is made may be modified to accommodate desirable variations in moduli and positions of the scales of the nomograph.

THREE-DIMENSIONAL HYPERBOLIC COORDINATES

13.1 XY-Plane

The XY-plane is identical with the hyperbolic plane coordinate system described in Chapter 12.

13.2 Vertical Axes

The principal vertical axis is the Z axis on the origin of the X and Y axes. Auxiliary vertical axes are the $\overline{Y}$ axis on the terminus of the X axis and the $\overline{X}$ axis on the terminus of the Y axis (Fig. 13:1). These axes are linear; their moduli have the relationships, $m_{\overline{y}} = r_x m_z$ and $m_{\overline{x}} = r_y m_z$. Points on these axes have the values q_z, $q_{\overline{y}}$, and $q_{\overline{x}}$, respectively.

13.3 Vertical Planes

The vertical axes and the axes of the XY plane together determine three vertical planes; the XZ plane, the YZ plane,

[1] Adapted from article published in Ind. Eng. Chem., 43:1823 (1951). The geometry of three-dimensional hyperbolic coordinates is discussed in Chapter 14.

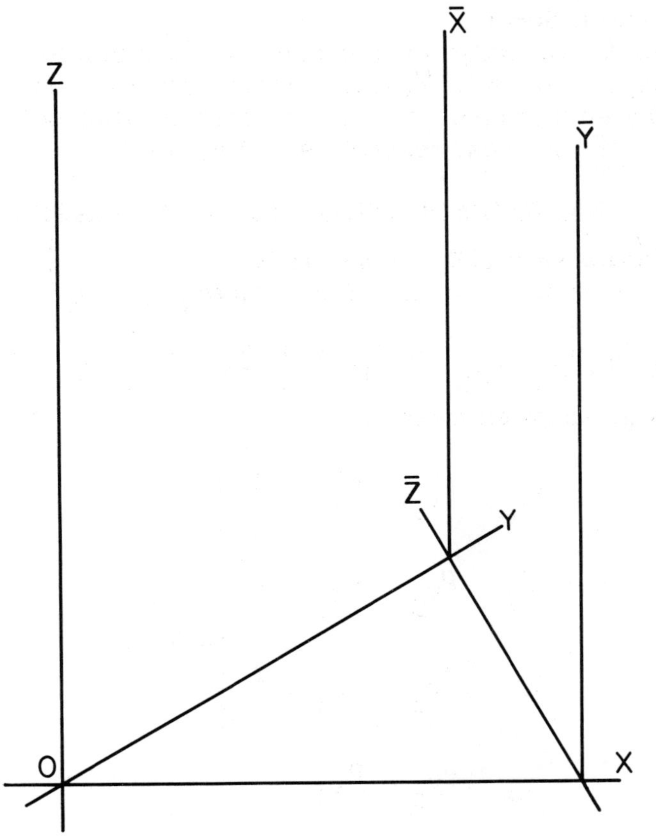

FIG. 13:1 SEMIHYPERBOLIC SOLID COORDINATES

and the $\overline{Z}\overline{Y}$ plane. Each of these planes is identical with the semihyperbolic planar coordinate system described in Chapter 10, which consists of a single hyperbolic axis and two linear axes. Points in these planes are located in the manner previously described: abscissas, by vertical coordinate lines; and ordinates, by coordinate lines drawn through the terminus of the hyperbolic axis.

13.4 Points in Space

A point not lying on the axes or axis planes is located with respect to its X, Y, and Z coordinates; the corresponding coordinate planes are determined by this point and by the $\overline{X}$, $\overline{Y}$, and $\overline{Z}$ axes, respectively (Fig. 13:2).

DEFINING EQUATION FOR NOMOGRAPHS

13.5 General Form of Defining Equation

The relationship of any four points $\left(p_{x_1}, p_{y_1}, q_{z_1}\right), \left(p_{x_2}, p_{y_2}, q_{z_2}\right), \left(p_{x_3}, p_{y_3}, q_{z_3}\right)$, and $\left(p_{x_4}, p_{y_4}, q_{z_4}\right)$ lying on a plane is given by expression

$$\begin{vmatrix} q_{z_1} & p_{y_1} & p_{x_1} & 1 \\ q_{z_2} & p_{y_2} & p_{x_2} & 1 \\ q_{z_3} & p_{y_3} & p_{x_3} & 1 \\ q_{z_4} & p_{y_4} & p_{x_4} & 1 \end{vmatrix} = 0 \qquad (13.1)$$

Any formula in four variables which can be expressed in this form, with the functions of any one variable confined to a single row of the determinant, can be represented as a three-dimensional nomograph. The elements of that determinant—cooresponding to the p's and q's of Equation 13.1—determine the values of the dimensions of the nomographic scales for the four corresponding functions.

13.6 Specific Types

Equation 13.1 is the form taken by nomographs in which no scales lie on the axes or axial planes of the coordinate system. Scales lying on axes or planes are indicated by modified rows in the determinant, as follows:

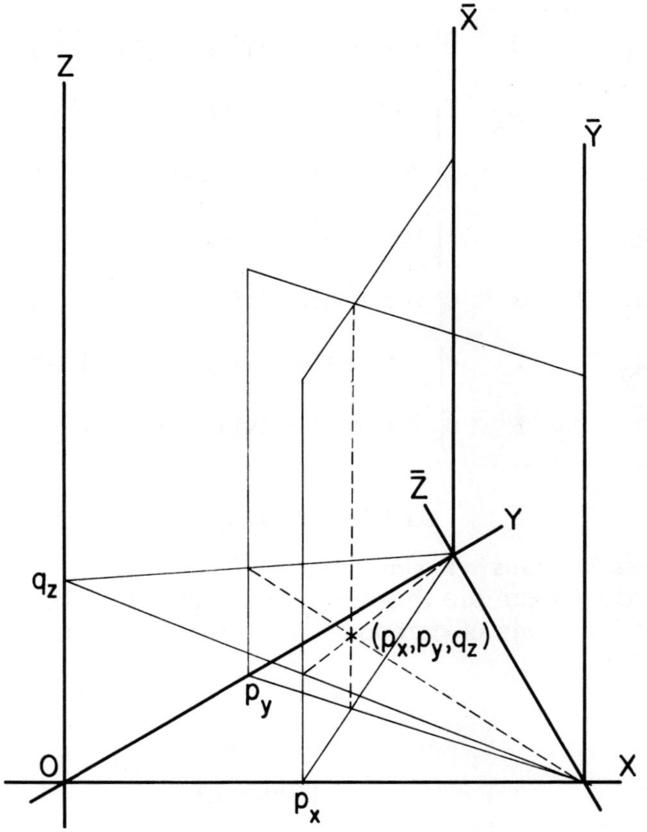

FIG. 13:2 METHOD OF LOCATING POINTS IN SEMI-HYPERBOLIC SPACE

$\begin{vmatrix} 0 & p_y & p_x & 1 \end{vmatrix}$ for scales lying on the XY plane

$\begin{vmatrix} q_z & 0 & p_x & 1 \end{vmatrix}$ for scales lying on the XZ plane

$\begin{vmatrix} q_z & p_y & 0 & 1 \end{vmatrix}$ for scales lying on the YZ plane

$\begin{vmatrix} q_{\overline{y}} & p_{\overline{z}} & 1 & 0 \end{vmatrix}$ for scales lying on the $\overline{ZY}$ plane

$\begin{vmatrix} 0 & 0 & p_x & 1 \end{vmatrix}$ for scales lying on the X axis

$\begin{vmatrix} 0 & p_y & 0 & 1 \end{vmatrix}$ for scales lying on the Y axis

$\begin{vmatrix} q_z & 0 & 0 & 1 \end{vmatrix}$ for scales lying on the Z axis

$\begin{vmatrix} q_{\overline{x}} & 1 & 0 & 0 \end{vmatrix}$ for scales lying on the $\overline{X}$ axis

$\begin{vmatrix} q_{\overline{y}} & 0 & 1 & 0 \end{vmatrix}$ for scales lying on the $\overline{Y}$ axis

$\begin{vmatrix} 0 & p_{\overline{z}} & 1 & 0 \end{vmatrix}$ for scales lying on the $\overline{Z}$ axis

APPLICATIONS

13.7 Mole Fractions in Residual Liquids

As a first example this method is applied to the equations for mole fractions of components in residual liquids.[2]

$$x_i = \frac{100\, x_{f_i}}{L + \frac{P_i}{P}(100 - L)}$$

where x_i is the mole fraction of component i in residual liquid (range 0 to 1);

x_{f_i}, mole fraction of component i in original liquid (0 to 1);

[2] Perry, John H., Chemical Engineers' Handbook, 3rd ed., p. 586: McGraw-Hill Book Company, New York, 1950.

L, total moles of liquid (0 to 100);
P_i, vapor pressure of component i (0 to 5000 mm.);
P, total pressure (0 to 7500 mm.).

The formula may be expressed in determinant form

$$\begin{vmatrix} -x_i & 0 & 1 & 0 \\ x_{f_i} & 0 & 0 & 1 \\ 0 & \dfrac{100 - L}{100} & 1 & 1 \\ 0 & P/(P - P_i) & 1 & 0 \end{vmatrix} = 0$$

The characteristics of the rows in this determinant indicate the following positions of the nomographic scales: x_i on the $\overline{Y}$ axis, x_{f_i} on the Z axis, L on the XY plane at $p_x = 1$, and P_i/P on the $\overline{Z}$ axis. From the ranges of these variables the desirable moduli of the vertical axes and the hyperbolic constants of the horizontal axes may be derived: $m_{\overline{y}} = 10$ inches, $m_z = 10$ inches, $r_{\overline{z}} = 1$; whence,

$$r_x = m_{\overline{y}}/m_z = 1,$$

$$r_y = r_x/r_{\overline{z}} = 1,$$

and

$$m_{\overline{x}} = m_z r_y = 10.$$

The nomograph so constructed is shown in Figure 13:3; the base of the figure is on the XZ plane.

13.8 Rayleigh Equation

A somewhat more involved example is the Rayleigh equation[2] for binary mixtures:

$$\ln \frac{L_1}{L_2} = \frac{1}{\alpha - 1}\left(\ln \frac{x_1}{x_2} + \alpha \ln \frac{1 - x_2}{1 - x_1}\right)$$

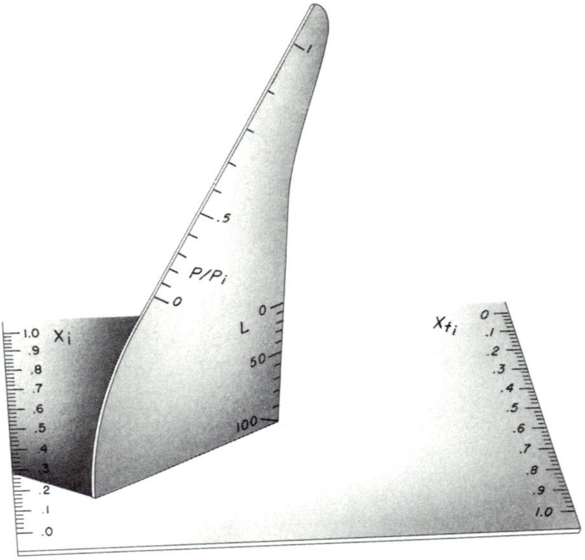

FIG. 13:3 NOMOGRAPH FOR MOLE FRACTIONS IN RESIDUAL LIQUIDS

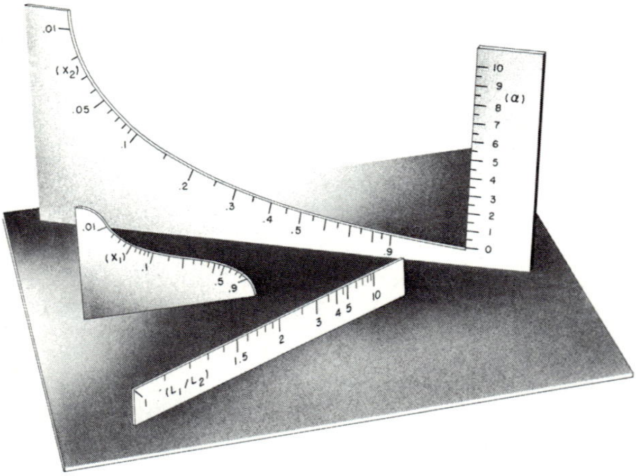

FIG. 13:4 NOMOGRAPH FOR RAYLEIGH EQUATION

where L_1 is the moles (or pounds) or original charge (25 to 250); L_2, moles (or pounds) of residual charge (25 to 250); x_1, mole fraction (or weight fraction) of more volatile component in original charge (0.01 to 1.0); x_2 mole fraction (or weight fraction) of more volatile component in residual charge (0.01 to 1.0); and α, relative volatility (0 to 10). From the ranges of L_1 and L_2, $\ln (L_1/L_2) = 0$ to 2.303.

The determinant form of this equation is

$$\begin{vmatrix} 0 & \ln \dfrac{L_1}{L_2} & 1 & 0 \\ \ln\left(\dfrac{1}{x_2} - 1\right) & -\ln(1 - x_2) & 0 & 1 \\ \ln\left(\dfrac{1}{x_1} - 1\right) & -\ln(1 - x_1) & 1 & 1 \\ \alpha - 1 & 1 & 0 & 0 \end{vmatrix} = 0$$

The rows of the determinant indicate for the nomographic scales the following positions: L_1/L_2 on the $\overline{Z}$ axis, x_2 on YZ plane, x_1 on the plane $p_x = 1$, and α on the $\overline{X}$ axis. The moduli and hyperbolic constants derived from the ranges of these variables are $m_z = 1$, $m_{\overline{y}} = 0.5$, $m_{\overline{x}} = 0.5$, $r_x = 0.5$, $r_y = 0.5$, $r_{\overline{z}} = 1$. Figure 13:4 shows this nomograph with XY plane as its base.

13.9 Castles Equation

In general, polynomials may be satisfactorily represented by planar nomographs, as shown in Chapter 11 for polynomials of three, four, and five parameters, By way of contrast, however, one of the formulas represented in that article by a planar nomograph is shown here in a three-dimensional nomograph. The formula is the Castles[3] equation for horizontal velocity of a helicopter in powered flight:

---

[3] Castles, W., Jr., J. Aeron. Sci., 12:477 (1945).

$$16F\left(\lambda^4_{hm} + 9.1\,\lambda^2_{hm}\right) + A\,\lambda^3_{hm} + C\,\lambda_{hm} + 20.8 = 0$$

in which λ_{hm} is horizontal velocity parameter (range 5 to 15), and A, C, and F are independent parameters with ranges: A = 0 to 0.1, C = -50 to 0, and F = 0 to 0.005.

A satisfactory determinant form of this equation is

$$\begin{vmatrix} \dfrac{20.8}{\left(\lambda^2_{hm} + 9.1\right)\lambda^2_{hm}} & \dfrac{1}{\left(\lambda^2_{hm} + 9.1\right)\lambda_{hm}} & \dfrac{\lambda_{hm}}{\lambda^2_{hm} + 9.1} & 1 \\ -C & 1 & 0 & 0 \\ -A & 0 & 1 & 0 \\ -16F & 0 & 0 & 1 \end{vmatrix} = 0$$

from which it is apparent that the A scale lies on the Y axis,

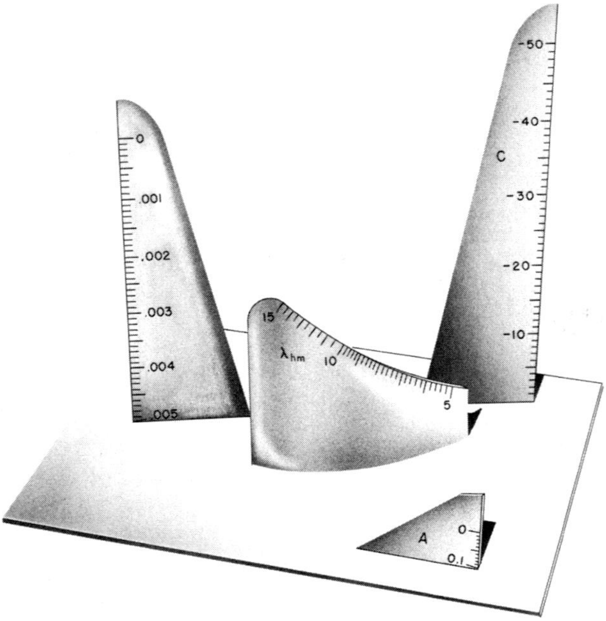

FIG. 13:5 NOMOGRAPH FOR CASTLES EQUATION

the C scale on the $\overline{X}$ axis, the F scale on the Z axis, and the λ_{hm} scale on a three-dimensional curve. From the ranges of the various parameters, it is desirable to select the values $m_{\overline{x}} = 0.2$, $m_{\overline{y}} = 10$, and $m_z = 100$. The resulting hyperbolic constants are $r_x = 0.1$, $r_y = 0.002$, and $r_{\overline{z}} = 0.02$. The nomograph constructed on these values is shown in Figure 13:5.

13.10 Methods of Reading

Reading three-dimensional nomographs is not so simple as reading planar nomographs, since constructing a plane through three given points presents more of a problem than drawing a straight line through two points. Nevertheless, several feasible methods present themselves, one of which is the method suggested by d'Ocagne.[4]

The nomograph is constructed in the form of a device having a peephole on one of the scales, through which the user sights to a line drawn between given points on two other scales. The point at which that line appears to intersect the fourth scale is the desired solution. I prefer to use two or three contacting straight edges, such as rulers or stretched wires. These are much simpler to construct and may be applied to any type of three-dimensional nomograph.

In most applications, the prime purpose of nomographs is computation. They are also useful, however, for the graphical representation of functionally related parameters, serving as aids to visualization of involved relationships. For this latter purpose the method of reading need not be precise. Hence, the three-dimensional nomograph is peculiarly well adapted to exhibits, and it presents a means of demonstrating relationships among four related parameters that is not easily achieved with three-dimensional graphs.

---

[4] d'Ocagne, M., "Traité de Nomographie," 1st ed., p. 348; Gauthier-Villars, Paris, 1921.

chapter 14

PROPERTIES OF HYPERBOLIC COORDINATE SYSTEMS

Previous chapters on the construction of nomographs with hyperbolic coordinates have dealt only briefly with the properties of the coordinate systems, attention being devoted primarily to practical applications. Extensive application of these methods, however, requires imagination and the freedom to select among, or depart from, the various types of coordinate systems suitable for the construction of nomographs. These requirements cannot be met without a clear mental picture of the properties of hyperbolic coordinate systems in general, and their relationships to cartesian coordinates. The purpose of this chapter is to close the many gaps left by the treatment given in previous chapters, and to provide a concise and systematic theoretical background for further application of these methods.[1]

14.1 Coordinate Systems and Projective Relationships

The coordinate system described here was devised as an aid in the construction of nomographs. In this application, it becomes necessary at times to employ the methods of projective geometry to produce a satisfactory arrangement of scales and index lines. The methods in common use, however, which are based on the analytical geometry of cartesian coordinates, are often cumbersome and tedious in this field of application. It has frequently been found preferable to define the scales and index lines in terms of a less specific coordinate system. In this way the coordinate system can be modified to produce the required arrangement of scales and

[1] Adapted from article published in J. Franklin Inst., 254-2:127 (1952).

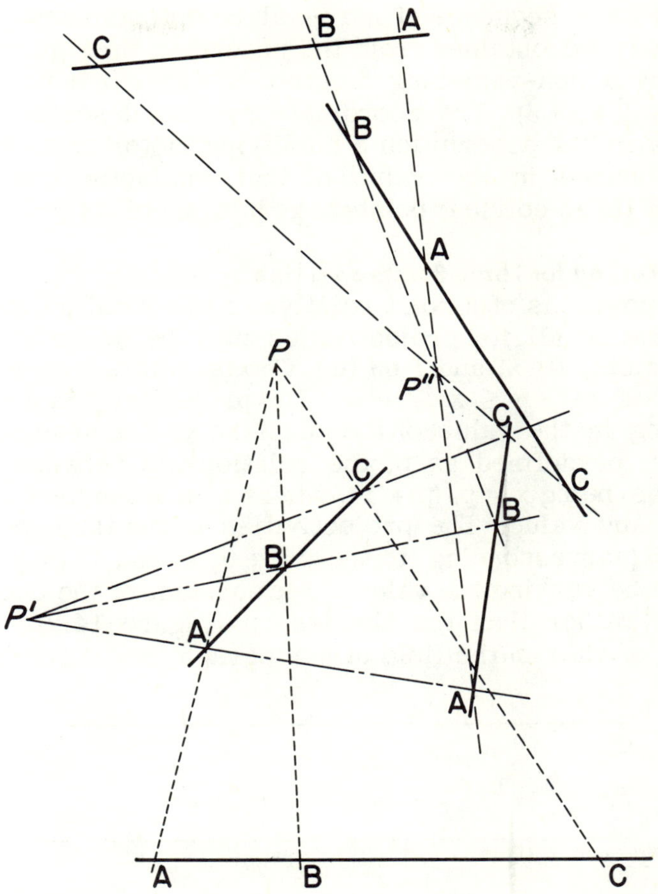

FIG. 14:1 RELATIONSHIPS OF THREE POINTS ON A LINE IN VARIOUS PROJECTIVITIES

index lines, and thereby avoid the extensive arithmetical treatment associated with cartesian projective transformations.

Figure 14:1 shows some of the infinite number of the projective relationships that exist among any three points on a line. These relationships are produced in synthetic geometry by the method of alternate projection and section, as shown in Figure 14:1. Analytically, this relationship may be expressed by a sequence of zero-valued determinants, each of which may be obtained from the preceding through multiplication by a non-vanishing "matrix of transformation." (see Sections 9.4, 9.5). The coordinate system described herein includes in its description a modifying factor of such nature that a change in the value of that one factor will cause a figure of three colinear points to go into any of its projections.

14.2 A Method for Three Points on a Line

The positions of a point relative to two fixed points on the same line in all its projectivities may be described by the three points, 0, x, and 1 on the X-axis, where x may take on all values of $-\infty \leq x \leq +\infty$. The projectivity is a result of modifying in the value of the abscissa **x**. The abscissa may, however, be defined as p, the relationship between the two abscissas being $x = p/(p + r)$, where r is a constant that may take on any value. The projectivities of the three points, 0, p, and ∞ (corresponding to the above 0, x, and 1) may then be realized by varying the value of r (a constant of the coordinate system) rather than the abscissa p (Figure 14:2). The advantage of this in the field of nomograph lies in the fact that

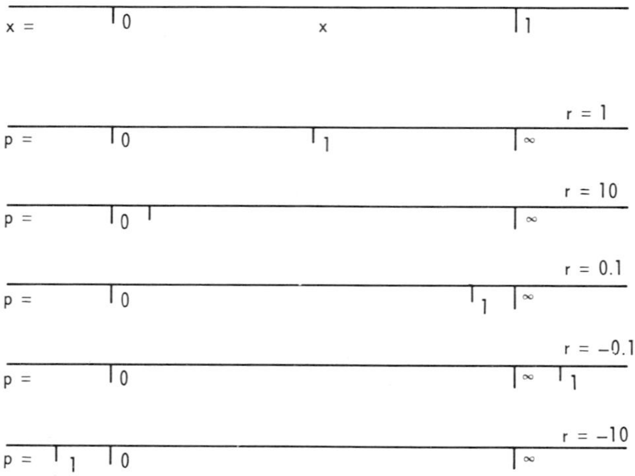

FIG. 14:2 PROJECTIVE RELATIONSHIPS REALIZED BY VARYING VALUE OF HYPERBOLIC CONSTANT, r

the abscissa (x or p) is frequently a complicated function whose evaluation is tedious, whereas r is a simple number, often an integer.

HYPERBOLIC PLANE COORDINATES

14.3 Points on the Axes

The axes of hyperbolic plane coordinates consist of two straight lines intersecting at a finite point, the origin, at an angle θ which is not equal to zero or any integral multiple of π. There is no advantage to be gained in the present discussion by considering special cases, such as $\theta = \pi/2$ (rectangular coordinates). These axes, are designated X and Y, and are superimposable upon the corresponding axes of a cartesian coordinate system whose axes form an angle equal to θ.

A point on the X-axis has the abscissa p_x, the relationship of this abscissa to its cartesian abscissa being

$$x = p_x/(p_x + r_x). \tag{14.1}$$

The ordinate of a point on the Y-axis is p_y, the relationship to the cartesian ordinate being

$$y = p_y/(p_y + r_y). \tag{14.2}$$

The values of r_x and r_y are independent. For all values of r_x and r_y, there are two fixed points on each axis: the origin ($p = 0$) and the point $p = \infty$ ($x = 1$ on the X-axis, $y = 1$ on the Y-axis) known as the terminus.

The relationship of Equations 14.1 and 14.2 to the cartesian hyperbola $y = x/(x + a)$ gives rise to the name "hyperbolic coordinates."

14.4 Points in the Plane

In cartesian coordinates, all points having a given abscissa lie on a line parallel to the Y-axis. All such lines converge at the point $y = \infty$. Ordinates are similarly related to a family of lines converging at the point $x = \infty$. These points may be regarded as the termini of the cartesian Y- and X-axes, respectively.

In the hyperbolic plane, all points having a given abscissa lie on a line that passes through the terminus of the Y-axis, and points having a given ordinate lie on a line that passes through the terminus of the X-axis (Figure 14:3). In cartesian

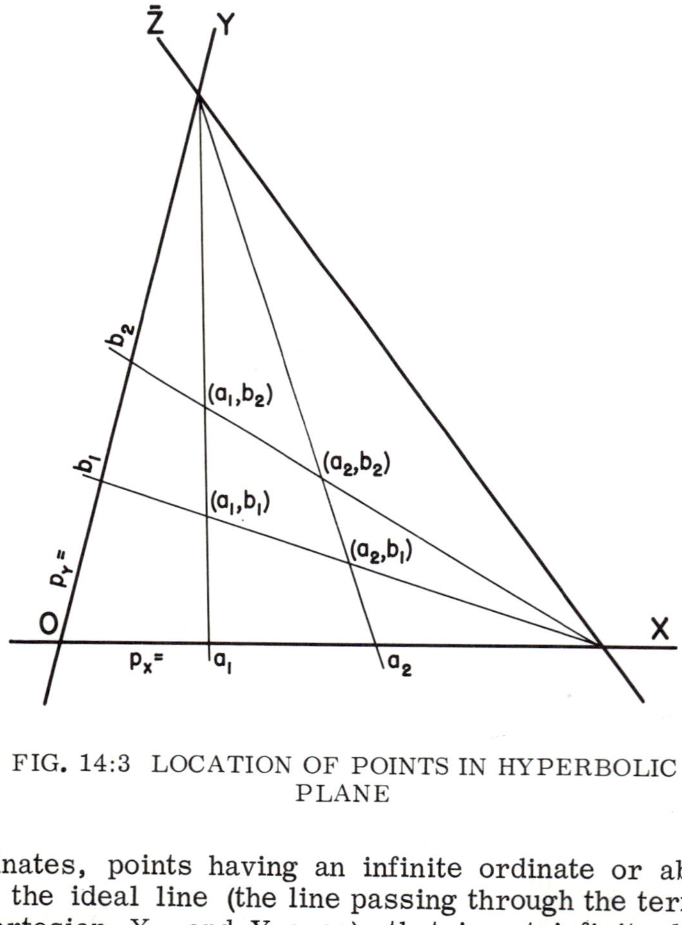

FIG. 14:3 LOCATION OF POINTS IN HYPERBOLIC PLANE

coordinates, points having an infinite ordinate or abscissa lie on the ideal line (the line passing through the termini of the cartesian X- and Y-axes), that is, at infinite distance from the origin. The association of infinite coordinate with infinite distance from the origin is not always a fortunate one. In hyperbolic coordinates, points having the finite coordinate $p_x = -r_x$, or $p_y = -r_y$, lie on the ideal line. Points that have infinite coordinates, $p_x = \infty$ or $p_y = \infty$, lie on the line that passes through the termini of the hyperbolic X- and Y-axes (a finite line).

All finite points may be located with respect to the hyperbolic X- and Y-axes, with the exception of points lying on the

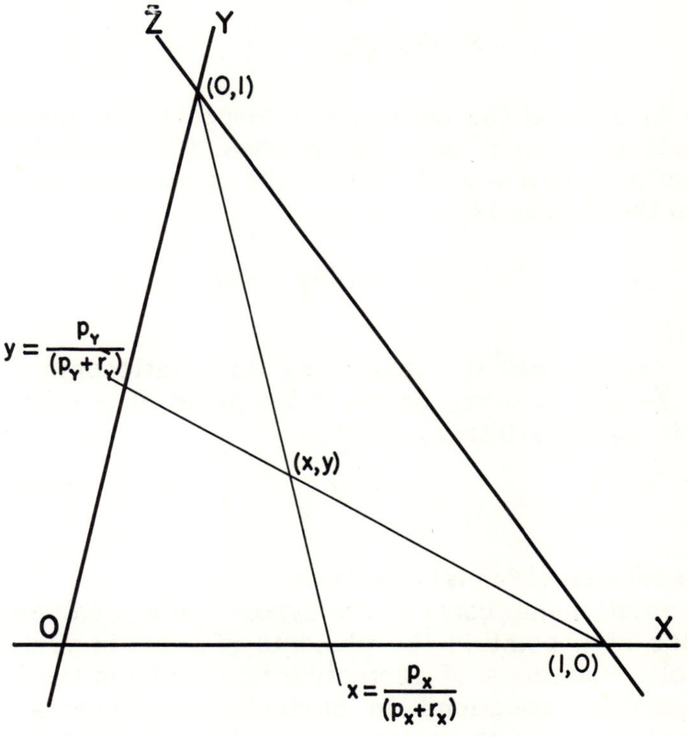

FIG. 14:4 CARTESIAN COORDINATES OF A POINT IN HYPERBOLIC PLANE

line that passes through the termini of those axes, because in this case both p_x and p_y are infinite. It is necessary, therefore, to define points on that line with respect to its fixed points, which are the termini of the X- and Y-axes. In so doing, it is desirable to consider the segment of that line lying between the two termini as being of unit length (as with the X- and Y-axes), unit length not being necessarily the same for any two of the three lines. The point at which this line intersects the X-axis is taken to be its origin, and its terminus is at the intersection with the Y-axis. A point on this line, known as the $\overline{Z}$-axis, is designated by the hyperbolic ordinate, $p_{\overline{z}}$, the relationship to its cartesian ordinate being

$$y = p_{\overline{z}}/(p_{\overline{z}} + r_{\overline{z}}). \tag{14.3}$$

The equation of the $\overline{Z}$-axis, with respect to the corresponding cartesian coordinates (as it passes through the points (1,0) and (0,1) is $x + y = 1$. Hence, the cartesian abscissa of a point on the $\overline{Z}$-axis is

$$x = 1 - y = r_{\overline{z}}/(p_{\overline{z}} + r_{\overline{z}}). \tag{14.4}$$

For the purpose of simplifying the relationship of points on the $\overline{Z}$-axis to other points in the plane, it is desirable to assign to $r_{\overline{z}}$ the arbitrary value,

$$r_{\overline{z}} = r_y/r_x. \tag{14.5}$$

14.5 Coordinates of Points in the Plane

The relationship between the hyperbolic and cartesian coordinates of a point in the plane is of value in deriving the hyperbolic equations of geometric figures from the familiar corresponding equations in cartesian coordinates. These relationships may be derived from the equations of the coordinate lines, as shown in Figure 14:4. The equation of the abscissa line, which passes through the points (0,1), (x,y), and $(p_x/(p_x + r_x), 0)$ is

$$x = \frac{p_x(1 - y)}{(p_x + r_x)}$$

and that of the ordinate line, which passes through the points (1,0), (x,y), and 0, $p_y/(p_y + r_y)$ is

$$y = \frac{p_y(1 - x)}{(p_y + r_y)}$$

Simultaneous solution of these equations gives the following values of x and y:

$$\left.\begin{aligned} x &= \frac{p_x r_y}{\left(p_x r_y + p_y r_x + r_x r_y\right)} \\ y &= \frac{p_y r_x}{\left(p_x r_y + p_y r_x + r_x r_y\right)} \end{aligned}\right\} \qquad (14.6)$$

At times it is desirable to give the position of a point in the plane in terms of the $\overline{Z}$-axis and one of the other axes. In so doing, the $\bar{z}$-coordinate line is taken as the line through the origin of the X- and Y- axes, the point in question, and a point $p_{\bar{z}}$ on the $\overline{Z}$-axis. The equation of this line, passing through the points (0,0), (x,y), and $(r_{\bar{z}}/(p_{\bar{z}} + r_{\bar{z}}), p_{\bar{z}}/(p_{\bar{z}} + r_{\bar{z}}))$, is $p_{\bar{z}} = yr_{\bar{z}}/x$. Substitution in this expression of the value of $r_{\bar{z}}$ (Eq. 14.5) and the values of x and y (Eq. 14.6) results in the relationship,

$$p_{\bar{z}} = p_y/p_x. \qquad (14.7)$$

The reciprocal relations are given by the equations (derived from Equations 14.6 and 14.7):

$$\left.\begin{aligned} p_x/r_x &= x/(1 - x - y) \\ p_y/r_y &= y/(1 - x - y) \\ p_{\bar{z}}/r_{\bar{z}} &= y/x. \end{aligned}\right\} \qquad (14.8)$$

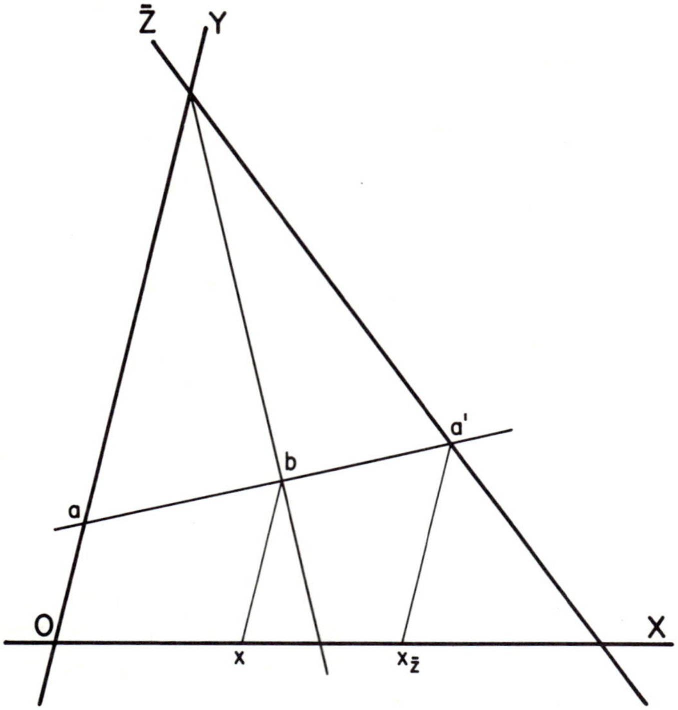

FIG. 14:5 DERIVATION OF HYPERBOLIC CONSTANTS OF A LINE

14.6 Equation of the Straight Line

The condition that three points (x_1,y_1), (x_2,y_2), and (x_3,y_3) lie on a straight line is that the following determinant vanish; that is

$$\begin{vmatrix} y_1 & x_1 & 1 \\ y_2 & x_2 & 1 \\ y_3 & x_3 & 1 \end{vmatrix} = 0 \qquad (14.9)$$

If the hyperbolic equivalents of the cartesian coordinates be substituted in Equation 14.9, the resulting determinant is

$$\begin{vmatrix} \dfrac{p_{y_1} r_x}{F_1} & \dfrac{p_{x_1} r_y}{F_1} & 1 \\[2ex] \dfrac{p_{y_2} r_x}{F_2} & \dfrac{p_{x_2} r_y}{F_2} & 1 \\[2ex] \dfrac{p_{y_3} r_x}{F_3} & \dfrac{p_{x_3} r_y}{F_3} & 1 \end{vmatrix} = 0$$

in which $F_i = p_{x_i} r_y + p_{y_i} r_x + r_x r_y$ $(i = 1, 2, 3)$.

This determinant is simplified by substracting both the first and second columns from the third column:

$$\begin{vmatrix} \dfrac{p_{y_1} r_x}{F_1} & \dfrac{p_{x_1} r_y}{F_1} & 1 - \dfrac{p_{x_1} r_y + p_{y_1} r_x}{F_1} \\[2ex] \dfrac{p_{y_2} r_x}{F_2} & \dfrac{p_{x_2} r_y}{F_2} & 1 - \dfrac{p_{x_2} r_y + p_{y_2} r_y}{F_2} \\[2ex] \dfrac{p_{y_3} r_x}{F_3} & \dfrac{p_{x_3} r_y}{F_3} & 1 - \dfrac{p_{x_3} r_y + p_{y_3} r_x}{F_3} \end{vmatrix} = 0$$

$$\text{whence} \qquad \begin{vmatrix} \dfrac{p_{y_1} r_x}{F_1} & \dfrac{p_{x_1} r_y}{F_1} & \dfrac{r_x r_y}{F_1} \\[2ex] \dfrac{p_{y_2} r_x}{F_2} & \dfrac{p_{x_2} r_y}{F_2} & \dfrac{r_x r_y}{F_2} \\[2ex] \dfrac{p_{y_3} r_x}{F_3} & \dfrac{p_{y_3} r_x}{F_3} & \dfrac{r_x r_y}{F_3} \end{vmatrix} = 0$$

The F's and r's may now be factored out:

$$\frac{r_x^2 r_y^2}{F_1 F_2 F_3} \begin{vmatrix} p_{y_1} & p_{x_1} & 1 \\ p_{y_2} & p_{x_2} & 1 \\ p_{y_3} & p_{x_3} & 1 \end{vmatrix} = 0 \qquad (14.10)$$

Except for a certain example to be discussed shortly, the factor that precedes the determinant does not vanish; whence, the expression

$$\begin{vmatrix} p_{y_1} & p_{x_1} & 1 \\ p_{y_2} & p_{x_2} & 1 \\ p_{y_3} & p_{x_3} & 1 \end{vmatrix} = 0 \qquad (14.11)$$

states the condition that the three points $\left(p_{x_1}, p_{y_1}\right)$, $\left(p_{x_2}, p_{y_2}\right)$, and $\left(p_{x_3}, p_{y_3}\right)$ lie on a straight line. In nomography, this expression is loosely termed as the equation of the straight line.

The determinant of Equation 14.11 is independent of the factors r_x and r_y; consequently, it is possible to produce on any line in the plane all those projections of three points shown in Figure 14:2 without changing the values of the hyperbolic coordinates of those points, the projection being accomplished by varying the values of r_x and r_y.

The only instance (except for the absurd case, $r_x = 0$, or $r_y = 0$, in which the factor preceding the determinant of Equation 14.10 vanishes) is that in which p_{x_i} or p_{y_i} becomes infinite, that is, when the point lies on the $\overline{Z}$-axis. In this case, convenient substitutions to be made in Equation 14.9 for that point may be derived from Equations 14.3, 14.4, and 14.5:

$$x_i = r_{\overline{z}} / \left(p_{\overline{z}_i} + r_{\overline{z}}\right) = r_y / \left(p_{\overline{z}_i} r_x + r_y\right)$$

$$y_i = p_{\overline{z}_i} / \left(p_{\overline{z}_i} + r_{\overline{z}}\right) = p_{\overline{z}_i} r_x / \left(p_{\overline{z}_i} r_x + r_y\right).$$

These expressions are substituted in Equation 14.10 for the point lying on the $\overline{Z}$-axis. The expressions in Equation 14.6 are substituted for the other two points. The resulting expression simplified as previously indicated, leads to a determinant preceded by a factor that does not vanish. The determinant form of the equation of the straight line with one point on the $\overline{Z}$-axis becomes

$$\begin{vmatrix} p_{y_1} & p_{x_1} & 1 \\ p_{y_2} & p_{x_2} & 1 \\ p_{\overline{z}} & 1 & 0 \end{vmatrix} = 0 \qquad (14.12)$$

The general equation of the straight line in hyperbolic coordinates is Equation 14.11. The appearance of a cipher in a row of the determinant indicates a point on one of the axes, as follows: first column, X-axis; second column, Y-axis; third column, $\overline{Z}$-axis. In the last example, the 1 appears in the second column instead of in the third, as it does in the other two examples.

14.7 The Hyperbolic Nucleus

The triangle formed by the X-, Y-, and $\overline{Z}$-axes may be regarded as the <u>nucleus</u> of a network formed by three pencils of lines. The X-pencil converges at the apex of the nucleus directly opposite the X-axis. The lines of this pencil intersect the X-axis at the points $x_i = p_{x_i} / \left(p_{x_i} + r_x\right)$. The Y- and $\overline{Z}$-pencils bear a similar relationship to the Y- and $\overline{Z}$-axes, respectively. Thus, each pencil may be regarded as intersecting one side (or axis) of the nucleus in a series of hyperbolic segments, the other two sides (or axes) of the nucleus being members of that same pencil.

It may now be shown that each of the pencils of lines also intersects every line that may be drawn in the plane in a series of hyperbolic segments, $x'_i = l_{x'} p_{x_i} / \left(p_{x_i} + r_{x'}\right)$, l_x being the length of the line segment bounded by the p_0 member and the p_∞ member of the intersecting pencil. The corresponding values of r are,

for lines intersected by the X-pencil,

$$r_{x'} = \frac{(p'_y + r_y)}{p'_{\overline{z}} + r_{\overline{z}}};$$

for lines intersected by the Y-pencil,

$$r_{y'} = \frac{(p'_x + r_x)}{\left(\frac{1}{p'_{\overline{z}}} + \frac{1}{r_z}\right)}; \qquad (14.13)$$

for lines intersected by the $\overline{Z}$-pencil,

$$r_{\overline{z}'} = \frac{\left(\frac{1}{p'_x} + \frac{1}{r_x}\right)}{\left(\frac{1}{p_y} + \frac{1}{r_y}\right)}$$

where p'_x, p'_y, and $p'_{\bar{z}}$ are the points of intersection of the line with the X-, Y-, and $\bar{Z}$-axes, respectively, while r_x, r_y, and $r_{\bar{z}}$ are the respective hyperbolic constants of those axes.

As every line in the plane, other than lines on the vertices of the nucleus, is intersected by all three pencils it is, consequently, subdivided by those pencils into hyperbolic segments whose constants have the values given in Equation 14.13. For any given line, the relationship of its three hyperbolic constants, from Equation 14.13 is

$$r_{\bar{z}'} = r_{y'}/r_{x'} \tag{14.14}$$

Derivation of Equation 14.13 for X-Pencil: Let aa' in Figure 14:5 be a line in the plane of the nucleus, intersecting the Y-axis at p'_y and the $\bar{Z}$-axis at $p'_{\bar{z}}$. From Equation 14.12, the equation of this line is

$$\begin{vmatrix} p'_y & 0 & 1 \\ p_y & p_x & 1 \\ p'_{\bar{z}} & 1 & 0 \end{vmatrix} = 0$$

or

$$p_y = p_x p'_{\bar{z}} + p'_y.$$

The ratio of the length of $\overline{ab}$ to that of $\overline{aa'}$ is equal to the ratio of x to $x_{\bar{z}}$ on the X-axis. If this ratio is designated x', then

$$x' = \overline{ab}/\overline{aa'} = x/x_{\bar{z}}.$$

But, by Equation 14.6,

$$x = p_x r_y/(p_x r_y + p_y r_x + r_x r_y)$$

and, by Equation 14.4

$$x_{\overline{Z}} = r_{\overline{Z}}/(p_{\overline{Z}} + r_{\overline{Z}})$$

Substitutions of these values, and of the value of p_y given above, leads to the expression

$$x' = p_x \Big/ \left[p_x + \frac{p'_y + r_y}{p'_{\overline{Z}} + r_{\overline{Z}}} \right].$$

For a given line in the plane, $(p'_y + r_y)/(p'_{\overline{Z}} + r_{\overline{Z}})$ is a constant. If this constant is designated $r_{x'}$, then

$$x' = p_x/(p_x + r_{x'}).$$

The values of $r_{y'}$ and $r_{\overline{Z}'}$ may be established in similar manner.

The relationship of points on the X-axis to points with equal hyperbolic abscissas on any line intersecting the X-pencil is a projective relationship. The X-pencil may be regarded as the projection; the intersecting line as the section. Projections of a figure of points on the X-axis may be realized by this method of projection and section. As has now been shown, however, this method of projection and section has the same effect as varying the value of r_x, since the distribution of points on the section is identical with that produced by substitution in Equation 14.1 of the value of $r_{x'}$ given in Equation 14.13.

14.8 Parallel Lines

The slope of a line in the hyperbolic place is $m = (y'_{\overline{Z}} - y')/x'_{\overline{Z}}$, where $x'_{\overline{Z}}$ and $y'_{\overline{Z}}$ are the cartesian coordinates of its point of intersection with the $\overline{Z}$-axis and y' is its y-intercept. Substituting the values of these given in Equations 14.2

through 14.5 and Equation 14.13 yields the relationship

$$m = 1 - r_x/r_{x'} \tag{14.15}$$

where r_x and $r_{x'}$ are hyperbolic constants of the X-axis and the line with slope m, respectively.

It follows from Equation 14.15 that the necessary and sufficient condition that two lines, a and b, be parallel is that $r_{x'}(a) = r_{x'}(b)$. It may be shown in similar manner that in two such lines $r_{y'}(a) = r_{y'}(b)$ and that $r_{\bar{z}}(a) = r_{\bar{z}'}(b)$. This principle is applied in rectifying nomographic scales; (see Section 12.4, ff.)

SEMIHYPERBOLIC COORDINATES

14.9 Relationship to Hyperbolic Coordinates

In the field of nomography a modified form of hyperbolic coordinates is extensively used. It is properly called semihyperbolic, because only the X-axis is hyperbolic, and the Y-axis is linear. This form of coordinates is derived as follows:

In the definition of hyperbolic coordinates, the length of the Y-axis was taken as unity, there being no loss of generality by so doing. In the modification now to be made, the length of the Y-axis, l_y, must be considered variable; whence, $y = l_y p_y/(p_y + r_y)$ gives the distance of a point p_y from the origin of the Y-axis. The values of l_y and r_y are now allowed to approach infinity simultaneously, while the ratio l_y/r_y retains a constant value, m_y, that is, $l_y = r_y m_y$; whence,

$$y = \lim_{\substack{l_y \to \infty \\ r_y \to \infty}} \frac{\dfrac{l_y p_y}{r_y}}{\dfrac{p_y}{r_y} + 1} = p_y m_y \tag{14.16}$$

Here m_y is the constant commonly known as the modulus of

the Y-axis, effectively the length of that axis representing an increase of 1 in the value of p_y. As the length of the $\overline{Z}$-axis and the value of $r_{\overline{z}}$ approach infinity simultaneously with l_y and r_y, the $\overline{Z}$-axis becomes linear and parallel to the Y-axis. The value of $l_{\overline{z}}/r_{\overline{z}}$ as l_y becomes infinite has the value $l_y r_x / r_y = r_x m_y = m_{\overline{z}}$, the modulus of the $\overline{Z}$-axis.

As the properties of semihyperbolic coordinates and their use in the construction of nomographs are described in Chapter 10, they will not be repeated here.

The X-axis might be modified in a manner similar to that given for the Y-axis. If both axes be modified in this manner, a coordinate system identical with cartesian coordinates is obtained. Hence, cartesian coordinates may be considered a special case of hyperbolic coordinates.

HYPERBOLIC SOLID COORDINATES

14.10 Axes of Hyperbolic Solid Coordinate System

Hyperbolic solid coordinates consist of three axes, X, Y, and Z, corresponding to the X-, Y-, and Z- axes of cartesian solid coordinates. Points on these axes are defined by the expressions given for x and y in Equations 14.1 and 14.2, and by $z = p_z/(p_z + r_z)$. Values of r_x, r_y, and r_z are independent. Origin and termini are defined as in plane coordinates.

Each pair of axes determines a plane in which those axes are the principal axes of a plane hyperbolic system (Fig. 14:6). Positions of points lying in the planes of the axes are determined with respect to their principal axes in the manner given for plane coordinates. Points having infinite abscissa and ordinate lie on auxiliary axes: in the XY-plane, this is the $\overline{Z}$-axis; in the XZ-plane, the $\overline{Y}$-axis; and in the YZ-plane, the $\overline{X}$-axis (Fig. 14:6). There are, then, three principal and three auxiliary axes for locating points in the coordinate planes. It will be noted that the symbol (–) over a letter denotes an auxiliary axis that is not concurrent with the principal axis having the corresponding letter. Thus, the $\overline{Z}$-axis is not concurrent with the Z-axis, although all the remaining axes (principal and auxiliary) are. This convention

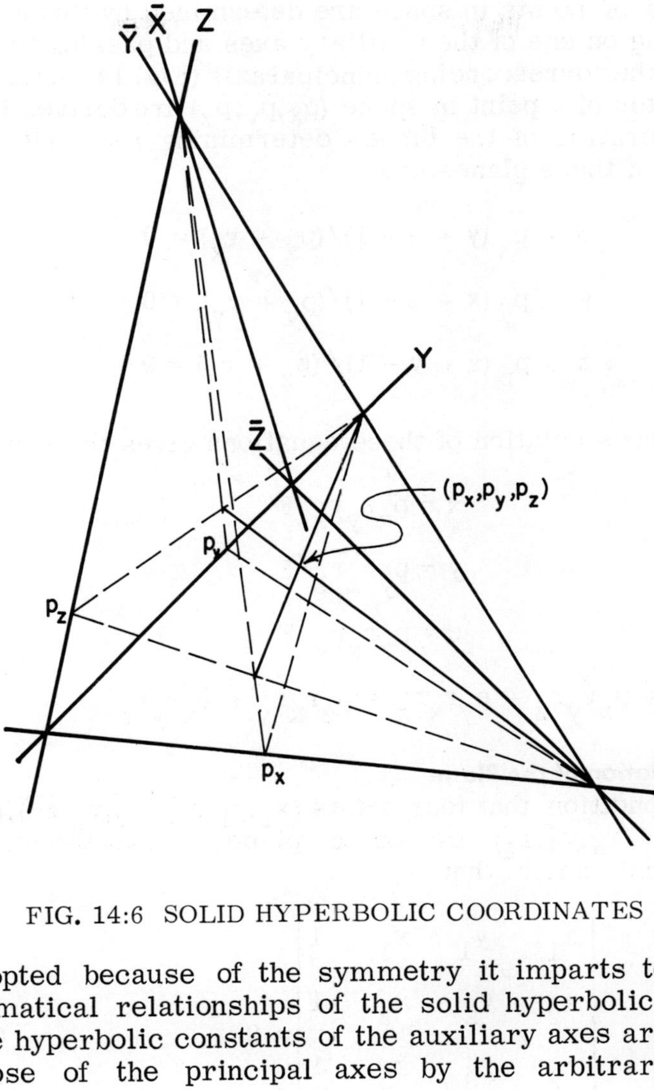

FIG. 14:6 SOLID HYPERBOLIC COORDINATES

is adopted because of the symmetry it imparts to certain mathematical relationships of the solid hyperbolic nucleus.

The hyperbolic constants of the auxiliary axes are related to those of the principal axes by the arbitrary values,

$$r_{\bar{z}} = r_y/r_x,$$

$$r_{\bar{y}} = r_z/r_x,$$

and

$$r_{\bar{x}} = r_z/r_y$$

Positions of points in space are determined by three planes, each lying on one of the auxiliary axes and passing through a point on the corresponding principal axis (Fig. 14:6). Cartesian coordinates of a point in space (p_x, p_y, p_z) are derived through a consideration of the planes determining its position. The equation of these planes are

$$x + p_x (y + z - 1)/(p_x + r_x) = 0$$

$$y + p_y (x + z - 1)/(p_y + r_y) = 0$$

$$z + p_z (x + y - 1)/(p_z + r_z) = 0$$

Simultaneous solution of these equations gives the values

$$x = p_x r_y r_z / F$$

$$y = p_y r_x r_z / F$$

$$z = p_z r_x r_y / F$$

where $F = p_x r_y r_z + p_y r_x r_z + p_z r_x r_y + r_x r_y r_z$.

14.11 Equation of the Plane

The condition that four points (x_1, y_1, z_1), (x_2, y_2, z_2), (x_3, y_3, z_3), and (x_4, y_4, z_4) lie on a plane is that the following determinant vanish: that is,

$$\begin{vmatrix} z_1 & y_1 & x_1 & 1 \\ z_2 & y_2 & x_2 & 1 \\ z_3 & y_3 & x_3 & 1 \\ z_4 & y_4 & x_4 & 1 \end{vmatrix} = 0 \qquad (14.17)$$

The hyperbolic values of these cartesian coordinates may be substituted in this determinant. Then subtraction of columns 1, 2, and 3 from column 4, followed by removal of certain factors, leaves the following relationship:

$$\frac{r_x^3 r_y^3 r_z^3}{F_1 F_2 F_3 F_4} \begin{vmatrix} p_{z_1} & p_{y_1} & p_{x_1} & 1 \\ p_{z_2} & p_{y_2} & p_{x_2} & 1 \\ p_{z_3} & p_{y_3} & p_{x_3} & 1 \\ p_{z_4} & p_{y_4} & p_{x_4} & 1 \end{vmatrix} = 0$$

Except for points lying on the auxiliary axes or on the plane through those axes, the F's are finite numbers; hence, the factor preceding the determinant does not vanish. Consequently, for points with finite coordinates, the expression

$$\begin{vmatrix} p_{z_1} & p_{y_1} & p_{x_1} & 1 \\ p_{z_2} & p_{y_2} & p_{x_2} & 1 \\ p_{z_3} & p_{y_3} & p_{x_3} & 1 \\ p_{z_4} & p_{y_4} & p_{x_4} & 1 \end{vmatrix} = 0 \qquad (14.18)$$

states the condition that four points $\left(p_{x_i}, p_{y_i}, p_{z_i}, (i = 1, 2, 3, \text{ and } 4)\right)$ lie on a plane.

Modifications of Equation 14.18 result when a point lies on one of the coordinate planes or on the plane through the three termini (the $\overline{XYZ}$-plane). These modifications arise through substitutions similar to those encountered in the equation of the straight line in plane hyperbolic coordinates, followed by factoring as just described. These modifications result in characteristic rows in the determinant, as follows:

$$\begin{vmatrix} 0 & p_{y_i} & p_{x_i} & 1 \end{vmatrix} \quad \text{for a point in the } XY\overline{Z}\text{-plane}$$

$$\begin{vmatrix} p_{z_i} & 0 & p_{x_i} & 1 \end{vmatrix} \quad \text{for a point in the } X\overline{Y}Z\text{-plane,}$$

$$\begin{vmatrix} p_{z_i} & p_{y_i} & 0 & 1 \end{vmatrix} \quad \text{for a point in the } \overline{X}YZ\text{-plane,}$$

$$\begin{vmatrix} p_{\overline{z}_i} & p_{\overline{y}_i} & 1 & 0 \end{vmatrix} \quad \text{for a point in the } \overline{X}\overline{Y}\overline{Z}\text{-plane}$$

For the values of x, y, and z of a point in the $\overline{X}\,\overline{Y}\,\overline{Z}$-plane, one may consider the equation of that plane, which is $x + y + z = 1$. But in that plane, $y = p_{\overline{y}}r_{\overline{z}}/(p_{\overline{y}}r_{\overline{z}} + p_{\overline{z}}r_{\overline{y}} + r_{\overline{y}}r_{\overline{z}})$ and $z = p_{\overline{z}}r_{\overline{y}}/(p_{\overline{y}}r_{\overline{z}} + p_{\overline{z}}r_{\overline{y}} + r_{\overline{y}}r_{\overline{z}})$; whence, $x = r_{\overline{y}}r_{\overline{z}}/(p_{\overline{y}}r_{\overline{z}} + p_{\overline{z}}r_{\overline{y}} + r_{\overline{y}}r_{\overline{z}})$. But $r_{\overline{z}} = r_y/r_x$ and $r_{\overline{y}} = r_z/r_x$; hence, $x = r_{\overline{y}}r_z/F'$, $y = p_{\overline{y}}/F'$ and $z = p_{\overline{z}}/F'$, where $F' = p_{\overline{y}}r_y + p_z r_{\overline{z}} + r_x r_y r_z$. Substitutions of these values in a row of Equation 14.17, followed by the steps employed above for simplification, leads to the characteristic row for a point in the $\overline{X}\,\overline{Y}\,\overline{Z}$-plane as given above.

Characteristic rows for points lying on the axes are derived in similar manner and are

$$\begin{vmatrix} 0 & 0 & p_{x_i} & 1 \end{vmatrix} \quad \text{for a point on the X-axis,}$$

$$\begin{vmatrix} 0 & p_{y_i} & 0 & 1 \end{vmatrix} \quad \text{for a point on the Y-axis,}$$

$$\begin{vmatrix} p_{z_i} & 0 & 0 & 1 \end{vmatrix} \quad \text{for a point on the Z-axis,}$$

$$\begin{vmatrix} p_{x_i} & 1 & 0 & 0 \end{vmatrix} \quad \text{for a point on the } \overline{X}\text{-axis,}$$

$$\begin{vmatrix} p_{\overline{y}_i} & 0 & 1 & 0 \end{vmatrix} \quad \text{for a point on the } \overline{Y}\text{-axis, and}$$

$$\begin{vmatrix} 0 & p_{\overline{z}_i} & 1 & 0 \end{vmatrix} \quad \text{for a point on the Z-axis.}$$

14.12 The Hyperbolic Solid Nucleus

The tetrahedron formed by the four planes bounded by the X-, Y-, Z-, $\overline{X}$-, $\overline{Y}$-, and $\overline{Z}$-axes may be regarded as the nucleus of a matrix formed by six bundles of planes. The X-bundle converges on the axis opposite the X-axis in the nucleus, that is, on the $\overline{X}$-axis; the Y-bundle on the $\overline{Y}$-axis.

It may now be shown that these bundles of planes intersect every plane in the hyperbolic space in such a manner as to form a plane hyperbolic nucleus whose principal axes are the lines of intersection of that plane with the planes of the solid hyberbolic nucleus.

Consider first a plane intersecting the $\overline{X}YZ$-, $X\overline{Y}Z$-, and $\overline{X}\overline{Y}\overline{Z}$-planes. Its intersection with the XYZ-plane is subdivided by the same bundle of planes which subdivides the X-axis; hence , it forms an X'-axis whose hyperbolic constant is $r_{x'} = (p'_z + r_z)/(p'_{\overline{y}} + r_{\overline{y}})$. Its intersection with the $\overline{X}YZ$-plane is subdivided by the same bundle of planes which subdivides the Y-axis; hence, it forms a Y'axis whose hyperbolic constant is $r_y = (p'_z + r_z)/(p'_{\overline{x}} + r_{\overline{x}})$. Its intersection with the $\overline{X}\overline{Y}Z$-plane is subdivided by the same bundle of planes which subdivides the $\overline{Z}$-axis, hence it forms a $\overline{Z}'$-axis whose hyperbolic constant is $r_{z'} = (p'_{\overline{y}} + r_{\overline{y}})/(p'_{\overline{x}} + r_{\overline{x}})$. From the above values of $r_{x'}$, $r_{y'}$ and $r_{\overline{z}'}$, it is obvious that $r_{\overline{z}'} = r_{y'}/r_{x'}$. Thus, the hyperbolic constants of the axes of the $X'Y'\overline{Z}'$-plane bear the same relationship to each other as do those of the $XY\overline{Z}$-plane.[2]

In similar manner it may be shown that a plane intersecting any three of the coordinate planes of the solid hyperbolic nucleus forms a plane nucleus whose hyperbolic constants are related in the same manner as are those of the fourth coordinate plane of the nucleus. As these values may be derived by inspection, using the relationships given by Equations 11.13, it is unnecessary to list them here.

SEMIHYPERBOLIC SOLID COORDINATES

14.13 Relationship to Hyperbolic Solid Coordinates

Hyperbolic solid coordinates may be modified after the same manner as hyperbolic plane coordinates by allowing

---

2 See section on Hyperbolic Plane Coordinates.

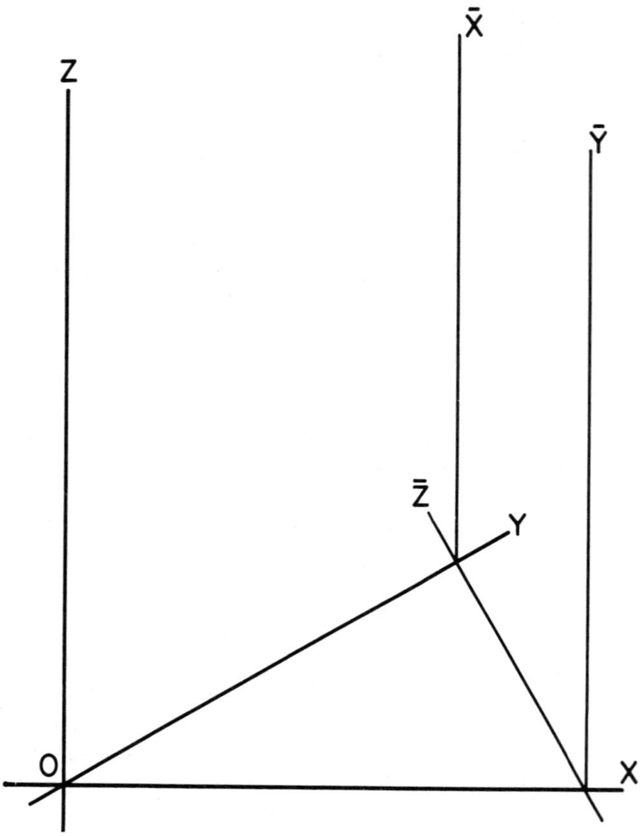

FIG. 14:7 SOLID SEMI-HYPERBOLIC COORDINATES

the length and hyperbolic constant of any one of the principal axes to approach infinity simultaneously, provided the ratio l_u/r_u(u = x,y,z) remains constant. For the construction of three-dimensional nomographs the form of coordinates most easily adaptable is that in which the length and hyperbolic constant of the Z-axis have been permitted to approach infinity. The resulting system (Fig. 14:7) consists of a triangular hyperbolic plane ($XY\bar{Z}$-plane with vertical axes arising from its three vertices. The vertical axes are parallel and linear, their moduli (derived in the same manner as with plane semihyperbolic coordinates) being related as follows: $m_{\overline{x}} = r_y m_z$ and $m_{\overline{y}} = r_x m_z$.

Any one, two, or three of the principal axes of solid coordinates may be allowed to approach infinite length (and infinite value of its hyperbolic constant). In the event that all three principal axes do so, the resulting coordinate system is identical with the cartesian solid coordinate system.

To summarize: By varying the value of r_x it is possible to produce within the X-axis (that is, in one dimension) all possible projections of a figure having two fixed points colinear with a third point, thereby obviating for such cases the method of projection and section that employs two dimensions. That similar projections of a three-point line figure may be produced on any line in the plane by varying r_x and/or r_y was shown in the discussion of the plane hyperbolic nucleus.

Likewise, every projection of the XYZ-plane is an X'Y'Z'-plane which differs from the XY$\overline{Z}$-plane only in the values of r_x and r_y. Hence, it follows that it is possible to produce within the XY$\overline{Z}$-plane (that is, in two dimensions) projections of a plane figure, merely by varying the values of r_x and r_y, and, thereby obviating the method of projection and section that employs three dimensions.

Similarly, varying the values of r_x, r_y, and r_z is a means of realizing projections of a solid figure. It is also possible to do this analytically, although, for many purposes, the analytical approach is much more difficult. In the field of nomography, for example, the construction of three-dimensional nomographs has been largely avoided because the application of analytical methods of projection, when applied to any but the simplest defining equations, is not only very laborious, but it also fails to provide the ease of visualization inherent in the hyperbolic coordinate system.

This ease of visualizing the projections of a figure (plane or solid) is enhanced by the fact that the mind readily considers variation in the value of the hyperbolic constant as a continuous process, with one-to-one correspondence between a figure and its projection. An analytical projection, being a discrete step, may be less easily visualized. In practical applications, this visualization is frequently valuable in avoiding unnecessary trials and leads more directly to the optimum projection for the purpose at hand.

chapter 15

NOMOGRAPHS FROM GRAPHS AND TABLES

In the preceding chapters, methods of constructing nomographs are confined to those examples in which the mathematical relationship to be represented by the nomograph is available in the form of an equation.

There are, however, two instances in which it is desirable to construct a nomograph directly from either a graph or a table. If the relationship is already in the form of a graph or table, it would be necessary to analyze the graphical data (see Sect. 6.17, 6.18) in order to convert the relationship to a formula from which a defining equation for the nomograph might be derived. This would be both circuitous and tedious. There are also numerous examples of formulas which do not lend themselves to expression as defining equations in any of the forms discussed in the preceding chapters. In these instances, a table of specific values might be computed from the formula and serve as a basis for the construction of the nomograph. Alternatively, the relationship might be plotted as a graph, which can subsequently be converted into a nomograph.[1]

RELATIONSHIP OF GRAPH TO NOMOGRAPH

15.1 Straight-line Graphs

Every three-variable system that can be represented by a graph consisting of a family of straight lines superimposed upon a graph-paper grid can also be represented by a nomo-

---

[1] Adapted from paper presented before the Division of Industrial and Engineering Chemistry during the 124th meeting of the American Chemical Society; Ind. Eng. Chem., 47:33 (1955).

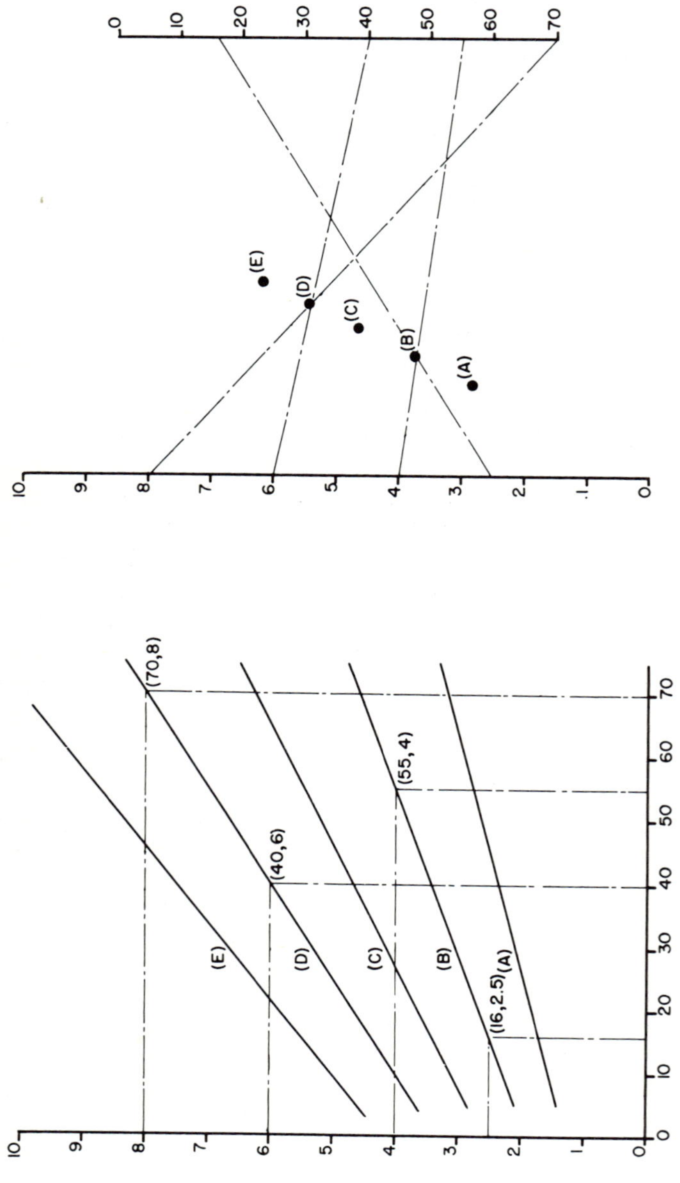

FIG. 15:1 METHOD OF PLOTTING THIRD SCALE OF NOMOGRAPH FROM POINTS ON GRAPH

graph having two parallel scales. These two scales may be identical with the X and Y axes of the graph-paper grid (except, possibly, for a change in modulus or sense). Indeed, the simplest method of constructing a nomograph from such a graph is: (1) to construct scales identical with the X and Y axes (except for sense), parallel to each another; and (2) to obtain points on the scale for the third variable by cross-plotting, as shown in Figure 15:1.

If the resulting nomograph lacks compactness, accuracy, or some other desirable characteristic, that defect may be remedied by a change in modulus of one or both of the outer scales and a replotting of the inner scale.

Figure 15:2 shows several types of straight-line graphs and the corresponding forms of the nomographs plotted by the method just described. In every example, the nomograph is superior to the graph in its simplicity and clarity. In view of the ease with which the graph may be converted to a nomograph, it is difficult to understand the large number of cross-section graphs that appear in the technical literature issued by manufacturers of equipment and materials. Without question, the nomograph has far greater appeal to the engineer and designer towards whom the technical information is directed.

15.2 Graphs With Curves

A large number of graphs found in the literature consist of families of curves superimposed upon a graph-paper grid. In many of them, these families of curves may be rectified by the proper selection of axes (Sec. 3.2). Where this is possible, the nomograph from the rectified graph may be constructed as shown in the preceding section.

If complete rectification is not possible, the method described in Section 15.7 for the construction of nomographs from tables should be followed. Readings for the cross-plot are obtained from the graph rather than from the table. The resulting nomograph contains more than one curved scale, as it is necessary to alter the positions of certain points on the exterior scales to accommodate the unrectified portions of the graphs. The relationship, how-

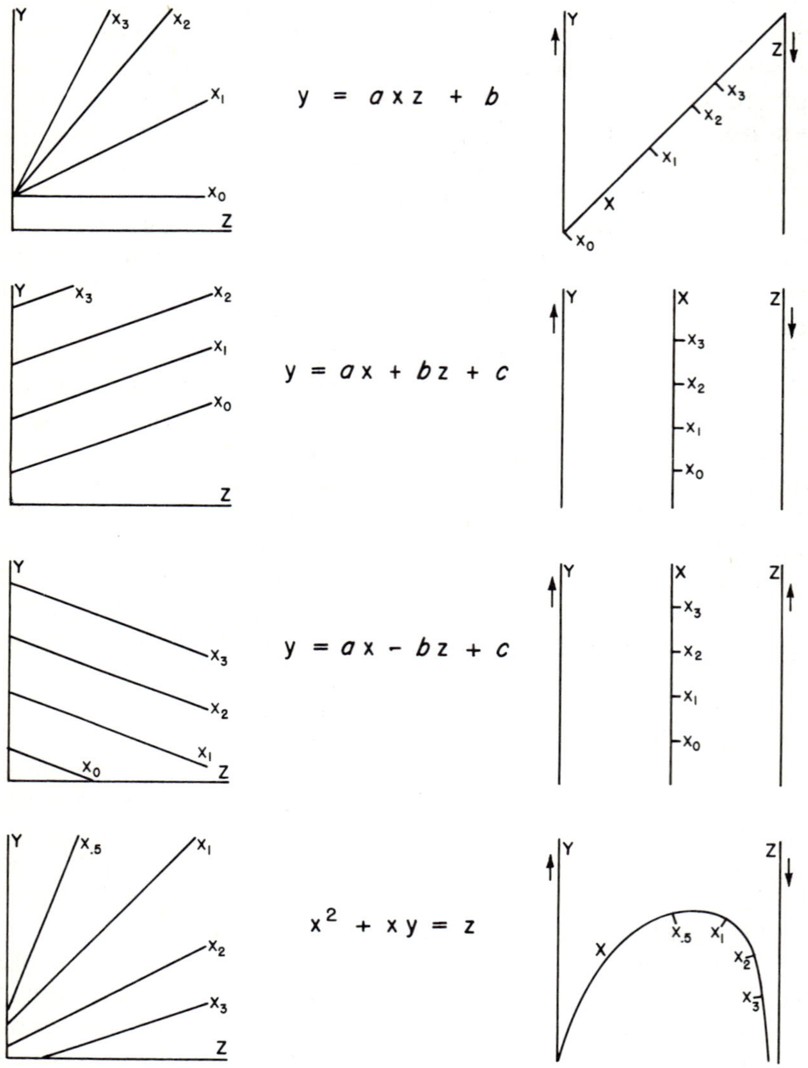

FIG. 15:2 RELATIONSHIPS OF SOME GRAPHS AND NOMOGRAPHS

ever, is more clearly understood in the light of the following discussion of construction of nomographs from tables.

CONSTRUCTING NOMOGRAPHS FROM TABLES: SYMMETRICAL

From the foregoing it is clear that the nature of the scales selected for the exterior scales of the nomograph (or the axes of the graph) affects profoundly the distribution of points comprising the third scale of the nomograph. The relationship is a complex one which must be visualized if it is to be used advantageously in the construction of nomographs. Toward this end, the various forms taken by nomographs that represent a symmetrical formula may be studied to reveal the effects of a change of scale on the nature of the other scales.

15.3 Logarithmic Mean Temperature Difference

An excellent example of a formula that yields an undesirable form of defining equation and for which it is desirable to construct a nomograph by other means is that for the logarithmic mean temperature difference,

$$\mathrm{MTD} = \frac{(\ln \Delta T_2 - \ln \Delta T_1)}{(\Delta T_2 - \Delta T_1)}$$

where ΔT_1 and ΔT_2 are the temperature changes experienced by two fluids as they pass simultaneously through a heat exchanger, and MTD is the logarithmic mean temperature difference.

This formula takes the determinant form,

$$\begin{vmatrix} \mathrm{MTD} & 1 & 0 \\ \Delta T_2 & \ln \Delta T_2 & 1 \\ \Delta T_1 & \ln \Delta T_1 & 1 \end{vmatrix} \qquad (15.1)$$

The nomographic representation of this defining equation

takes the form shown in Figure 15:3. The ΔT_1 and ΔT_2 scales, it will be noticed, are identical, making readings of MTD difficult and uncertain for cases in which ΔT_1 and

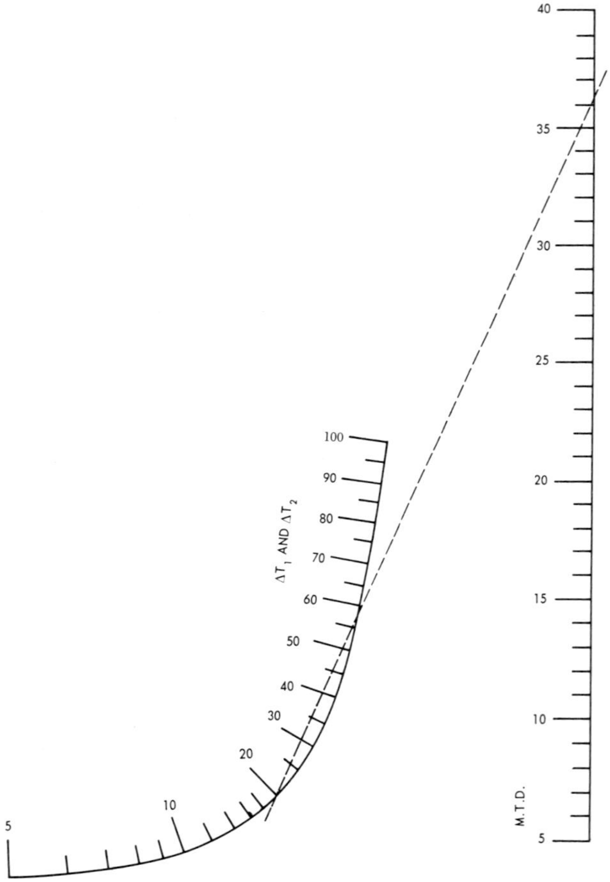

FIG. 15:3 NOMOGRAPH FOR LOGARITHMIC MTD CONSTRUCTED FROM DEFINING EQUATION

ΔT_2 have nearly the same value. Aside from this, however, the unsymmetrical form of this nomograph makes adjustment of scale moduli and positions especially unwieldly.

Consequently, the construction of an accurate nomograph by the use of this defining equation is very difficult.

A much more desirable form would be one in which the ΔT_1 and ΔT_2 scales were separate, with the MTD scale between them. However, to convert Equation 15.1 into a defining equation for such a form involves determining the nature of the functions that define the ΔT_1 and ΔT_2 scales. No general deductive process is available for this determination, and it is usually a matter of chance that such a conversion can be effected inductively.

Whether a nomograph of the type desired is compatible with the formula is generally indicated by examination of the formula. In the case of the logarithmic mean temperature difference the following aspects indicate the form of nomograph that might be constructed:

1. In the formula ΔT_1 and ΔT_2 may be interchanged without altering the value of MTD. This indicates the possibility of a symmetrical arrangement with identical or mirror-image scales for ΔT_1 and ΔT_2 on opposite sides of the MTD scale.

2. For the case $\Delta T_1 = \Delta T_2$, the value of MTD is equal to that of ΔT_1 and ΔT_2. Consequently, if the symmetrical form can be constructed, ordinates on the three scales will be equal for equal values of ΔT_1, ΔT_2, and MTD.

15.4 Method For Parallel Scales

In attempting to construct a symmetrical nomograph of this type, the obvious procedure is as follows (Fig. 15:4).

1) Locate suitable points (A, A', B, B') for the extremes of the exterior scales.
2) Cross-plot (I) between opposite points to obtain a point on the middle scale, and compute the value of this point.
3) Draw a horizontal line (II) through this point intersecting the exterior scales and compute the values of these points (C, C') on the exterior scales.
4) Cross-plot (III) between these points and the terminal points of the scales to obtain two more points on the middle scale. Compute the values of these points.

5) Draw horizontal lines (IV) through these points intersecting the exterior scales. Repeat this process of subdivision until enough points have been located that the remainder of each scale may be plotted by graphical interpolation.

In actual practice the values (Table 15:1) of the points obtained by this cross-plot are calculated in advance from the formula and need not be inserted at the plotted points until the scale is to be subdivided by graphical interpolation. In Figure 15:4 this method is illustrated for the logarithmic mean

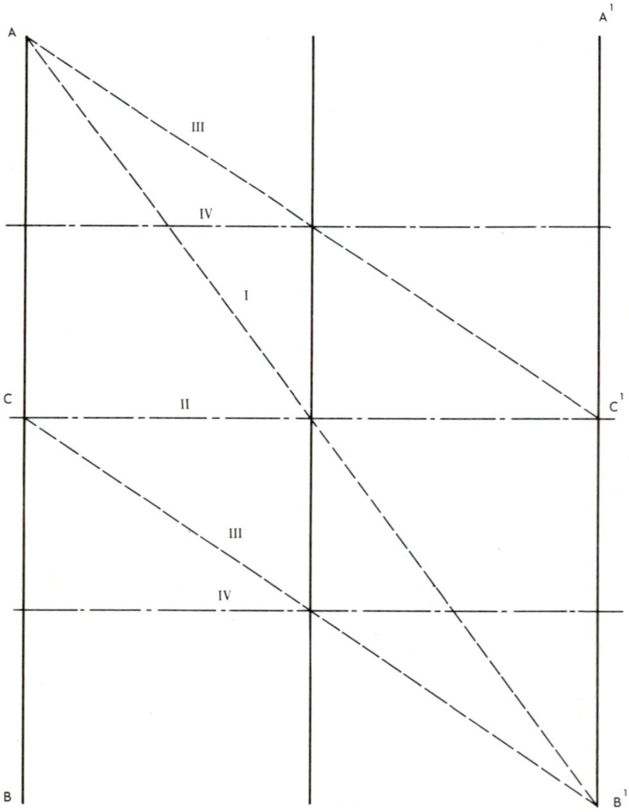

FIG. 15:4 CONSTRUCTION OF SYMMETRICAL NOMOGRAPH FROM DATA BY SUCCESSIVE SUB-DIVISION

Table 15:1
ORDINATES OBTAINED BY SUCCESSIVE CROSS PLOTTING OF MTD BETWEEN ΔT_1 AND ΔT_2 EXTREMES 10 TO 100

	Cross Plot Number 1	2	3	4
100.00				
				90.22
			81.13	
				72.66
		64.83		
				57.54
			50.86	
				44.70
	39.08			
				33.96
			29.31	
				25.10
		21.33		
				17.95
			14.95	
				12.31
10.00				

temperature chart with parallel scales. Table 15:1 shows the values of the ordinates for all three scales for the points of subdivision obtained by this method of cross-plotting. Figure 15:5 shows the nomograph, with scales completed by graphical interpolation.

The defining equation (Cartesian coordinates) of this nomograph is

$$\begin{vmatrix} f_1\ (\text{MTD}) & 1/2 & 1 \\ f_1\ (\Delta T_1) & 0 & 1 \\ f_1\ (\Delta T_2) & 1 & 1 \end{vmatrix} = 0 \qquad (15.2)$$

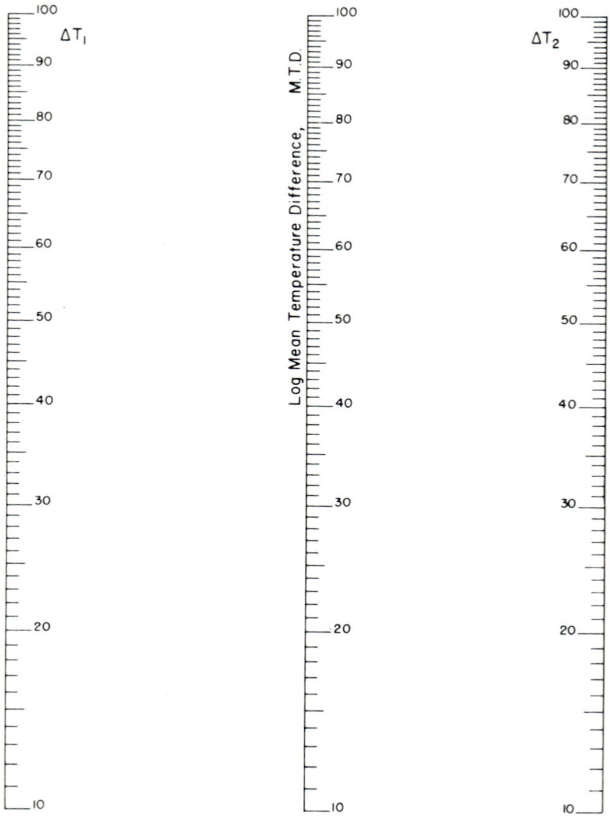

FIG. 15:5 NOMOGRAPH FOR LOGARITHMIC MTD CONSTRUCTED FROM DATA

The f_1's are identical for the three scales, although the nature of f_1 is not specified.

15.5 Method for Special Scales

Occasionally it is desirable that the MTD scale be linear or logarithmic so that the MTD nomograph may be combined with other scales in the construction of a more complex nomograph. In this event the middle scale is first laid off, and a point P_1 is selected for one of the termini of an exterior scale. Directly opposite this point (specifically, on a line perpendicular to the MTD scale and equally distant from the MTD scale) a point P_2 is placed for the corresponding terminus of the other exterior scale. Ordinates corresponding to the values in Table 15.1 are located on the MTD scale, and horizontal lines A, B, C, are drawn to establish the corresponding ordinates of points on the exterior scales (Fig. 15:6). Abscissas of those points are located by diagonal I, II, III lines corresponding to the method of cross-plotting by which the values of the points in Table 15:1 were derived.

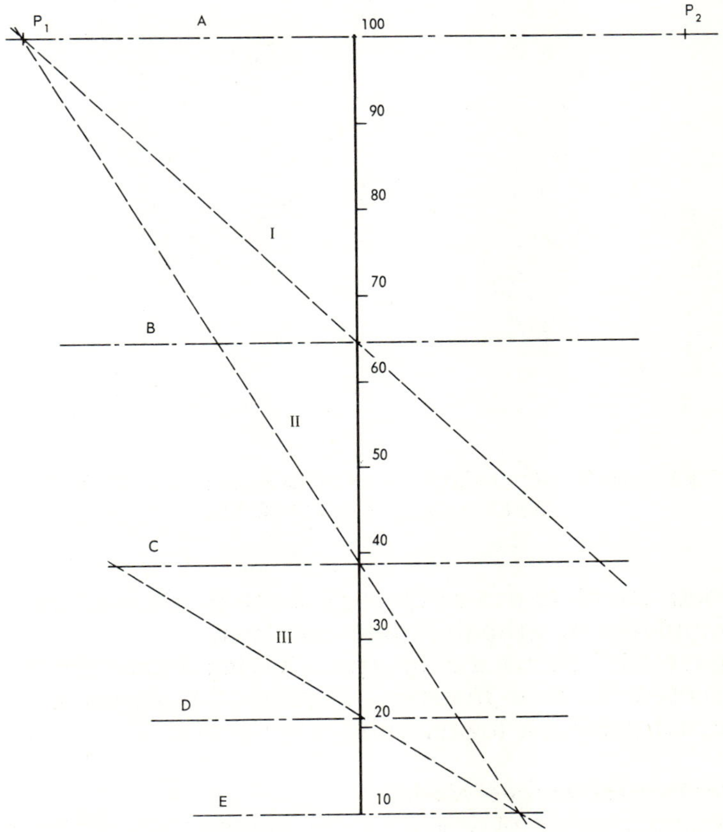

FIG. 15:6 METHOD OF CROSS-PLOTTING TO OBTAIN ABSCISSAS OF EXTERIOR SCALES

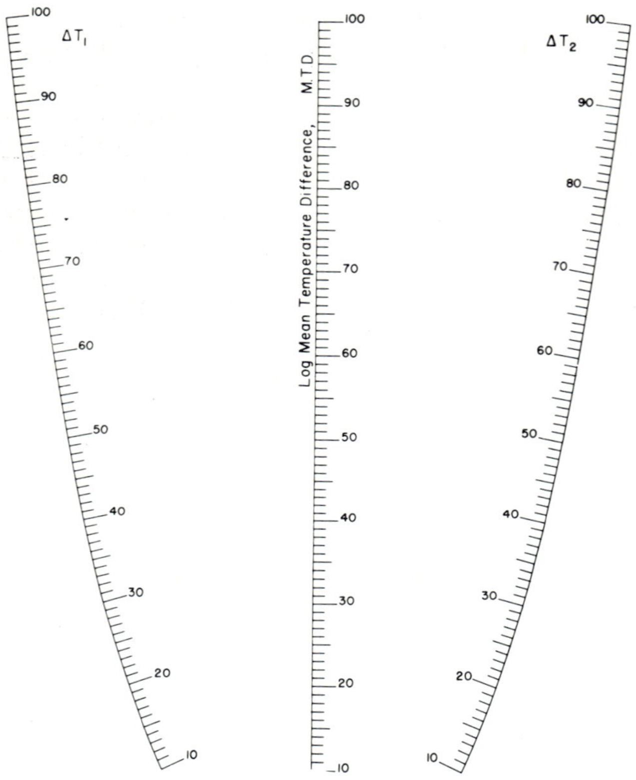

FIG. 15:7 NOMOGRAPH FOR LOGARITHMIC MTD WITH LINEAR MTD SCALE

A smooth curve is drawn through these points and the scales are completed by graphical interpolation.

Figure 15:7 shows a nomograph having a linear MTD scale constructed in this manner. Figure 15:8 shows a similar nomograph having a logarithmic MTD scale.

15.6 Method for Unequal Moduli

Unequal moduli of the exterior scales may be desired as a means of presenting a wider range on one scale than on the other without departing from rectangularity, or as a means of fitting the nomograph to an irregularly shaped space.

This requires only a slight modification of the method just described. A suitable point of projection P is employed, and points on the three scales having equal ordinates are located on lines that pass through this point of projection, rather than on horizontal lines. Instead of equidistance from the MTD scale as measured on a linear scale, one substitutes "equidistance" on a hyperbolic scale (Chapter 10). Figure 15:9 shows the method applied to an MTD nomograph with parallel scales.

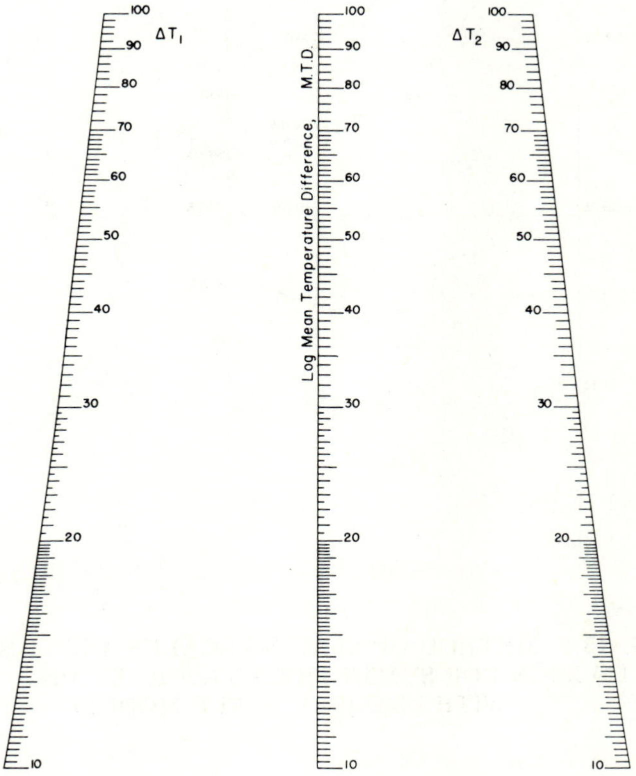

FIG. 15:8 NOMOGRAPH FOR LOGARITHMIC MTD WITH LOGARITHMIC MTD SCALE

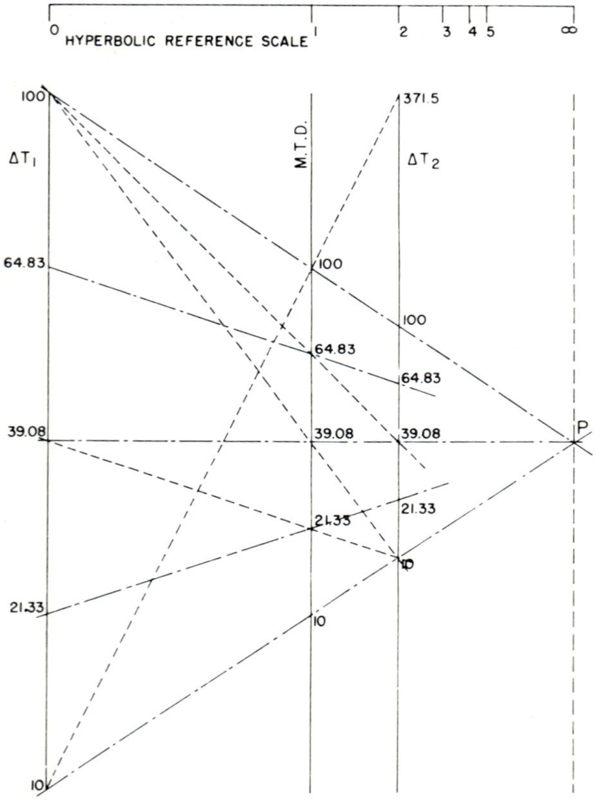

FIG. 15:9 METHOD OF SPACING SCALES AND SUBDIVISION FOR SYMMETRICAL NOMOGRAPH WITH UNEQUAL SCALE MODULI

CONSTRUCTING NOMOGRAPHS FROM TABLES: GENERAL CASE

In the foregoing example, a great deal of freedom could be exercised in the choice of a scale for the MTD; the scale function could be varied to accommodate various demands other than providing a nomograph having certain degrees of accuracy and compatibility of its scales. Most tables of data, however, are unsymmetrical and leave far less choice in the selection of scale functions. A poorly selected function for one of the scales may result in a nomograph so distorted that it would be virtually illegible over portions of certain of its scales. On the other hand, with well-selected scale functions it is usually possible to construct a useful nomograph, even for a table that presents seemingly great irregularities. It is the purpose of the remainder of this chapter to present (1) a method for converting tables of asymmetrical data into nomographic form; and (2) some feasible steps leading to a proper selection of the most critical scale functions.

15.7 Method of Construction: Arbitrary Features

The features of the nomograph (which may be selected at will, subject to certain reservations to be pointed out later,) are:

1) A scale for the dependent variable (Scale I), usually taken as one of the extreme scales;
2) The extreme points of one of the other scales, usually Scale II opposite to the dependent variable scale.

Figure 14:10 shows this stage in the constructing of a nomograph. As the construction proceeds, it is often necessary to revise the features originally selected. The required revisions are usually clearly indicated by the shapes of the scales that result from cross-plotting between the original points of the nomograph.

Cross-plotting: Using the points originally laid out, it is possible to locate the points of the Scale III by cross-plotting, as shown in Figure 15:11. Enough of these points should be located to determine whether the resulting scale will be sufficiently accurate and well-proportioned before proceed-

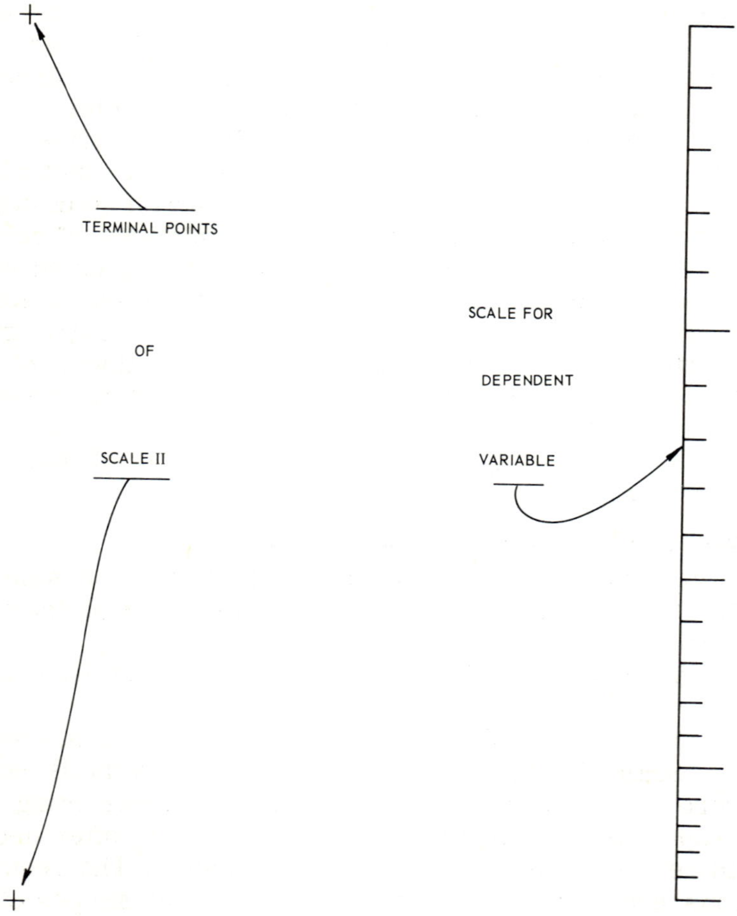

FIG. 15:10 ARBITRARY FEATURES OF GENERAL NOMOGRAPH CONSTRUCTED FROM DATA

ing to the next step; enough, also, to serve for the location of the remaining points of the Scale III. The intermediate points of the Scale II are located by a similar manner of cross-plotting, as shown in Figure 15:12. Portions of the

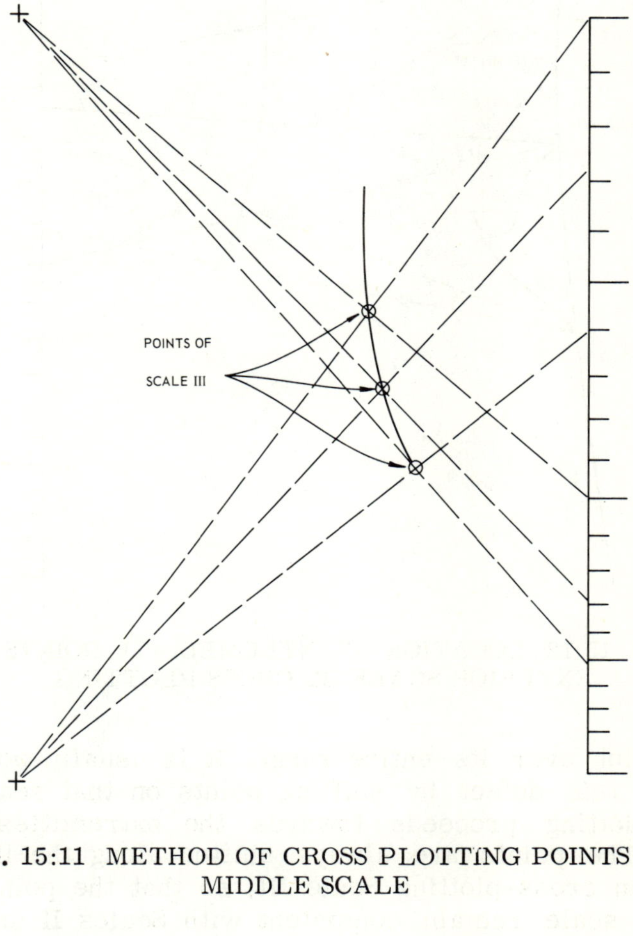

FIG. 15:11 METHOD OF CROSS PLOTTING POINTS OF MIDDLE SCALE

Scale III that could not be plotted from the terminal points of Scale II are now plotted.

At times it develops that the scale for the dependent variable, as originally laid out, fails to yield a satisfactory

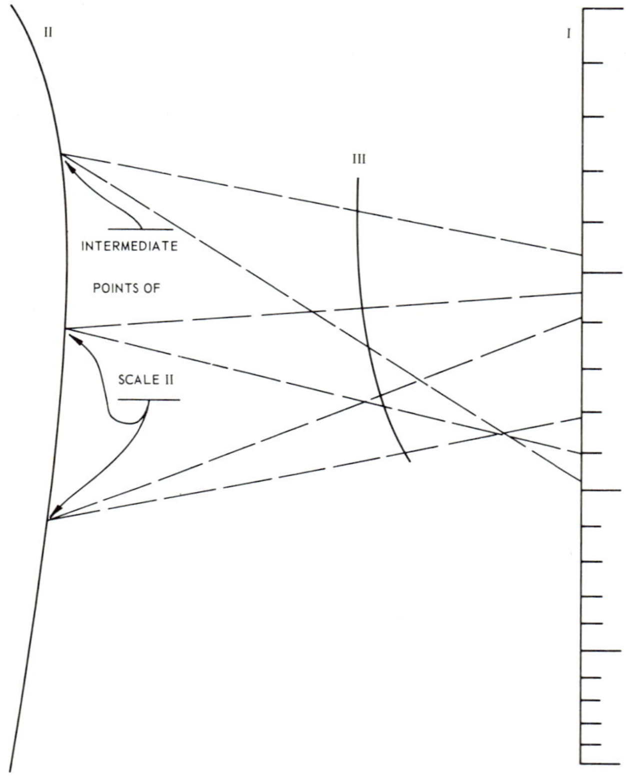

FIG. 15:12 LOCATION OF INTERMEDIATE POINTS ON EXTERIOR SCALE BY CROSS PLOTTING

cross-plot over its entire range. It is usually possible to remedy this defect by shifting points on that scale as the cross-plotting proceeds towards the extremities of that scale. The points are always shifted along the lines employed in cross-plotting Scale III, so that the points on the revised scale remain consistent with Scales II and III, as shown in Figure 15:13. The result is a curved scale; it is through this device that empirical nomographs having three curved scales sometimes make their appearance.

15.8 Selection of Scale for Dependent Variable

In the selection of a scale for the dependent variable, three points deserve primary consideration:

1) The range of the scale must be such as to include all the values of the dependent variable given in the table;
2) The subdivision of the scale must be such as to provide the degree of accuracy required in each of these values,
3) The scale must be free of discontinuities foreign to the dependent variable.

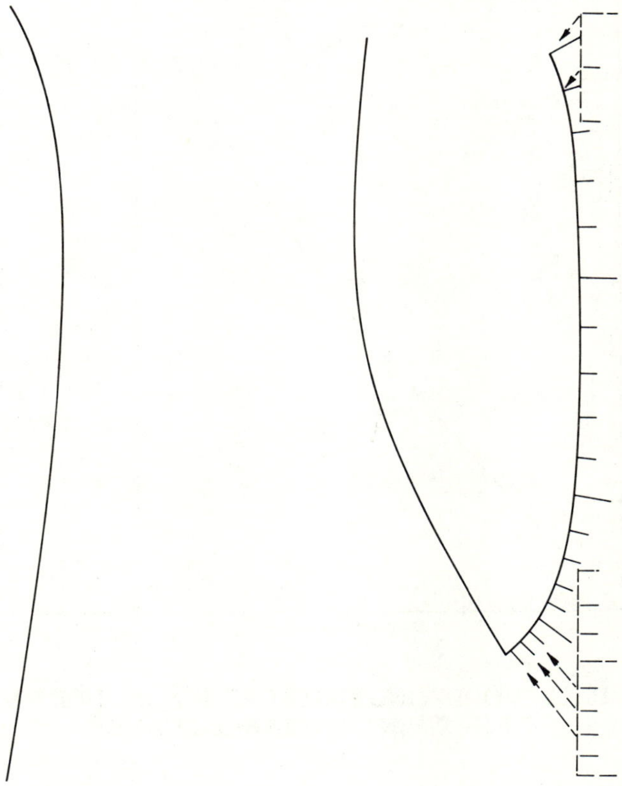

FIG. 15:13 MODIFYING SHAPE OF DEPENDENT VARIABLE SCALE

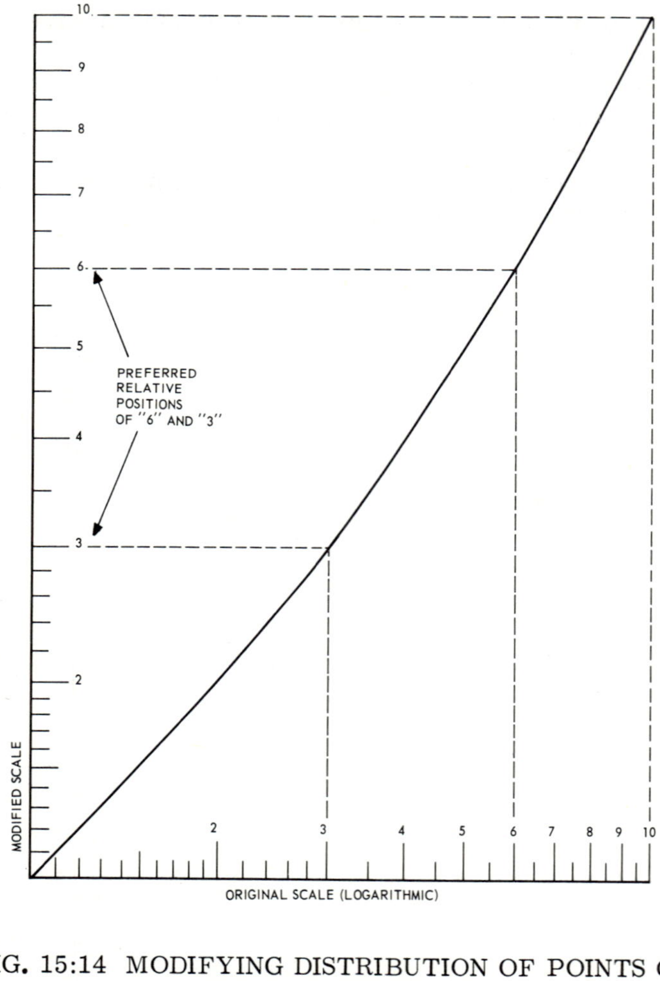

FIG. 15:14 MODIFYING DISTRIBUTION OF POINTS ON DEPENDENT VARIABLE SCALE

A good starting point in the selection of the scale is one of the common functions, or a function known to be generic with the variable to be represented. Some of the common functions that should be considered as to range and dis-

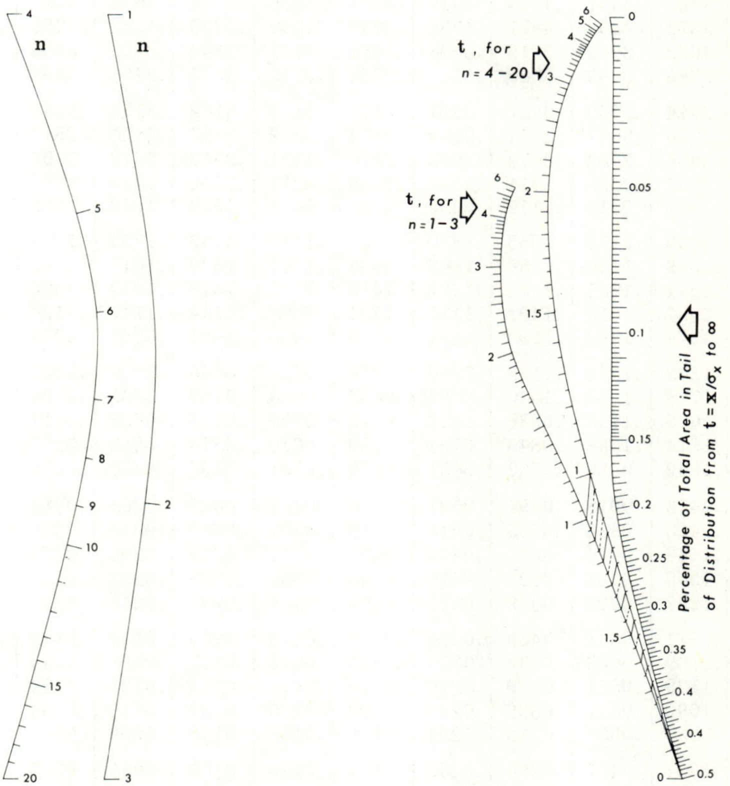

FIG. 15:15 NOMOGRAPH SHOWING "STUDENT'S" t DISTRIBUTION

tribution of points are $y = x^n$, $y = e^x$, $y = \log x$, $y = x/(x + r)$, $y = 1/x$, $y = ax^2 + bx + c$, $y = \sin x$, $y = \tan x$, and so on. It is desirable to be familiar with the distribution of points on scales representing the above functions over the entire range of x from $-\infty$ to 0 to 1 to $+\infty$.

Table 15:2

DISTRIBUTION OF STUDENT'S t

(Percentage of the Total Area in Tail of Distribution from $t = x/\sigma_x$ to ∞)

t	n = 1 n' = 2	2 3	3 4	4 5	5 6	6 7	7 8	8 9	9 10	10 11
0	.5000	.5000	.5000	.5000	.5000	.5000	.5000	.5000	.5000	.5000
0.1	.4683	.4647	.4633	.4626	.4621	.4618	.4616	.4614	.4613	.4612
0.2	.4372	.4300	.4271	.4256	.4247	.4240	.4236	.4232	.4230	.4227
0.3	.4072	.3962	.3919	.3896	.3881	.3871	.3864	.3859	.3855	.3852
0.4	.3789	.3639	.3580	.3548	.3528	.3515	.3505	.3498	.3492	.3488
0.5	.3524	.3333	.3257	.3217	.3191	.3174	.3162	.3153	.3145	.3139
0.6	.3280	.3047	.2954	.2904	.2873	.2852	.2837	.2826	.2817	.2809
0.7	.3056	.2782	.2672	.2613	.2576	.2551	.2533	.2519	.2508	.2499
0.8	.2852	.2538	.2411	.2342	.2300	.2271	.2250	.2234	.2222	.2212
0.9	.2667	.2316	.2172	.2095	.2047	.2014	.1990	.1972	.1958	.1946
1.0	.2500	.2113	.1955	.1870	.1816	.1780	.1753	.1733	.1717	.1704
1.1	.2349	.1930	.1758	.1665	.1607	.1567	.1539	.1517	.1499	.1486
1.2	.2211	.1765	.1581	.1482	.1419	.1377	.1346	.1322	.1304	.1289
1.3	.2087	.1616	.1422	.1317	.1252	.1207	.1174	.1149	.1130	.1114
1.4	.1974	.1482	.1280	.1171	.1102	.1055	.1021	.0995	.0975	.0959
1.5	.1872	.1362	.1153	.1040	.0970	.0921	.0886	.0860	.0839	.0823
1.6	.1778	.1254	.1040	.0924	.0852	.0804	.0768	.0741	.0720	.0703
1.7	.1693	.1156	.0938	.0822	.0749	.0700	.0665	.0638	.0617	.0600
1.8	.1614	.1068	.0848	.0731	.0659	.0610	.0574	.0548	.0527	.0510
1.9	.1542	.0989	.0768	.0651	.0579	.0531	.0496	.0470	.0449	.0433
2.0	.1476	.0918	.0697	.0581	.0510	.0462	.0428	.0403	.0383	.0367
2.1	.1415	.0853	.0633	.0518	.0449	.0402	.0369	.0345	.0326	.0310
2.2	.1358	.0794	.0576	.0463	.0395	.0351	.0319	.0295	.0277	.0262
2.3	.1305	.0741	.0525	.0415	.0349	.0306	.0275	.0252	.0235	.0221
2.4	.1257	.0692	.0479	.0372	.0308	.0266	.0237	.0216	.0199	.0187
2.5	.1211	.0648	.0439	.0334	.0272	.0233	.0205	.0185	.0169	.0157
2.6	.1169	.0608	.0402	.0300	.0242	.0203	.0177	.0158	.0144	.0132
2.7	.1129	.0571	.0369	.0270	.0214	.0178	.0153	.0135	.0122	.0112
2.8	.1092	.0537	.0339	.0244	.0190	.0156	.0133	.0116	.0104	.0094
2.9	.1057	.0506	.0313	.0221	.0169	.0137	.0115	.0099	.0088	.0079
3.0	.1024	.0477	.0288	.0200	.0150	.0120	.0100	.0085	.0075	.0067
3.1	.0993	.0451	.0266	.0181	.0134	.0106	.0087	.0073	.0064	.0056
3.2	.0964	.0427	.0247	.0165	.0120	.0093	.0075	.0063	.0054	.0047
3.3	.0937	.0404	.0229	.0150	.0107	.0082	.0066	.0054	.0046	.0040
3.4	.0911	.0383	.0212	.0136	.0096	.0072	.0057	.0047	.0039	.0034

If one of these functions gives a scale that approaches but does not quite reach the required standards, it may be modified by the method shown in Figure 15:14. The scale is laid out on the abscissa of coordinate paper, and the proposed extremes of the modified scale are marked off on the ordinate.

Corresponding extremes of the two scales determine points in the plane, as shown. Approximate positions of additional ordinates are indicated to provide the distribution of points desired. These ordinates, with the corresponding abscissas from the original scale, also determine points in the plane. A smooth curve is then drawn through the approximate positions of these points, it being much more important that the curve be smooth and regular than that it pass exactly through the points. The points of the original scale are then projected upward to the curve and across to the ordinate scale, thus producing the modified scale. This method is used in constructing the nomograph shown in Figure 15:15, derived from the values of Table 15:2.[2]

15.9 Variations in Plotting

The method of plotting may be varied, producing inversions of the scale positions. One variation is the substitution of the second independent variable for the one first plotted opposite the dependent variable scale. The other variation is transposition of the terminal points of that scale. In the nomographs shown here, these variations would have contributed little to the appearance or legibility of the nomograph. Such modifications should be considered, however, if it is desirable that certain scale points lie in predetermined positions, or if the length of a scale is out of proportion to the degree of accuracy required of the variable it represents.

This method may be applied to other formulas for which a suitable defining equation may not be obtained or to empirical tables for which no formula has been derived. In the latter case, the nomograph scales may be analyzed by standard methods to obtain general expressions for the functions defining their ordinates and abscissas. The values may then be inserted in a skeleton defining equation, thereby giving rise to an empirical formula for the relationships inherent in the table and in the nomograph.

---

[2]Table 15:2 shows one corner of the table used to prepare the nomograph of Figure 15:15. For complete table, see Statistical Procedures and their Mathematical Bases (Peters, Charles C., Van Voorhis, Walter R.), 2nd Ed.; McGraw-Hill Book Company, Inc., New York, 1940.

APPENDIX

TABLE

Values of $x = \frac{p}{p + r}$ and 1 - x

p/r	x	1 - x	p/r	x	1 - x
.000	.00000	1.00000	.046	.04398	.95602
.001	.00100	.99900	.047	.04489	.95511
.002	.00200	.99800	.048	.04580	.95420
.003	.00299	.99701	.049	.04671	.95329
.004	.00398	.99602	.050	.04762	.95238
.005	.00498	.99502	.051	.04853	.95147
.006	.00596	.99404	.052	.04943	.95057
.007	.00695	.99305	.053	.05033	.94967
.008	.00794	.99206	.054	.05123	.94877
.009	.00892	.99108	.055	.05213	.94787
.010	.00990	.99010	.056	.05303	.94697
.011	.01088	.98912	.057	.05393	.94607
.012	.01186	.98814	.058	.05482	.94518
.013	.01283	.98717	.059	.05571	.94429
.014	.01381	.98619	.060	.05660	.94340
.015	.01478	.98522	.061	.05749	.94251
.016	.01575	.98425	.062	.05838	.94162
.017	.01672	.98328	.063	.05927	.94073
.018	.01768	.98232	.064	.06015	.93985
.019	.01865	.98135	.065	.06103	.93897
.020	.01961	.98039	.066	.06191	.93809
.021	.02057	.97943	.067	.06279	.93721
.022	.02153	.97847	.068	.06367	.93633
.023	.02248	.97752	.069	.06455	.93545
.024	.02344	.97656	.070	.06542	.93458
.025	.02439	.97561	.071	.06629	.93371
.026	.02534	.97466	.072	.06716	.93284
.027	.02629	.97371	.073	.06803	.93197
.028	.02724	.97276	.074	.06890	.93110
.029	.02818	.97182	.075	.06977	.93023
.030	.02913	.97087	.076	.07063	.92937
.031	.03007	.96993	.077	.07150	.92850
.032	.03101	.96899	.078	.07236	.92764
.033	.03195	.96805	.079	.07322	.92678
.034	.03288	.96712	.080	.07407	.92593
.035	.03382	.96618	.081	.07493	.92507
.036	.03475	.96525	.082	.07579	.92421
.037	.03568	.96432	.083	.07664	.92336
.038	.03661	.96139	.084	.07749	.92251
.039	.03754	.96246	.085	.07834	.92161
.040	.03846	.96152	.086	.07919	.92081
.041	.03939	.96061	.087	.08004	.91996
.042	.04031	.95969	.088	.08088	.91912
.043	.04123	.95877	.089	.08173	.91827
.044	.04215	.95785	.090	.08257	.91743
.045	.0430[illegible]	.95694	.091	.08341	.91659

p/r	x	x - x	p/r	x	1 - x
.092	.08425	.91575	.143	.12511	.87489
.093	.08509	.91491	.144	.12587	.87413
.094	.08592	.91408	.145	.12664	.87336
.095	.08676	.91324	.146	.12740	.87260
.096	.08759	.91241	.147	.12816	.87184
.097	.08842	.91158	.148	.12892	.87108
.098	.08925	.91075	.149	.12968	.87032
.099	.09008	.90992	.150	.13044	.86956
.100	.09091	.90909	.151	.13119	.86881
.101	.09174	.90826	.152	.13194	.86806
.102	.09256	.90744	.153	.13270	.86730
.103	.09338	.90662	.154	.13345	.86655
.104	.09420	.90580	.155	.13420	.86580
.105	.09502	.90498	.156	.13495	.86505
.106	.09584	.90416	.157	.13570	.86430
.107	.09666	.90334	.158	.13644	.86356
.108	.09747	.90253	.159	.13719	.86281
.109	.09829	.90171	.160	.13793	.86207
.110	.09910	.90090	.161	.13876	.86133
.111	.09991	.90009	.162	.13942	.86058
.112	.10072	.89928	.163	.14016	.85984
.113	.10153	.89847	.164	.14089	.85911
.114	.10233	.89767	.165	.14163	.85837
.115	.10314	.89686	.166	.14237	.85763
.116	.10394	.89606	.167	.14310	.85690
.117	.10475	.89525	.168	.14384	.85616
.118	.10555	.89445	.169	.14457	.85543
.119	.10635	.89365	.170	.14530	.85470
.120	.10714	.89286	.171	.14603	.85397
.121	.10794	.89206	.172	.14676	.85324
.122	.10873	.89127	.173	.14749	.85251
.123	.10953	.89047	.174	.14821	.85179
.124	.11032	.88968	.175	.14894	.85106
.125	.11111	.88889	.176	.14966	.85034
.126	.11190	.88810	.177	.15038	.84962
.127	.11269	.88731	.178	.15110	.84890
.128	.11348	.88652	.179	.15182	.84818
.129	.11426	.88574	.180	.15254	.84746
.130	.11504	.88496	.181	.15326	.84674
.131	.11583	.88417	.182	.15398	.84602
.132	.11661	.88339	.183	.15469	.84531
.133	.11739	.88261	.184	.15541	.84459
.134	.11817	.88183	.185	.15612	.84388
.135	.11894	.88106	.186	.15683	.84317
.136	.11972	.88028	.187	.15754	.84246
.137	.12049	.87951	.188	.15825	.84175
.138	.12127	.87873	.189	.15896	.84104
.139	.12204	.87796	.190	.15966	.84034
.140	.12281	.87719	.191	.16037	.83963
.141	.12358	.87642	.192	.16107	.83893
.142	.12434	.87566	.193	.16178	.83822

p/r	x	1 - x	p/r	x	1 - x
.194	.16248	.83752	.245	.19679	.80321
.195	.16318	.83682	.246	.19743	.80257
.196	.16388	.83612	.247	.19808	.80192
.197	.16458	.83542	.248	.19872	.80128
.198	.16528	.83472	.249	.19936	.80064
.199	.16597	.83403	.250	.20000	.80000
.200	.16667	.83333	.251	.20064	.79936
.201	.16736	.83264	.252	.20128	.79872
.202	.16805	.83195	.253	.20192	.79808
.203	.16875	.83125	.254	.20255	.79745
.204	.16944	.83056	.255	.20319	.79681
.205	.17012	.82988	.256	.20382	.79618
.206	.17081	.82919	.257	.20446	.79554
.207	.17150	.82850	.258	.20509	.79491
.208	.17219	.82781	.259	.20572	.79428
.209	.17287	.82713	.260	.20635	.79365
.210	.17355	.82645	.261	.20698	.79302
.211	.17424	.82576	.262	.20761	.79239
.212	.17492	.82508	.263	.20823	.79177
.213	.17560	.82440	.264	.20886	.79114
.214	17628	.82372	.265	.20949	.79051
.215	.17696	.82304	.266	.21011	.78989
.216	.17763	.82237	.267	.21073	.78927
.217	.17831	.82169	.268	.21136	.78864
.218	.17898	.82102	.269	.21198	.78802
.219	.17966	.82034	.270	.21260	.78740
.220	.18033	.81967	.271	.21322	.78678
.221	.18100	.81900	.272	.21384	.78616
.222	.18167	.81833	.273	.21445	.78555
.223	.18234	.81766	.274	.21507	.78493
.224	.18301	.81699	.275	.21569	.78431
.225	.18367	.81633	.276	.21630	.78370
.226	.18434	.81566	.277	.21692	.78308
227	.18500	.81500	.278	.21753	.78247
.228	.18567	.81433	.279	.21814	.78186
.229	.18633	.81367	.280	.21875	.78125
.230	.18699	.81301	.281	.21936	.78064
.231	.18765	.81235	.282	.21997	.78003
.232	.18831	.81169	.283	.22058	.77942
.233	.18897	.81103	.284	.22118	.77882
.234	.18963	.81037	.285	.22179	.77821
.235	.19028	.80972	.286	.22240	.77760
.236	.19094	.80906	.287	.22300	.77700
.237	.19159	.80841	.288	.22360	.77640
.238	.19225	.80775	.289	.22421	.77579
.239	.19290	.80710	.290	.22481	.77519
.240	.19355	.80645	.291	.22541	.77459
.241	.19420	.80580	.292	.22601	.77399
.242	.19485	.80515	.293	.22661	.77339
.243	.19550	.80450	.294	.22720	.77280
.244	.19614	.80386	.295	.22780	.77220

p/r	x	1 - x	p/r	x	1 - x
.296	.22840	.77160	.396	.28367	.71633
.297	.22900	.77100	.398	.28469	.71531
.298	.22958	.77042	.400	.28571	.71429
.299	.23018	.76982	.402	.28673	.71327
.300	.23077	.76923	.404	.28775	.71226
.302	.23195	.76805	.406	.28876	.71124
.304	.23313	.76687	.408	.28977	.71023
.306	.23430	.76570	.410	.29078	.70922
.308	.23547	.76453	.412	.29178	.70822
.310	.23664	.76336	.414	.29279	.70721
.312	.23781	.76219	.416	.29379	.70621
.314	.23897	.76103	.418	.29478	.70522
.316	.24012	.75988	.420	.29577	.70423
.318	.24127	.75873	.422	.29677	.70323
.320	.24242	.75758	.424	.29775	.70225
.322	.24357	.75643	.426	.29874	.70126
.324	.24471	.75529	.428	.29972	.70028
.326	.24585	.75415	.430	.30070	.69930
.328	.24699	.75301	.432	.30168	.69832
.330	.24812	.75188	.434	.30265	.69735
.332	.24925	.75075	.436	.30362	.69638
.334	.25038	.74962	.438	.30459	.69541
.336	.25150	.74850	.440	.30556	.69444
.338	.25262	.74738	.442	.30652	.69348
.340	.25373	.74627	.444	.30748	.69252
.342	.25484	.74516	.446	.30844	.69156
.344	.25595	.74405	.448	.30939	.69061
.346	.25706	.74294	.450	.31034	.68966
.348	.25816	.74184	.452	.31129	.68871
.350	.25926	.74074	.454	.31224	.68776
.352	.26036	.73964	.456	.31319	.68681
.354	.26145	.73855	.458	.31413	.68587
.356	.26254	.73746	.460	.31507	.68493
.358	.26362	.73638	.462	.31601	.68399
.360	.26471	.73529	.464	.31694	.68306
.362	.26579	.73421	.466	.31787	.68213
.364	.26686	.73314	.468	.31880	.68120
.366	.26794	.73206	.470	.31973	.68027
.368	.26901	.73099	.472	.32068	.67935
.370	.27007	.72993	.474	.32157	.67843
.372	.27114	.72886	.476	.32249	.67751
.374	.27220	.72780	.478	.32341	.67659
.376	.27326	.72674	.480	.32432	.67568
.378	.27431	.72569	.482	.32524	.67476
.380	.27536	.72464	.484	.32615	.67385
.382	.27641	.72359	.486	.32705	.67295
.384	.27746	.72254	.488	.32796	.67204
.386	.27850	.72150	.490	.32886	.67114
.388	.27954	.72046	.492	.32976	.67024
.390	.28058	.71942	.494	.33066	.66934
.392	.28161	.71839	.496	.33155	.66845
.394	.28264	.71736	.498	.33244	.66756

p/r	x	1 - x	p/r	x	1 - x
.500	.33333	.66667	.602	.37578	.62422
.502	.33422	.66578	.604	.37656	.62344
.504	.33511	.66489	.606	.37733	.62267
.506	.33599	.66401	.608	.37811	.62189
.508	.33687	.66313	.610	.37888	.62112
.510	.33775	.66225	.612	.37965	.62035
.512	.33862	.66138	.614	.38042	.61958
.514	.33950	.66050	.616	.38119	.61881
.516	.34037	.65963	.618	.38195	.61805
.518	.34124	.65876	.620	.38272	.61728
.520	.34211	.65789	.622	.38347	.61652
.522	.34297	.65703	.624	.38424	.61576
.524	.34383	.65617	.626	.38499	.61501
.526	.34469	.65531	.628	.38575	.61425
.528	.34555	.65445	.630	.38650	.61350
.530	.34641	.65359	.632	.38725	.61275
.532	.34726	.65274	.634	.38800	.61200
.534	.34811	.65189	.636	.38875	.61125
.536	.34896	.65104	.638	.38950	.61050
.538	.34980	.65020	.640	.39024	.60976
.540	.35065	.64935	.642	.39099	.60901
.542	.35149	.64851	.644	.39173	.60827
.544	.35233	.64767	.646	.39247	.60753
.546	.35317	.64683	.648	.39320	.60680
.548	.35401	.64599	.650	.39394	.60606
.550	.35484	.64516	.652	.39467	.60533
.552	.35567	.64433	.654	.39541	.60459
.554	.35650	.64350	.656	.39614	.60386
.556	.35733	.64267	.658	.39686	.60314
.558	.35815	.64185	.660	.39759	.60241
.560	.35897	.64103	.662	.39832	.60168
.562	.35980	.64020	.664	.39904	.80096
.564	.36061	.63939	.666	.39976	.60024
.566	.36143	.63857	.668	.40048	.59952
.568	.36224	.63776	.670	.40120	.59880
.570	.36306	.63694	.672	.40191	.59809
.572	.36387	.63613	.674	.40263	.59737
.574	.36468	.63532	.676	.40334	59666
.576	.36548	.63452	.678	.40405	.59595
.578	.36629	.63371	.680	.40476	.59524
.580	.36709	.63291	.682	.40547	.59453
.582	.36789	.63211	.684	.40618	.59382
.584	.36869	.63131	.686	.40688	.59312
.586	.36948	.63052	.688	.40758	.59242
.588	.37028	.62972	.690	.40828	.59172
.590	.37107	.62893	.692	.40898	.59102
.592	.37186	.62814	.694	.40968	.59033
.594	.37265	.62735	.696	.41038	.58962
.596	.37343	.62657	.698	.41107	.58893
.598	.37422	.62578	.700	.41176	.58824
.600	.37500	.62500	.702	.41246	.58754

p/r	x	1 - x	p/r	x	1 - x
.704	.41315	.58685	.806	.44629	.55371
.706	.41383	.58617	.808	.44690	.55310
.708	.41452	.58548	.810	.44751	.55249
.710	.41520	.58480	.812	.44812	.55188
.712	.41589	.58411	.814	.44873	.55127
.714	.41657	.58343	.816	.44934	.55066
.716	.41725	.58275	.818	.44994	.55006
.718	.41793	.58207	.820	.45055	.54945
.720	.41860	.58140	.822	.45115	.54885
.722	.41928	.58072	.824	.45175	.54825
.724	.41995	.58005	.826	.45235	.54765
.726	.42063	.57937	.828	.45295	.54705
.728	.42130	.57870	.830	.45355	.54645
.730	.42197	.57803	.832	.45415	.54585
.732	.42263	.57737	.834	.45474	.54526
.734	.42330	.57670	.836	.45534	.54466
.736	.42396	.57604	.838	.45593	.54407
.738	.42463	.57537	.840	.45652	.54348
.740	.42529	.57471	.842	.45711	.54289
.742	.42595	.57405	.844	.45770	.54230
.744	.42661	.57339	.846	.45829	.54171
.746	.42726	.57274	.848	.45887	.54113
.748	.42792	.57208	.850	.45946	.54054
.750	.42857	.57143	.852	.46004	.53996
.752	.42922	.57078	.854	.46063	.53937
.754	.42987	.57013	.856	.46121	.53879
.756	.43052	.56948	.858	.46179	.53821
.758	.43117	.56883	.860	.46237	.53763
.760	.43182	.56818	.862	.46294	.53706
.762	.43246	.56754	.864	.46352	.53648
.764	.43311	.56689	.866	.46409	.53591
.766	.43375	.56625	.868	.46467	.53533
.768	.43439	.56561	.870	.46524	.53476
.770	.43503	.56497	.872	.46581	.53419
.772	.43567	.56433	.874	.46638	.53362
.774	.43630	.56370	.876	.46695	.53305
.776	.43696	.56304	.878	.46752	.53248
.778	.43757	.56243	.880	.46809	.53191
.780	.43820	.56180	.882	.46865	.53135
.782	.43883	.56117	.884	.46921	.53079
.784	.43946	.56054	.886	.46978	.53022
.786	.44009	.55991	.888	.47034	.52966
.788	.44072	.55928	.890	.47090	.52910
.790	.44134	.55866	.892	.47146	.52854
.792	.44196	.55804	.894	.47202	.52798
.794	.44259	.55741	.896	.47257	.52743
.796	.44321	.55679	.898	.47313	.52687
.798	.44383	.55617	.900	.47368	.52632
.800	.44444	.55556	.902	.47424	.52576
.802	.44506	.55494	.904	.47479	.52521
.804	.44568	.55432	.906	.47534	.52466

p/r	x	1 - x	p/r	x	1 - x
.908	.47589	.52411	1.025	.50617	.49383
.910	.47644	.52356	1.030	.50739	.49261
.912	.47699	.52301	1.035	.50860	.49140
.914	.47753	.52247	1.040	.50980	.49020
.916	.47808	.52192	1.045	.51100	.48900
.918	.47862	.52138	1.050	.51220	.48780
.920	.47917	.52083	1.055	.51338	.48662
.922	.47971	.52029	1.060	.51456	.48544
.924	.48025	.51975	1.065	.51574	.48426
.926	.48079	.51921	1.070	.51691	.48309
.928	.48133	.51867	1.075	.51807	.48193
.930	.48187	.51813	1.080	.51923	.48077
.932	.48240	.51760	1.085	.52038	.47962
.934	.48294	.51706	1.090	.52153	.47847
.936	.48347	.51653	1.095	.52267	.47733
.938	.48400	.51600	1.100	.52381	.47619
.940	.48454	.51546	1.105	.52494	.47505
.942	.48507	.51493	1.110	.52607	.47393
.944	.48560	.51440	1.115	.52719	.47281
.946	.48613	.51387	1.120	.52830	.47170
.948	.48665	.51335	1.125	.52941	.47059
.950	.48718	.51282	1.130	.53052	.46948
.952	.48770	.51230	1.135	.53162	.46638
.954	.48823	.51177	1.140	.53271	.46729
.956	.48875	.51125	1.145	.53380	.46620
.958	.48927	.51073	1.150	.53488	.46512
.960	.48980	.51020	1.155	.53596	.46404
.962	.49032	.50968	1.160	.53704	.46296
.964	.49084	.50917	1.165	.53811	.46189
.966	.49135	.50865	1.170	.53917	.46083
.968	.49187	.50813	1.175	.54023	.45977
.970	.49239	.50761	1.180	.54128	.45872
.972	.49290	.50710	1.185	.54233	.45767
.974	.49341	.50659	1.190	.54338	.45662
.976	.49393	.50607	1.195	.54442	.45558
.978	.49444	.50556	1.200	.54545	.45455
.980	.49495	.50505	1.205	.54649	.45351
.982	.49546	.50454	1.210	.54751	.45249
.984	.49597	.50403	1.215	.54853	.45147
.986	.49648	.50352	1.220	.54955	.45045
.988	.49698	.50302	1.225	.55056	.44944
.990	.49749	.50251	1.230	.55157	.44843
.992	.49799	.50201	1.235	.55257	.44743
.994	.49850	.50150	1.240	.55357	.44643
.996	.49900	.50100	1.245	.55457	.44543
.998	.49950	.50050	1.250	.55556	.44444
1.000	.50000	.50000	1.255	.55654	.44346
1.005	.50125	.49875	1.260	.55752	.44248
1.010	.50249	.49751	1.265	.55850	.44150
1.015	.50372	.49628	1.270	.55947	.44053
1.020	.50495	.49505	1.275	.56044	.43956

p/r	x	1 - x	p/r	x	1 - x
1.280	.56140	.43860	1.535	.60552	.39448
1.285	.56236	.43764	1.540	.60630	.39370
1.290	.56332	.43668	1.545	.60707	.39293
1.295	.56427	.43573	1.550	.60784	.39216
1.300	.56522	.43478	1.555	.60861	.39139
1.305	.56616	.43384	1.560	.60937	.39063
1.310	.56710	.43290	1.565	.61014	.38986
1.315	.56803	.43197	1.570	.61089	.38911
1.320	.56897	.43103	1.575	.61165	.38835
1.325	.56989	.43011	1.580	.61240	.38760
1.330	.57081	.42919	1.585	.61315	.38685
1.335	.57173	.42827	1.590	.61390	.38610
1.340	.57265	.42735	1.595	.61464	.38536
1.345	.57356	.42644	1.600	.61538	.38462
1.350	.57447	.42553	1.605	.61612	.38388
1.355	.57537	.42463	1.610	.61686	.38314
1.360	.57627	.42373	1.615	.61759	.38241
1.365	.57717	.42283	1.620	.61832	.38168
1.370	.57806	.42194	1.625	.61905	.38095
1.375	.57895	.42105	1.630	.61977	.38023
1.380	.57983	.42017	1.635	.62049	.37951
1.385	.58071	.41929	1.640	.62121	.37879
1.390	.58159	.41841	1.645	.62193	.37807
1.395	.58246	.41754	1.650	.62264	.37736
1.400	.58333	.41667	1.655	.62335	.37665
1.405	.58420	.41580	1.660	.62406	.37594
1.410	.58506	.41494	1.665	.62477	.37523
1.415	.58592	.41408	1.670	.62547	.37453
1.420	.58678	.41322	1.675	.62617	.37383
1.425	.58763	.41237	1.680	.62687	.37313
1.430	.58848	.41152	1.685	.62756	.37244
1.435	.58932	.41068	1.690	.62825	.37175
1.440	.59016	.40984	1.695	.62894	.47106
1.445	.59100	.40900	1.700	.62963	.37037
1.450	.59184	.40816	1.705	.63031	.36969
1.455	.59267	.40733	1.710	.63100	.36900
1.460	.59350	.40650	1.715	.63168	.36832
1.465	.59432	.40568	1.720	.63235	.36765
1.470	.59514	.40486	1.725	.63303	.36697
1.475	.59596	.40404	1.730	.63370	.36630
1.480	.59677	.40323	1.735	.63437	.36563
1.485	.59759	.40241	1.740	.63504	.36496
1.490	.59839	.40161	1.745	.63570	.36430
1.495	.59920	.40080	1.750	.63636	.36364
1.500	.60000	.40000	1.755	.63702	.36298
1.505	.60080	.39920	1.760	.63768	.36232
1.510	.60159	.39841	1.765	.63834	.36166
1.515	.60239	.39761	1.770	.63899	.36101
1.520	.60317	.39683	1.775	.63964	.36036
1.525	.60396	.39604	1.780	.64029	.35971
1.530	.60474	.39526	1.785	.64093	.35907

p/r	x	1 - x	p/r	x	1 - x
1.790	.64158	.35842	2.090	.67638	.32362
1.795	.64222	.35778	2.10	.67742	.32258
1.800	.64286	.35714	2.11	.67846	.32154
1.805	.64349	.35651	2.12	.67949	.32051
1.810	.64413	.35587	2.13	.68051	.31949
1.815	.64476	.35524	2.14	.68153	.31847
1.820	.64539	.35461	2.15	.68254	.31746
1.825	.64602	.35398	2.16	.68354	.31646
1.830	.64664	.35336	2.17	.68454	.31546
1.835	.64727	.35273	2.18	.68553	.31447
1.840	.64789	.35211	2.19	.68652	.31348
1.845	.64851	.35149	2.20	.68750	.31250
1.850	.64912	.35088	2.21	.68847	.31153
1.855	.64973	.35027	2.22	.68944	.31056
1.860	.65035	.34965	2.23	.69040	.30960
1.865	.65096	.34904	2.24	.69136	.30864
1.870	.65157	.34843	2.25	.69231	.30769
1.875	.65217	.34783	2.26	.69325	.30675
1.880	.65278	.34722	2.27	.69419	.30581
1.885	.65338	.34662	2.28	.69512	.30488
1.890	.65398	.34602	2.29	.69605	.30395
1.895	.65458	.34542	2.30	.69697	.30303
1.900	.65517	.34483	2.31	.69789	.30211
1.905	.65577	.34423	2.32	.69880	.30120
1.910	.65636	.34364	2.33	.69970	.30030
1.915	.65695	.34305	2.34	.70060	.29940
1.920	.65753	.34247	2.35	.70149	.29851
1.925	.65812	.34188	2.36	.70238	.29762
1.930	.65870	.34130	2.37	.70326	.29674
1.935	.65928	.34072	2.38	.70414	.29586
1.940	.65986	.34014	2.39	.70501	.29499
1.945	.66044	.33956	2.40	.70588	.29412
1.950	.66102	.33898	2.41	.70674	.29326
1.955	.66159	.33841	2.42	.70760	.29240
1.960	.66216	.33784	2.43	.70845	.29155
1.965	.66273	.33727	2.44	.70930	.29070
1.970	.66330	.33670	2.45	.71014	.28986
1.975	.66387	.33613	2.46	.71098	.28902
1.980	.66443	.33557	2.47	.71181	.28819
1.985	.66499	.33501	2.48	.71264	.28736
1.990	.66555	.33445	2.49	.71347	.28653
1.995	.66611	.33389	2.50	.71429	.28571
2.000	.66667	.33333	2.51	.71510	.28490
2.010	.66777	.33223	2.52	.71591	.28409
2.020	.66887	.33113	2.53	.71671	.28329
2,030	.66997	.33003	2.54	.71751	.28249
2.040	.67105	.32895	2.55	.71831	.28169
2.050	.67213	.32787	2.56	.71910	.28090
2.060	.67320	.32680	2.57	.71989	.28011
2.070	.67427	.32573	2.58	.72067	.27933
2.080	.67532	.32468	2.59	.72145	.27855

p/r	x	1 - x	p/r	x	1 - x
2.60	.72222	.27778	3.11	.75669	.24331
2.61	.72299	.27701	3.12	.75728	.24272
2.62	.72376	.27624	3.13	.75787	.24213
2.63	.72452	.27548	3.14	.75845	.24155
2.64	.72527	.27473	3.15	.75904	.24096
2.65	.72603	.27397	3.16	.75962	.24038
2.66	.72678	.27322	3.17	.76019	.23981
2.67	.72752	.27248	3.18	.76077	.23923
2.68	.72826	.27174	3.19	.76134	.23866
2.69	.72900	.27100	3.20	.76190	.23810
2.70	.72973	.27027	3.21	.76247	.23753
2.71	.73046	.26954	3.22	.76303	.23697
2.72	.73118	.26882	3.23	.76359	.23641
2.73	.73190	.26810	3.24	.76415	.23585
2.74	.73262	.26738	3.25	.76471	.23529
2.75	.73333	.26667	3.26	.76526	.23474
2.76	.73404	.26596	3.27	.76581	.23419
2.77	.73475	.26525	3.28	.76636	.23364
2.78	.73545	.26455	3.29	.76690	.23310
2.79	.73615	.26385	3.30	.76744	.23256
2.80	.73684	.26316	3.31	.76798	.23202
2.81	.73753	.26247	3.32	.76852	.23148
2.82	.73822	.26178	3.33	.76905	.23095
2.83	.73890	.26110	3.34	.76959	.23041
2.84	.73958	.26042	3.35	.77011	.22988
2.85	.74026	.26974	3.36	.77064	.22936
2.86	.74093	.25907	3.37	.77117	.22883
2.87	.74160	.25840	3.38	.77169	.22831
2.88	.74227	.25773	3.39	.77221	.22779
2.89	.74293	.25707	3.40	.77273	.22727
2.90	.74359	.25641	3.41	.77324	.22676
2.91	.74425	.25575	3.42	.77376	.22624
2.92	.74490	.25510	3.43	.77427	.22573
2.93	.74555	.25445	3.44	.77477	.22523
2.94	.74619	.25381	3.45	.77528	.22472
2.95	.74684	.25317	3.46	.77578	.22422
2.96	.74747	.25253	3.47	.77629	.22371
2.97	.74811	.25189	3.48	.77679	.22321
2.98	.74874	.25126	3.49	.77728	.22272
2.99	.74937	.25063	3.50	.77778	.22222
3.00	.75000	.25000	3.51	.77827	.22173
3.01	.75062	.24938	3.52	.77876	.22124
3.02	.75124	.24876	3.53	.77925	.22075
3.03	.75186	.24814	3.54	.77974	.22026
3.04	.75248	.24752	3.55	.78022	.21978
3.05	.75309	.24691	3.56	.78070	.21930
3.06	.75369	.24631	3.57	.78118	.21882
3.07	.75430	.24570	3.58	.78166	.21834
3.08	.75490	.24510	3.59	.78214	.21786
3.09	.75550	.24450	3.60	.78251	.21739
3.10	.75610	.24390	3.61	.78308	.21692

p/r	x	1 - x	p/r	x	1 - x
3.62	.78355	.21645	4.26	.80989	,19011
3.63	.78402	.21598	4.28	.81061	.18939
3.64	.78448	.21552	4.30	.81132	.18839
3.65	.78495	.21505	4.32	.81203	.18797
3.66	.78541	.21459	4.34	.81273	.18727
3.67	.78587	.21413	4.36	.81343	.18657
3.68	.78632	.21368	4.38	.81413	.18587
3.69	.78678	.21322	4.40	.81481	.18519
3.70	.78723	.21277	4.42	.81550	.18450
3.71	.78769	.21231	4.44	.81618	.18382
3.72	.78814	.21186	4.46	.81685	.18315
3.73	.78858	.21142	4.48	.81752	.18248
3.74	.78903	.21097	4.50	.81818	.18182
3.75	.78947	.21053	4.52	.81884	.18116
3.76	.78992	.21008	4.54	.81949	.18051
3.77	.79036	.20964	4.56	.82014	.17986
3.78	.79079	.20921	4.58	.82079	.17921
3.79	.79123	.20877	4.60	.82143	.17857
3.80	.79167	.20833	4.62	.82206	.17794
3.81	.79210	.20790	4.64	.82270	.17730
3.82	.79253	.20747	4.66	.82332	.17668
3.83	.79296	.20704	4.68	.82394	.17606
3.84	.79339	.20661	4.70	.82456	.17544
3.85	.79381	.20619	4.72	.82517	.17483
3.86	.77424	.20576	4.74	.82578	.17422
3.87	.79466	.20534	4.76	.82639	.17361
3.88	.79508	.20492	4.78	.82699	.17301
3.89	.79550	.20450	4.80	.82759	.17241
3.90	.79592	.20408	4.82	.82818	.17182
3.91	.79633	.20367	4.84	.82877	.17123
3.92	.79675	.20325	4.86	.82935	.17065
3.93	.79716	.20284	4.88	.82993	.17007
3.94	.79757	.20243	4.90	.83051	.16949
3.95	.79798	.20202	4.92	.83108	.16892
3.96	.79839	.20161	4.94	.83165	.16835
3.97	.79879	.20121	4.96	.83221	.16779
3.98	.79920	.20080	4.98	.83278	.16722
3.99	.79960	.20040	5.00	.83333	.16667
4.00	.80000	.20000	5.02	.83389	.16611
4.02	.80080	.19920	5.04	.83444	.16556
4.04	.80159	.19841	5.06	.83498	.16502
4.06	.80237	.19763	5.08	.83553	.16447
4.08	.80315	.19685	5.10	.83607	.16393
4.10	.80392	.19608	5.12	.83660	.16340
4.12	.80469	.19531	5.14	.83713	.16287
4.14	.80545	.19455	5.16	.83766	.16234
4.16	.80620	.19380	5.18	.83819	.16181
4.18	.80695	.19305	5.20	.83871	.16129
4.20	.80769	.19231	5.22	.83923	.16077
4.22	.80843	.19157	5.24	.83974	.16026
4.24	.80916	.19084	5.26	.84026	.15974

p/r	x	1 - x	p/r	x	1 - x
5.28	.84076	.15924	6.75	.87097	.12903
5.30	.84127	.15873	6.80	.87179	.12821
5.32	.84177	.15823	6.85	.87261	.12939
5.34	.84227	.15773	6.90	.87342	.12658
5.36	.84277	.15723	6.95	.87421	.12579
5.38	.84326	.15674	7.00	.87500	.12500
5.40	.84375	.15625	7.05	.87578	.12422
5.42	.84424	.15576	7.10	.87654	.12346
5.44	.84472	.15528	7.15	.87730	.12270
5.46	.84520	.15480	7.20	.87805	.12195
5.48	.84568	.15432	7.25	.87879	.12121
5.50	.84615	.15385	7.30	.87952	.12048
5.52	.84663	.15337	7.35	.88024	.11976
5.54	.84709	.15291	7.40	.88095	.11905
5.56	.84756	.15244	7.45	.88166	.11834
5.58	.84802	.15198	7.50	.88235	.11765
5.60	.84848	.15152	7.55	.88304	.11696
5.62	.84894	.15106	7.60	.88372	.11628
5.64	.84940	.15060	7.65	.88439	.11561
5.66	.84985	.15015	7.70	.88506	.11494
5.68	.85030	.14970	7.75	.88571	.11429
5.70	.85075	.14925	7.80	.88636	.11364
5.72	.85119	.14881	7.85	.88701	.11299
5.74	.85163	.14837	7.90	.88764	.11236
5.76	.85207	.14793	7.95	.88827	.11173
5.78	.85251	.14749	8.00	.88889	.11111
5.80	.85294	.14706	8.05	.88950	.11050
5.82	.85337	.14663	8.10	.89011	.10989
5.84	.85380	.14620	8.15	.89071	.10929
5.86	.85423	.14577	8.20	.89130	.10870
5.88	.85465	.14535	8.25	.89189	.10811
5.90	.85507	.14493	8.30	.89247	.10753
5.92	.85549	.14451	8.35	.89305	.10695
5.94	.85591	.14409	8.40	.89362	.10638
5.96	.85632	.14368	8.45	.89418	.10582
5.98	.85673	.14327	8.50	.89474	.10526
6.00	.85714	.14286	8.55	.89529	.10471
6.05	.85816	.14184	8.60	.89583	.10417
6.10	.85915	.14085	8.65	.89637	.10363
6.15	.86014	.13986	8.70	.89691	.10309
6.20	.86111	.13889	8.75	.89744	.10256
6.25	.86209	.13793	8.80	.89796	.10204
6.30	.86301	.13699	8.85	.89848	.10152
6.35	.86395	.13605	8.90	.89899	.10101
6.40	.86486	.13514	8.95	.89950	.10050
6.45	.86577	.13423	9.00	.90000	.10000
6.50	.86667	.13333	9.05	.90050	.09950
6.55	.86755	.13245	9.10	.90099	.09901
6.60	.86842	.13158	9.15	.90148	.09852
6.65	.86928	.13072	9.20	.90196	.09804
6.70	.87013	.12987	9.25	.90244	.09756

p/r	x	1 - x	p/r	x	1 - x
9.30	.90291	.09709	15.4	.93902	.06098
9.35	.90338	.09662	15.6	.93976	.06024
9.40	.90385	.09615	15.8	.94048	.05952
9.45	.90431	.09569	16.0	.94118	.05882
9.50	.90476	.09524	16.5	.94286	.05714
9.55	.90521	.09479	17.0	.94444	.05556
9.60	.90566	.09434	17.5	.94595	.05405
9.65	.90610	.09390	18.0	.94737	.05263
9.70	.90654	.09346	18.5	.94872	.05128
9.75	.90698	.09302	19.0	.95000	.05000
9.80	.90741	.09259	19.5	.95122	.04878
9.85	.90783	.09217	20.0	.95238	.04762
9.90	.90826	.09174	20.5	.95349	.04651
9.95	.90868	.09132	21.0	.95465	.04546
10.00	.90909	.09091	21.5	.95556	.04444
10.1	.90991	.09009	22.0	.95652	.04348
10.2	.91071	.08929	22.5	.95745	.04255
10.3	.91150	.08850	23.0	.95833	.04167
10.4	.91228	.08772	23.5	.95918	.04082
10.5	.91304	.08696	24.0	.96000	.04000
10.6	.91379	.08621	24.5	.96078	.03922
10.7	.91453	.08547	25.0	.96154	.03846
10.8	.91525	.08475	26	.96296	.03704
10.9	.91597	.08403	27	.96429	.03571
11.0	.91667	.08333	28	.96552	.03448
11.1	.91736	.08264	29	.96667	.03333
11.2	.91803	.08197	30	.96774	.03226
11.3	.91870	.08130	31	.96875	.03125
11.4	.91935	.08065	32	.96970	.03030
11.5	.92000	.08000	33	.97059	.02941
11.6	.92063	.07937	34	.97143	.02857
11.7	.92126	.07874	35	.97222	.02778
11.8	.92187	.07813	36	.97297	.02703
11.9	.92248	.07752	37	.97368	.02632
12.0	.92308	.07692	38	.97436	.02564
12.2	.92424	.07576	39	.97500	.02500
12.4	.92537	.07463	40	.97561	.02439
12.6	.92647	.07353	42	.97674	.02326
12.8	.92754	.07246	44	.97778	.02222
13.0	.92857	.07143	46	.97872	.02128
13.2	.92958	.07042	48	.97959	.02041
13.4	.93056	.06944	50	.98039	.01961
13.6	.93151	.06849	55	.98214	.01786
13.8	.93243	.06757	60	.98361	.01639
14.0	.93333	.06667	65	.98485	.01515
14.2	.93421	.06579	70	.98592	.01408
14.4	.93506	.06494	75	.98684	.01316
14.6	.93590	.06410	80	.98765	.01235
14.8	.93671	.06329	85	.98837	.01163
15.0	.93750	.06250	90	.98901	.01099
15.2	.93827	.06173	95	.98958	.01042
			100	.99010	.00990

INDEX